A LAND
ON FIRE

A LAND ON FIRE

The Environmental Consequences of the Southeast Asian Boom

James David Fahn

A Member of the Perseus Books Group

Westview Press books are available at special discounts for bulk purchases in the United States by corporations, institutions, and other organizations. For more information, please contact the Special Markets Department at the Perseus Books Group, 11 Cambridge Center, Cambridge MA 02142, or call (617) 252-5298, (800) 255-1514 or email j.mccrary@perseusbooks.com.

Published in the United States of America by Westview Press, 5500 Central Avenue, Boulder, Colorado 80301–2877, and in the United Kingdom by Westview Press, 12 Hid's Copse Road, Cumnor Hill, Oxford OX2 9JJ

Find us on the World Wide Web at www.westviewpress.com

Designed by Reginald R. Thompson
Set in 11-point AGaramond by the Perseus Books Group

Library of Congress Cataloging-in-Publication Data
Fahn, James
A land on fire : the environmental consequences of the Southeast Asian boom / James Fahn
p. cm
Includes bibliographical references and index.
ISBN 0-8133-4053-5 (hardcover : alk. paper)
1. Environmental policy—Asia, Southeastern. 2. Asia, Southeastern—Economic policy. 3. Pollution—Asia, Southeastern. I. Title.
HC441.Z9 E538 2003
333.7'0959—dc21

2002153121

The paper used in this publication meets the requirements of the American National Standard for Permanence of Paper for Printed Library Materials Z39.48–1984.

10 9 8 7 6 5 4 3 2 1

This book is dedicated to
Ping
Because you can't say her name without smiling!

CONTENTS

ACRONYMS AND ABBREVIATIONS

ABS	access and benefits sharing
ABSDF	All-Burma Students Democratic Front
ADB	Asian Development Bank
API	Air Pollutants Index
ASEAN	Association of Southeast Asian Nations
BPKP	Bolisat Pattana Khet Pudoi
BPP	Border Patrol Police
BMA	Bangkok Metropolitan Administration
CBD	Convention on Biological Diversity
CCC	Counter Corruption Commission
CDM	Clean Development Mechanism
CGIAR	Consultative Group on International Agricultural Research
DEQP	Department of Environmental Quality Promotion
DIW	Department of Industrial Works
DMR	Department of Mineral Resources
DSM	demand-side management
EGAT	Electricity Generating Authority of Thailand
EIA	environmental impact assessment
EII	Earth Island Institute
EPA	Environmental Protection Agency
ERI	EarthRights International
EU	European Union

FAO	UN Food and Agriculture Organization
FCCC	Framework Convention on Climate Change
FIO	Forest Industry Organization
GATT	General Agreement on Tariffs and Trade
GDP	Gross Domestic Product
GEF	Global Environmental Facility
GHGs	greenhouse gases
GM	genitically modified
GMOs	genetically modified organisms
IEAT	Industrial Estates Authority of Thailand
IFREMER	French Research Institute for the Exploitation of the Sea
IIEC	International Institute for Energy Conservation
ILO	International Labor Organization
IMF	International Monetary Fund
IPCC	Inter-Governmental Panel on Climate Change
ITQs	Individual Transferable Quotas
IUCN	World Conservation Union
JI	Joint Implementation
KNPP	Karenni National Progressive Party
LST	Law Society of Thailand
MAP	Mangrove Action Project
MEAs	multilateral environmental agreements
mg/l	micrograms per liter
MOGE	Myanmar Oil and Gas Enterprise
MoU	Memorandum of Understanding
MRC	Mekong River Commission
M.P.	member of Parliament
MW	megawatts
NACA	Network of Aquaculture Centres in Asia-Pacific
NAFTA	North American Free Trade Agreement
NASA	U.S. National Aeronautics and Space Administration
NEPO	National Energy Policy Office
NGO	nongovernmental organization
NPD	National Parks Division
NIMBY	"Not in My Backyard"
NORMs	naturally occurring radioactive materials
NPD	National Parks Division
NPKC	National Peacekeeping Council

PCD	Pollution Control Department
PDA	Population and Community Development Association
PER	Project for Ecological Recovery
POPs	persistent organic pollutants
ppb	parts per billion
ppm	parts per million
PTT	Petroleum Authority of Thailand
PTTEP	PTT Exploration & Production
R&D	research and development
RFD	Royal Forest Department
RID	Royal Irrigation Department
SAIN	Southeast Asia Information Network
SLORC	State Law and Order Restoration Council
TAT	Tourism Authority of Thailand
TBT	tributyl tin
TDRI	Thailand Development Research Institute
TEDs	turtle escape devices
TERRA	Towards Ecological Recovery and Regional Alliance
TPI	Thai Petrochemical Industry Co. Ltd.
TRIPs	Trade-Related Intellectual Property Rights
UN	United Nations
UNEP	UN Environment Program
UNESCAP	UN Economic and Social Commission for Asia-Pacific
UNESCO	UN Economic, Social and Cultural Organization
U.K.	United Kingdom
U.S.	United States
WCD	Wildlife Conservation Division
WCS	Wildlife Conservation Society
WFT	Wildlife Fund Thailand
WHO	World Health Organization
WTO	World Trade Organization
WWF	Worldwide Fund for Nature

ACKNOWLEDGMENTS

I OWE MY START IN journalism to Thepchai Yong, who hired me to work at the *Nation* (an English-language newspaper in Bangkok) and was my first editor there. I am equally indebted to the current editor, Pana Janviroj, who gave his full backing to the environment desk, and gave our stories prominence. They and other editors at the *Nation* provided us with rare freedom to pursue the stories we were interested in, and always backed us up—even in the face of powerful external pressure—when our investigative stories proved controversial.

I am also grateful to have worked with so many skillful and friendly colleagues at the *Nation,* particularly on the environment desk: Nantiya Tangwisutijit, Klomjit Chandrapanya, Kamol Sukin, Walakkamon Eamwiwatkit, and Pennapa Hongthong. They made work seem like fun. Ann Danaiya Usher, a pioneer of environmental coverage in Thailand, provided me with invaluable advice when I was starting out, and Pichaya Changsorn was always helpful in covering industry. Termsak Palanupap and Sonny Inbaraj encouraged me to write editorials. Tulsathit Taptim and Boonrawd Krasaesin guided my stories onto the news desk, and Sandy Barron, Dsylan Jones, and Nithinand Yorsaengrat likewise helped my feature pieces along. Scot Donaldson and David Squires taught me that, yes, design is important, too. Finally, Kwanruan Thaworntaweewong at Nation TV was an amazingly good-natured boss amid all the turmoil we faced.

There were so many other people who helped me during my stay in Southeast Asia that it would be impossible to mention them all here. But special thanks go to Ing Kanjanavanit, who taught me so much about Thailand; to Sarah MacLean and Denis Gray of the Indochina Media Memorial Foundation; and to all my colleagues at the Thai Society of Environmental Journalists. Others who deserve mention are Belinda Stewart-Cox, Paul Handley, Kasem Snidvongse, Dhira Phantumvanit, Wasant Techawongtham, Faith Doherty, Dr. Oy Kanjanavanit, David Lazarus, Peter Du Pont, John Baker, Jack Hurd, Steve Wilkins and the folks at Wildlife Fund Thailand (WFT), Terra/Project for Ecological Recovery (PER), UN Environment Program (UNEP), PDA (Population and Community Development Association), and the Seub Nakhasathien Foundation. Michael Rivers, Simon Johnstone, Alan Parkhouse, Chris Burslem, Roger Beaumont, and the rest of the gang who gathered at Ping's place provided camaraderie far from home. Bob Ingall deserves special thanks for rescuing my cat, Scratchy, from the clutches of The Swamp. Many others, particularly those who served as sources for my stories, must remain anonymous, but you know who you are.

Then there are the people who helped me in preparing and selling this manuscript. I am especially grateful to Steve Rayner, my advisor at Columbia University, who read it in its entirety and helped me figure out what the book is actually about. Chris Baker, an academic and author based in Thailand, also took time out to read the manuscript and offered invaluable advice. Columbia's Arthur Small, Ann-Marie Murphy, and the ever-encouraging Marguerite Holloway each read sections and gave me useful advice. So did Dick and Helen Lynn, John Heller, Klomjit Chandrapanya, and Ted Bardacke, a fellow traveler in Bangkok and Manhattan who was always ready to lend an ear and a hand. Allen Zelon provided his invaluable legal services gratis, and the ever-generous Stephen Moskowitz was a constant source of help and encouragement.

This book would never have been published if Karl Yambert and Holly Hodder at Westview Press had not taken a chance, in the face of adversity, on a new author. And since I wrote it while being treated and monitored for cancer over the last three years, it wouldn't exist if not for the medical care of Dr. Franklin Lowe, who performed my orchiectomy; Dr. Mitchell Benson, who carried out my surveillance program; Dr. Joel Sheinfeld, who performed my retroperitaneal lymph node dissection once the cancer had metastasized; and Dr. Robert Motzer, who is currently overseeing my chemotherapy. The nurses at Memorial Sloan-Kettering Hospital helped the healing tremendously with their friendly and patient care.

Most of all, I'd like to thank my family for all their support and love. I owe everything to my parents, Dr. Stanley and Charlotte Fahn, who have provided a foundation for my life in so many ways. My brother Paul has been there whenever I needed him. And finally there is my wife, Ping, who made the maps for this book, but, most important, stuck with me through all the dark days of illness and dark nights of noisy typing, bringing joy into my life. I love them all dearly.

James David Fahn
Manhattan, August 2002

INTRODUCTION
THE FIRE OF KNOWLEDGE

"How can people live in such filth?!" The woman sitting next to me, well-dressed and European, was staring out of the bus window as we passed one of Bangkok's many canal-side slums. Litter lay everywhere, under and around the decrepit shacks, lining the fetid stream like flotsam. She shook her head. Her disdain was palpable.

It's a question many have asked, for it's not just the slums where you see lots of litter, but also the streets and even, sadly, in Thailand's national parks. This time I was with a group of foreign visitors (or *farang,* as Thais call people of European and American descent) in town to attend a conference. Now let's face it, stories about trash are not terribly sexy. The really hot topics on the environment beat, and in this book, are issues such as climate change, deforestation, the battle over national parks, and toxic waste. You don't become a journalist to expose the truth about litterbugs. But sometimes the mundane subjects speak most profoundly about us, reflecting deep-rooted concerns intimately bound up with our values. So I told her my theory about littering, that it is linked more closely to poverty than nationality, but most of all it depends on the strength (or lack) of community feeling.

"If you venture into one of these shacks, I bet you'd find it spotless," I suggested. "But they don't own the land they're living on and they probably don't feel any responsibility for what's not theirs, no sense of community, no respect for public land. So when it comes to tossing away their garbage, maybe they just decide if it's out of their house, it's out of mind."

Western countries were just as garbage strewn at one time, and their poorer neighborhoods still are. In the United States, Iron Eyes Cody helped change that. A Native American actor who performed in a commercial that first appeared on Earth Day in 1971, he was shown crossing a landscape befouled by trash; finally, he shed a single eloquent tear. It had a profound effect on America. Certainly, everyone from my generation remembers it, for Native Americans serve as the ecological conscience of the United States. I wondered who would serve a similar role in Thailand. An aged rice farmer? A Karen hill tribesman? A member of the royal family?

But the woman on the bus didn't buy my explanation. "Surely they must see the garbage, so why do they continue to throw it?" she kept asking. It was a typical example of environmental culture clash, and I wouldn't have thought much more about it save for what happened that evening.

I had returned to the newsroom to write up a story, and was joking with some of my Thai colleagues. Thais love to tease, and they can be mercilessly personal about it. So, noting my haggard appearance at the end of a long day, one of them asked when I had last had a shower. *"Farang sokaprok jang loy,"* she needled, "*Farang* are so dirty."

Of course, she only meant to rib me, but many Thais secretly do question the personal hygiene habits of foreigners. And they may have a point. Because they live in the tropics, Thais realize that you have to shower at least twice a day in such a hot and moist climate. Embarrassing as it may seem, it often takes visitors from temperate countries a while to learn that key lesson, simply because people living in colder climates generally shower less often (especially if they don't have access to hot water). Also, although foreigners generally begin to drip with sweat the minute they step outside into Bangkok's humid air, Thais seem able to spend the entire day on the grimy streets, or marching through primeval forest for that matter, with nary a stain blotting their perfectly creased clothes. It's really quite an amazing achievement.

The contrasting attitudes towards cleanliness expressed by the European visitor and my Thai colleague are a beautiful (or perhaps ugly) illustration of how Western and Asian societies can fail to understand each other in regard to environmental issues. Each person discretely looked on the other as being "dirty": the *farang* reproach Thais because of how they treat their public spaces, the Thais disapprove of how *farang* look after their personal space.

And for each there are reasonable explanations that make it possible to understand the behavior in question, even if you don't excuse it.

This anecdote offers a glimpse of what I've found to be so enthralling about my experience in Southeast Asia, and what I hope to convey in this book. Observing how environmental issues are handled turns out to be a novel way to learn about a region and its people. At the same time, examining how a region that's growing as quickly as Southeast Asia responds to its environmental crises can help us understand the issues facing us all. Sometimes, they play out in surprisingly different ways than in the West; but often, the differences are largely superficial, and the fundamental issues turn out to be surprisingly similar.

Either way, coming to grips with the environmental challenges facing developing countries is vital to the future of our planet. The events of the previous decade—culminating with the protests over free trade at the 1999 World Trade Organization (WTO) summit in Seattle—demonstrated with alarming clarity that in this age of globalization our economic and environmental destinies are intertwined. Asia is home to 60 percent of the world's population. Considering its rich biodiversity, its ever-increasing demand for natural resources, and its vast potential to emit greenhouse gases, the future of the world's environment will depend on how it develops. So it's critical to understand how people in Asia and developing countries elsewhere view environmental issues, and how their environmental movements are evolving. Some Westerners may not even be aware that Asia has active green groups, but, in the long run, local environmental movements are likely to have a greater impact on national policies than foreign pressure.

As a journalist who was born and raised in the United States but who covered environmental issues for a Thai newspaper and learned to speak the native language, I was in a unique position to see daily how the people and institutions of Southeast Asia deal with environmental issues—to explore their concerns and motivations, their strengths and weaknesses—and compare them to our own in the West. Beginning in 1990, I worked for nearly a decade at the *Nation,* a daily newspaper based in Bangkok and owned by one of Thailand's leading media conglomerates. Although published in English, most of its readers are Thai, and it's a newspaper of record for the country, having received numerous international press freedom awards thanks in part to its bold coverage of Thailand's democracy uprising in 1992.

After serving as the newspaper's science and technology editor for a couple of years, I ended up as the *Nation*'s environment editor, and I put together a team of Thai reporters to form an environment desk. We set about covering a wide variety of issues related to natural resources around the Southeast Asian region. Our brief was writ large. We not only reported on current events but also tried to examine the social changes and political factors that lay behind them. Our award-winning investigative stories broke new ground in the region by uncovering some serious domestic and international scandals. To cite one example of the impact we had, following an exposé in 1993 about a Japanese aid package to Cambodia that included forty tons of outdated pesticides,[1] Japan's then–prime minister, Kiichi Miyazawa, held up my article at a press conference and vowed to halt the shipment of pesticides.

Starting in mid-1998, I worked as a scriptwriter, reporter, and part-time host of a television program called *Rayngan Si-khhiow* (Green Report)—a weekly feature program about the environment that aired nationally on Thai television. It was a fascinating experience, and a worthwhile one, but also frustrating at times because we did suffer from censorship. So a year later, I decided to take a break from my career and return to the United States to pursue a master's degree in international affairs at Columbia University, where I put this book together.

The 1990s were an amazing time to be in Southeast Asia. There was so much growth, so much hope and energy in the air, that it felt as if the entire region were on fire. Certainly the economies were. The "Asian tigers" were in the midst of a raging boom that was rapidly transforming them into industrial societies. And Thailand was outpacing them all: From 1986 to 1996, its economy expanded by an average of 8 percent every year. With an open and vibrant society, a free press (at least for the written word; television is much less free), and a government actively engaged not only with its neighbors but the world at large, Thailand, in the words of one UN Development Program officer, was "a great success story." It's also the country with which I am most familiar, so it takes center stage in *A Land on Fire,* although I shall examine issues in neighboring countries, as well, and survey the response to key issues throughout Southeast Asia.

If you've never been to Thailand but happen to be familiar with Europe, think of it as "the Italy of Asia." There are some startling similarities between the two countries, superficial and profound. Like the Italians, Thais are famous for their beauty and elegance, for their delicious cuisine and their insatiable appetite for fun and passion, for the splendor of their ancient traditions and the depths of their corruption, the continual changing of their governments and their disparities in income and power, their religious devotion and reckless debauchery, their grace and glitz, style and squalor, and, of course, their cities' ceaseless traffic jams.

During the Southeast Asian boom, it was in the cities where Thailand's rapid development was concentrated, particularly in the chaotic capital of Bangkok. This sprawling megalopolis grew so fast that its skies were filled with cranes—not the avian kind, but the towering metallic sort—which flocked to the country in droves; indeed, early in the decade, some civil engineers estimated that around 15 to 20 percent of the world's cranes could be found in Thailand.[2] All that growth, combined with the opening up of neighboring Indochina after decades of strife and isolation, gave the entire region an incredible dynamism. The surge of excitement in Bangkok was almost palpable, as if the roaring economic inferno had ignited a flame of unbridled optimism, as if an entire city had come to feel there were no limits to the possibilities before it.

Inevitably, however, there was also a dark side to this dash for prosperity. For while Bangkok's skyline soared during the boom, its streets became choked with cars, its canals turned into sewers, and its air transformed into a nasty, steaming soup of dust and toxins. And those are only the most visible manifestation of the environmental catastrophe that is Thailand today. Virtually every resource the country owns has been squandered with too little thought for the future. The forests which forty years ago covered 60 percent of the country now probably cover only 15 percent, and their amazingly rich biodiversity became endangered before it could even be fully recorded. Wanton use of pesticides has poisoned the countryside, shrimp farms have ravaged the coasts, and the seas have been plundered of their fish.

These disastrous side effects of industrialization and urbanization are not unprecedented. The situation in Bangkok (or Jakarta or Shanghai or Mexico City) today is probably quite similar to that of Dickensian England or turn-of-the-twentieth-century New York, and overall the developed countries (the "global North") have done far more to transform the global environment than those in the developing world (the "global South"). There are even some positive environmental signs in Southeast Asia; for instance, a rapid decline in population growth, thanks largely to improved educational opportunities for women and a competent public health service, supported by a thriving community of nongovernmental organizations (NGOs).

But in general, the region's rampant growth has been unsustainable, and although the financial crisis that rocked Southeast Asia beginning in 1997 was harsher than generally expected, many observers felt that a pause in the rapid pace of development was not only predictable but probably necessary.

One of my goals in writing this book is to document how the resources of Southeast Asia in general, and Thailand in particular, have been abused and often squandered during the region's rapid rise toward industrialization. To do so, it will be necessary to examine the social and political roots of the mismanagement and corruption that have contributed to the environmental crisis. Just as important is the reaction to this crisis: how governments and other institutions have responded; how an indigenous environmental movement has emerged from the grass roots and, when allowed a modicum of freedom, evolved into a potent force. Examined closely, many of these green groups turn out to be part of a broader non-governmental participatory democracy movement. In Thailand at least, this is the closest thing there is to a true left-wing opposition.

Another goal is to compare Southeast Asia's environmental issues, and the way they play out, with those in the West. It's particularly fascinating to look at the contrasts, to examine, for instance, why golf courses and national parks are so much more controversial in Thailand than in the United States. Sometimes these differences can be explained by the vagaries of culture and history and sometimes by discrepancies in wealth and education levels. But one of the biggest factors—and one that receives surprisingly scarce mention in connection to preserving resources—is the prevalence of corruption in the global South. Although corruption certainly occurs among governments in the global North, it exists on an entirely different scale in the developing world. A major feature of this book, therefore, is to reveal how my colleagues and I exposed various scandals, and to point out how such systemic abuse has made protecting the environment more difficult than it need be.

Home to half a billion people and some of the most biodiverse ecosystems on Earth, Southeast Asia's growing demand for resources is already having a direct impact on the global environment. More broadly, exploring the region's environmental crisis and its responses helps us understand the challenges being faced throughout the developing world, including the region's giant neighbors, China and India. Such extrapolation is always dangerous, of course, since ultimately every locale faces its own unique situation, but general lessons can be learned.

For one thing, environmentalism turns out to be something of a different beast in the South. Unlike the North, where the movement began to flourish in wealthy democracies where people had already migrated to the cities, in developing countries it is emerging in societies that are still mostly rural and poor but are moving toward democracy as they industrialize. So whereas the green movement in the North tends to focus on the middle class, in the South not only is it centered more on the farmers and fishermen who rely on natural resources for their livelihoods but it's also concerned more with who gets to use resources, not just with how they are used. This mixture of social and green activism—what I call the "environmental democracy" movement—is closest in spirit to the environmental justice movement in the United States, whose activists generally focus on urban and minority populations but have less clout than the mainstream green groups. In Asia, such activists *are* the mainstream because the majority of the population there is still underprivileged, and the middle class remains a minority.

Given the nature of societies in the developing world, and the endemic problem of corruption, the strategies needed to protect forests, fisheries, and other resources are much more basic in the South. When it comes to the environment, most Asian governments have failed their citizens. Even those that have passed good laws tend to do a poor job of enforcing them. Improving governance therefore requires the development of institutions—particularly the media, the legal system, NGOs, and an environmentally responsible business sector—that make up a healthy and vibrant civil society.

For concerned citizens in the West, it should be of great comfort to know that the urge towards environmentalism is found in all societies, although it may take different forms. It also suggests ways in which the North, which has been quicker to pass on its destructive technologies than the laws and customs needed to constrain them, can help the South develop more sustainably. Asia's rapid growth is occurring during a period of equally rapid globalization: There is now a great deal of scrutiny from abroad, and more opportunity to learn from other countries' experiences. On a practical level, this means the technology to make development cleaner already exists; the problem lies in determining how to pay for its installment. More fundamentally, it's now possible to establish all kinds of linkages—public and private, official and informal—to help societies in the South find a suitable path toward sustainable development.

Such collaboration is crucial because yet another difference between North and South is the huge difference in time scales of development. What Europe and the United States took many generations to achieve, Thailand and its neighbors are trying to do in one generation. So not only are the effects of industrialization more concentrated but much of the populace grew up in a very different world from that of today. And while it may be possible to "leapfrog" technologically, it is far more difficult to accelerate changes in the way people think. Most leaders currently in power are ill equipped to handle certain modern problems, and the development of a civil society to push those leaders toward greener policies has not kept pace with economic growth.

Finally, it is my hope that this book will refute one of the most common misconceptions about protecting the environment in developing countries: that it is a luxury for the benefit of the wealthy, whether at home or abroad. Almost inevitably, the people who suffer the most from environmental disasters—whether it's a landslide caused by denuded hillsides, illness caused by exposure to toxic waste, or dislocation to make way for a dam reservoir—are the poor. Developing countries need to address their environmental problems not because the rich world tells them to, but for the benefit of their own people.

By mixing in some thoughtful analysis of Southeast Asia's environmental situation with stories of the characters I've met and the adventures I've had as an investigative journalist, I've tried to make *A Land on Fire* appealing both to the serious and the casual reader. Generally, I present the facts as they occurred and then add some insights and synthesis. I prefer to let others provide commentary, and I have made a conscious effort to convey Asian views regarding the environment—even if I privately disagree with them—

since they aren't often heard in the West. But as an editorial writer for the *Nation,* and as a human being, I felt a responsibility at times to go beyond reporting and suggest viable alternatives to current policies and practices that seemed damaging. So I have done that here, too, when appropriate.

The most difficult thing in writing this book was deciding what to leave out. Some environmental issues stem from too little development, but for the most part I focus on those resulting from too much (or unplanned) development. I have avoided some issues that justifiably receive a great deal of attention in Southeast Asia—such as the spread of HIV/AIDS, the sex industry, and drugs—simply because they are off-topic. But some critical environmental issues also did not make the final cut even though we covered them at the *Nation.* We wrote extensively about wildlife conservation and the wildlife trade, currently thriving in Burma and Indochina, forest fires, and workers' health and safety, particularly concerning the many industrial disasters the region has faced, but they receive scarce mention in this book. Other important issues—such as agriculture, (non-toxic) industrial pollution, population growth, and the spread of genetically modified organisms—are discussed only briefly or tangentially because I simply ran out of space.

In Southeast Asia, it was not only the economies that were on fire in the 1990s. The countryside also burns, quite literally. Every year during the dry season, forests and farmland are set ablaze. The fires of Borneo and Sumatra are already infamous for spreading an annual haze across the region. But in Thailand, as well, the central plains are set alight to burn off the stubble from the previous harvest, and highland forests are scorched by hunters seeking to smoke out game or slashed by farmers trying to clear more land. A dwindling supply of nutrients is left behind in the ash, and a heavy smog collects in the leaden sky.

One scene seared into my memory came in early 1998, following a trip to the site where the Burma gas pipeline was being laid down in western Thailand. It had been a long and emotionally draining day. Along with visiting protestors who had camped out in the hills but seemed to know their cause was doomed, I had watched the route of the pipeline take shape as a swathe of dusty red clay twisting its way into the jungle. Now it was night-time, and a fellow journalist and I were making the long drive back to Bangkok. But the hills were streaked spectacularly with light. Hunters had set off blazes to spook out their prey, and the lines of fire had turned into eerie snakes of flame crawling across the dark, brush-covered hillsides. Occasionally we passed through a great strand of fire that had leapt the highway and imagined we were feeling the hot breath of dragons lurking by the side of the road.

In Buddhism, fire often represents our insatiable desires for physical, mental, and emotional fulfillment. It's said to be a destructive force, scorching the world around us as we "crackle, snap, and blaze away in our endless pursuit of . . . ultimate satisfaction."[3] Viewed in this light, Thailand's manic push for modernity can be seen as a society simply giving in to its baser instincts. But that seems unduly harsh. There was a brighter side to the boom, as well.

A Buddhist parable, called appropriately enough "The Fire of Knowledge," hints at this ambiguity. It's the story of a monk struggling to understand the Buddha's teachings. He sets out for a monastery and along the way comes across a major forest fire. Watching from atop a mountain as the blaze spreads, he suddenly realizes that just as the fire burns everything up, the development of insight within him can destroy the fetters of his life. The Buddha appears to him, and tells him he is on the right track. Eventually, he makes it to a kind of saintliness.

The monk achieves the knowledge he is seeking, but it does not come without a cost. The fire has consumed the land around him. In the same way, Thailand has achieved some real gains from its rapid development—gains in education, in wealth, in dreams realized and opportunities opened up. My great fear is that much of its potential gain was squandered through corruption and excess—symbolized by the empty real estate projects that now dot Bangkok's skyline—that it didn't use those gains to make the necessary investment in its people, particularly in education, which is vital if the countries of Southeast Asia are to follow in the footsteps of Taiwan and South Korea and become more prosperous and progressive societies. That, however, is not what this book is about.

Rather, *A Land on Fire* is about how Southeast Asia, and Thailand in particular, simultaneously seemed to consume itself through its breakneck pursuit of progress. It can serve as a role model, a positive one for its openness and its vibrant society, and a negative one for its corruption and mismanagement. Thailand has squandered many of its resources and destroyed many of its communities, but it has also made significant democratic advances and served as an island of peace in a region torn apart by war and civil strife. It's my hope, of course, that other countries will learn from these mistakes and achievements.

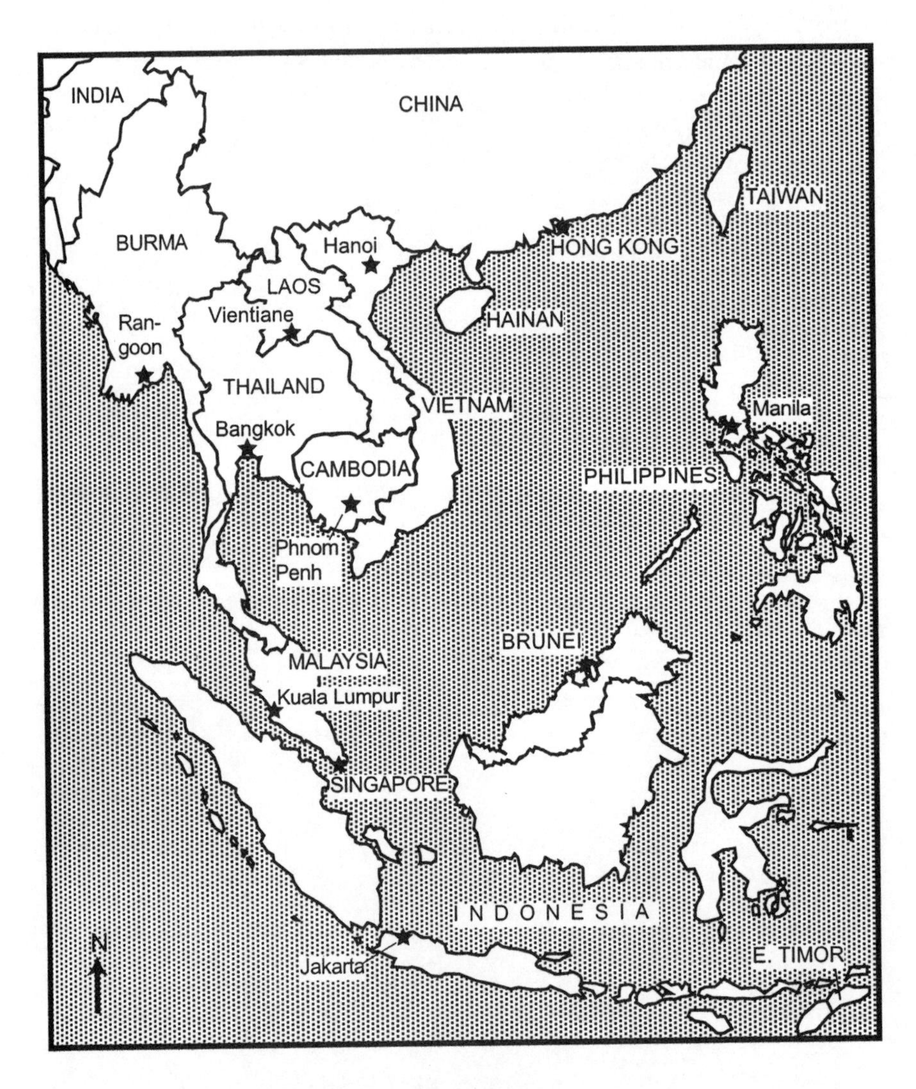

Southeast Asia

1

THE CITY

MANGO MADNESS

SWAMPED BY DEVELOPMENT

If living in Bangkok taught me nothing else, I learned the importance of a green vista. In this chaotic, sprawling capital, having a room with a pleasant view—along with air conditioning and a decent commute—was vital to maintaining my sanity. Now, those who've visited Bangkok only briefly probably think of it as just another gray, polluted Asian mega-city. And it is, on one level. But behind the exterior, it's also a garden city. To see for yourself, simply head to the top of a skyscraper and take a look outside. Amazingly, there's green everywhere, and trees generally seem to outnumber buildings.

Soon after arriving in Bangkok, I took advantage of this hidden trait and found a sanctuary in a small lane off Sukhumvit Road, a major thoroughfare that dissects the eastern part of the city. The apartment, a cramped studio built onto the roof of a shop house, seemed at first glance to have little to recommend it. The sun blasted the roof and walls, cooking everything inside. The parquet floor had a noticeable slope, and as you walked across it in bare feet, some of the little wooden pieces would stick to your soles and come popping out. In fact, the apartment had probably been an illegal addition to the building.

But it had one special feature: a balcony overlooking what my friends and I came to know simply as . . . The Swamp. A patch of wilderness hidden away in the thicket of Bangkok's concrete jungle, The Swamp actually had pretensions of being a proper wetland, an acre or two in size. In the center was a marsh covered with vines and low-lying vegetation. Scattered around the edges were major trees: cotton trees, whose drooping pods contain down soft enough to put in your pillow; a coconut palm, its shiny trunk so vertical that it looked artificial; a flame tree bursting in scarlet; a banana palm that had grown into a luxuriant fan; a thin but heavy-fruited papaya tree; and crowning the scene was a luxuriant rain tree, its stately trunk arched into a symmetrical sweep of thick, green glory.

The Swamp had wildlife, too, particularly birds: long-legged waterfowl that hunted around the marsh; tiny, flitting swallows that loved to soar and dive on the air vents at sunset; and a pair of auburn-winged Greater Coucal, darting from bush to bush in an eternal mating game. Sometimes, they would disappear into the lush undergrowth, but we could follow their trail as they played hide-and-seek among the bobbing lotus leaves. The Swamp even served as a home away from home for itinerant monks who camped beneath the trees when they passed through Bangkok. I visited them once, hoping for some kind of spiritual experience. Instead, they offered to sell me holy wax. If I daubed a little on my lips, they assured me, I'd be a big hit with the ladies.

At night, when Bangkok is at its best, its grime hidden by the shadows, The Swamp came alive. After a rainfall, an orchestra of invisible toads would start up, their croaks drowning out even the roar of nearby traffic. The crooked silhouettes of bats would wing by as my friends and I sat on the balcony, enjoying the cool breeze and drinking in the good times.

It wasn't just the greenery that made the view from my balcony so impressive; for above The Swamp rose the mountains of Bangkok. Not real mountains, of course—Bangkok is so flat that water doesn't even know which way to flow—but the concrete towers of a brightly-lit boomtown lunging skyward, grasping eagerly for that modern-day manna from heaven, foreign investment, growing day by day as countless cranes bobbed and weaved, adding to their build. It was a mountain range made of cement and glass, an artificial watershed from whose slopes pour forth a mighty torrent of metal, gleaming and honking. Just to the right of The Swamp, a cascade of cars roared down the asphalt canyon that was Sukhuvmit Road—its flow so mighty that even in this most gridlocked of cities, its legendary traffic jams made intrepid motorists tremble.

We lived in a kind of symbiosis with The Swamp. We would flush our wastewater into it, and it would give us back cool air and a beautiful view. But The Swamp was far kinder to us than we were to it. My ruminations of the scenery would occasionally be disturbed by the rustling sound of a plastic bag full of trash whooshing through the air, then landing with a thud. Some of the people living in the shop houses that partially ringed The Swamp considered it to be a garbage dump. Amma, the lady who looked after my apartment block, was one of the worst offenders. I'd taken the matter up with her several times. But she either denied using our backyard as an impromptu landfill or, faced with irrefutable proof of a growing rubbish heap peeking out through the undergrowth next to our building, laughed it off, explaining that she was too tired to take the garbage downstairs.

Anyway, The Swamp was on private land, so we always figured that its days were numbered. Rumor had long had it that a skyscraping hotel would be built on the land. But The Swamp had beaten back civilization before. Smack in the center sat the skeletal remains of an old house, its roof caved in and replaced by a cool canopy of vegetation. Nearby stood several lampposts also smothered in green, as if a couple of arbored ribs had sprouted from the earth. Workers had once set up shacks in The Swamp to live in while they did construction work nearby. We feared these residents would stay permanently, but eventually they moved away, leaving behind as the only reminder of their existence a most appropriate totem—a cement squat toilet. It was soon covered by weeds.

The Swamp, however, finally met its match. One day, a neighbor rushed up to me, eyes gaping in alarm: "Did you see it?!"

"See what?"

"The Swamp! Didn't you look out the back this morning?"

"No. What happened?" But a sickening feeling was already taking hold in my gut.

"The bulldozers, man. They came last night. It's all gone!"

Not quite all, as it turned out. Bulldozers had turned about half The Swamp into mud, and the small building in its midst had been stripped of its green canopy. But who knew what would come next? Rumors flew. It was going to become a parking lot, or perhaps that long-awaited hotel.

We soon found out what came next—a slum. Well, not exactly a slum, but a tin-roofed shanty town inhabited by construction workers building a new hospital wing down the road. At least they weren't erecting a condo on The Swamp itself; that construction would have kept us awake every night for the next year. Nevertheless, we sat down one night on my balcony to drum up strategies for driving the intruders out.

"How about starting up a protest campaign? Bring in the media, student activists . . ."

"Nah. Too many other environmental crises going on."

"I know. We'll get those monks back in here to ordain the trees. We can set up a kind of forest temple or something."

"Yeah. Or we can dress up as ghosts and go down there and spook them out. We'll say we used to live in that old house and were murdered. Now we're coming back to reclaim our land!"

One of the gang got a little overzealous and chucked a rock onto the tin roofs that were now sprouting below. It landed with a thunk, and made a tremendous clatter as it rolled off.

The racket brought us to our senses. It was wrong to take out our frustration on these laborers, who after all were just looking for some honest work. The Swamp's demise wasn't their fault. Whose fault was it? No one's? Everyone's? We realized there was nothing we could do. We were caught in the grip of Bangkok's development and there was no way out. There would be no happy ending.

The birds soon flew the coop. The monks no longer came. We were left with a village of migrant workers who tossed their garbage into what was left of The Swamp. Some of the residents around the area started to move away, and I did too, eventually. The apartment I moved to, off a different stretch of Sukhumvit Road, was also pleasant, quiet, and surrounded by a canopy of trees. It was almost like living in a tree house.

But it couldn't match The Swamp. I recently went back to my old street to see what had become of it, only to find it had been turned into, of all things, a golf driving range. It is a fate so poetically unjust that all I could do was sit back and laugh.

TERRORIZED BY TRAFFIC

There's a reason why Bangkok looks so gray on the outside and green on the inside: Just as they were in The Swamp, most of the city's trees are hidden when you're sitting in one of Bangkok's endless traffic jams, surrounded by clouds of smog and all manner of noisy vehicles—cars, trucks, rickshaws, motorcycles, bicycles, *tuk-tuks,* even the occasional elephant—seemingly moving in every direction. The city's main streets tend to be lined with rows of monotonous gray shop houses interspersed with colorful shop fronts, roadside vendors, open-air markets, and, in more upscale sections, steel- and glass-enclosed office buildings. The Bangkok Metropolitan Administration (BMA), the city government, has made a valiant effort to plant trees along the roads, but even they look gray and forlorn, wilted and weighed down as they are by several layers of dust and soot.

Head off the main streets into a small *soi* (side street), however, and the situation is often completely different. Take the place I moved to following the demise of The Swamp: Soi 27, off Sukhumvit Road. A few hundred meters in, you find yourself on a pleasant, almost suburban-like, road. Even better, it's a dead-end *soi,* and thus surprisingly quiet (and especially prized) because the interminable queue of cars and ear-splitting motorcycles desperately seeking shortcuts know they must search elsewhere. Stroll around these back streets and you'll catch glimpses not just of majestic trees but of manicured lawns and magnificent tropical gardens. Glimpses, however, are all you'll get. Because throughout Bangkok, affectionately known as the Big Mango, greenery is hoarded within private compounds, hidden behind walls designed to keep out prying eyes and would-be burglars. If white picket fences are the emblem of Western suburbia, then the symbols of Third World cities are tall, spiked gates.

For it is the wall between people that prevents Bangkok from becoming a more livable place. Without a sense of community, people won't cooperate to improve their environment, whether that means carpooling, curbing their wastes, accepting zoning rules, or simply disposing of trash properly. And communities are fragile. Difficult to create, easy to destroy, they are developed when humans share a common history or a common area: time and space. Bangkok, in other words, needs to develop its civil society, a key element the world over in achieving a better quality of life. Otherwise, it is left solely to the government—sluggish, corrupt, and much derided—to protect the common environment. Unless pushed by an active civic sector, authorities are slow to take necessary steps, and when they finally do so, they are often accused of acting on behalf of special interests.

Bangkok is a classic boomtown. It is the center of the Thai universe, in some ways a nation apart from the rest of the country. It has Thailand's best schools and its highest-paying jobs; the most powerful officials and the most popular pop singers; the most exciting nightlife and the wealthiest tycoons. People from all over the country, and the world, pour into the city to seek their fortunes. Officially, Bangkok is home to 6 million people, but the population of the greater metropolitan area is probably closer to twice that figure, especially when migrants are considered. The next biggest Thai city, Chiang Mai, contains more than a million people, but it still has the (pleasant) feel of a provincial capital.

And Bangkok is just one of dozens of mega-cities that are springing up all over the global South. They form the destination for the greatest mass migration in human history: the urbanization of the developing world. Half a century ago, two-thirds of the earth's population lived in the countryside, and only greater New York had a population larger than 10 million. Now there are twenty such cities, and in thirty years, two-thirds of the world's population is expected to be urban. Essentially, that means a new Bangkok (at its official population) is being created every two months.[1] The environmental woes of these new mega-cities are legion, and their infrastructure demands enormous. So much so that the problems posed by the lack of a civil society are usually overlooked.

Creating communal feelings is often difficult in Bangkok, and in other rapidly growing cities, too, because many residents simply haven't lived there for long. With economic change so rapid, city dwellers throughout the developing world are even more transient than in the West. Many, perhaps most, of the people living in the Big Mango have migrated there within the last generation or two, predominantly from two places: China and the Thai countryside. You don't find many high walls around dwellings in rural Thailand; instead, they're typically surrounded by modest wooden fences, not just because the countryside is safer but because it traditionally has a greater sense of communal spirit. Farmers had private land to till, but they also benefited from common areas, and from rules on how common resources (water, in particular) could be used. In Bangkok, however, although there are some open-access lands—notably along the banks of canals, where slums have sprung up—there are few common areas to help develop a sense of mutual obligation. Even government land is deemed to belong to a certain agency—the city parks department, perhaps, or the Treasury Department—and not to "the public." Bangkokians have therefore responded to urbanization by building walls, not only because of the insecurity bred by a transient, disparate society, but also because city people simply have a different way of looking at the land.

Without all these walls, it might be possible to imagine Bangkok as it was originally: a massive flood plain for the Chao Phraya River—a wetland. It used to be largely under water during the rainy season, and parts of it still are. When the Thai kings moved their capital downriver to this location more than two centuries ago, they modeled the city after their previous, storied capital of Ayutthaya, which had been ransacked by the Burmese. Because it was so much easier to get around by boat, it was only natural that the city's founders created a city laced with canals, an international entrepot dubbed the "Venice of the East." The canals still exist, but they have been turned into sewers, and many are covered by pavement.

That, too, is a symbol of the one object that more than anything or anyone else—more than any king, or prime minister, or mayor, or urban planner—has utterly transformed the face of Bangkok: the automobile. Philip Blenkinsop, an Australian photographer who has lived in Bangkok for many years, has published a brutally graphic book called *The Cars That Ate Bangkok.*[2] The title is appropriate. This human artifact has had such an enormous impact on the formerly bucolic city that it's tempting to describe it as a rampaging monster, able to command an entire society to do its bidding: be it building roads, demolishing communities to make way for expressways, or waking up at four in the morning to get your kids to school ahead of the morning traffic jam.

Of course, environmentalists have long demonized the private car, but all the shouting has done little to dissuade people that for comfort and convenience, the automobile is an unrivalled form of transport. Thais absolutely love their cars (and motorcycles, and pickup trucks). They adorn them inside with garlands, pillows, and assorted decorations. They keep the outside spotless, washing it repeatedly. Wealthy families aren't just satisfied with two cars; they may have half a dozen, some for the kids and relatives and servants. And these aren't little urban runabouts; the sedans you see on the street are all four-doors. Pickup trucks are even more popular—Thailand is the second largest producer in the world, after the United States.[3] And cars are important status symbols; despite Thailand's being a developing country, before the 1997 financial crisis, Thais purchased as many as 13,000 Mercedes-Benzes a year.[4]

The car has not been kind to Bangkok in return. All semblance of Venice has now gone; in its place is a sprawling megalopolis closer in spirit to Los Angeles or Houston. In pictures taken from orbiting satellites, the city appears unnervingly like a giant tumor spreading out remorselessly into the surrounding green paddy fields. It has become dominated by its gridlock, indeed, famous for it. Traffic is the number one topic of conversations. Strangers getting to know each other over cocktails chat about how bad a certain intersection is, or about the recent horrendous accident they witnessed, in much the same way people elsewhere discuss the weather. The situation was particularly bad in the early 1990s, when foreign journalists looking for a story would inevitably turn to the old standby about "Life in the Slow Lane," pointing out that the average speed of a car traveling Bangkok's streets was 12 miles per hour,[5] and less than half that during peak hours (incredibly, that is still better than the average traffic speed in Hong Kong, Taipei, Bombay, and Manila during rush hours, or rather, crawl hours).[6] More than 70 percent of commuters spend more than two hours a day mired in traffic jams, and a fifth of all drivers say they spend at least three hours commuting. Altogether, the average Bangkokian is estimated to spend the equivalent of around forty-four days a year on the road.[7]

The gridlock has an enormous environmental impact. The stalled cars, along with the stop-and-go traffic, contribute significantly to Bangkok's serious air pollution problems. But it also affects city residents in ways we rarely think about. Take noise pollution, for instance. The din from vehicles is so loud that it's often impossible to carry on any kind of conversation while walking down a major Bangkok thoroughfare. In 1996, Thailand's Pollution Control Department (PCD) listed a dozen major roads where noise levels regularly exceed the U.S. Environmental Protection Agency (EPA) standard of seventy decibels. And this cacophony has a serious effect on people's health: A survey of traffic cops, for example, found that more than a quarter of them had suffered hearing loss.[8] The economic impact of Thailand's traffic problems is also enormous: Bangkokians lose an estimated $2.4 billion annually in wasted fuel, time, and medical costs due to the traffic.[9]

But mere numbers don't do justice to the insidious social impact of the city's gridlock. Imagine what life is like for schoolchildren, having to eat your breakfast in the car so you can get to school on time, or having to sit on a hot, crowded bus and inhale clouds of smog for hours as you make your way home. For adults, the traffic robs life of its spontaneity; you can't suddenly decide to meet your friend somewhere in fifteen minutes; you have to plan everything—how far you have to go, your route, the time of the meeting—well in advance to avoid agonizing delays. Crossing the city, even escaping it altogether, becomes an enormous hassle. In effect, your world becomes radically smaller, and the prime advantage of urban life—being in close proximity to interesting people and places—becomes negated.

And yet, life goes on. When discussing all the problems of Bangkok and other mega-cities, it's tempting to speak in apocalyptic terms. But people in general, and Thais in particular, are amazingly practical. They've found ways to cope with the traffic. Mobile phones, for instance, have become a vital appliance for commuters; they can while away the time in traffic by flirting, doing business, or chatting with friends and relatives. Major intersections, furthermore, are now home to troupes of young motorcyclists who ferry passengers around for a reasonable fee. It can be a hair-raising ride, weaving in and out of stalled traffic, sucking in noxious fumes spewed out of ramshackle buses a few feet away; but if you want to get somewhere during rush hour in a reasonable amount of time, it's the only way to travel. The service has proved incredibly popular. It has spawned an industry that in 1994 was worth $200 million, and a motorcycle livery fleet estimated at 40,000 vehicles.[10]

As if helping to keep Bangkok moving were not enough, one motorcycle taxi driver actually deserves credit for having thwarted an act of global terrorism. In March 1994, on a typically crowded afternoon along Soi Chidlom, right in the middle of town, a truck was making its way out of the Central Department Store parking lot when it smacked into a motorcycle taxi parked on the corner. The truck driver, described as an "Arab-looking man,"[11] was prevented from departing the scene by the irate owner of the bike, a young lad who proceeded to demand compensation. Heated negotiations ensued for nearly an hour, but apparently to no avail. The truck driver grew increasingly agitated and finally fled the scene on foot. The truck, a rental hire, was eventually towed to a police impound, where it remained untouched for more than a week until the owner finally came to claim it. Someone then noticed a peculiar smell emanating from the rear of the truck. Investigation revealed the stench came from the body of the truck's original Thai driver, who had been murdered and stuffed in the back, along with some explosives, a tank filled with a hundred liters of diesel fuel, and large amounts of fertilizer—a similar concoction to that which had exploded under New York's World Trade Center a year earlier. The bomb makers had been planning to pay an unannounced visit to the Israeli embassy just three hundred meters from the department store. A couple of Iranian suspects were eventually arrested. The terrorists had been foiled, thanks not to Interpol or the Thai police, but to Bangkok's insurmountable traffic.

THERE'S NO PLACE LIKE A MOBILE HOME

How did Bangkok get itself into such a terrible mess? Automotive technology and its support system of roads, expressways, petrol stations, and refineries has transformed landscapes and cultures the world over, but somehow the impact seems especially dramatic on Thailand.

Part of the reason is that Bangkok was ill suited for the automobile. The yearly floods—made worse by land subsidence, a result of the pumping up of groundwater by industry—cause havoc with the traffic flow, creating jams that can extend as far as 50 kilometers. The city's traditional road system also aggravates matters. Roads account for only about 7 to 8 percent of the city's terrain,[12] roughly one-half to one-third the average figure in modern Western cities. And although road capacity is increasing at a rate of 1.5 percent every year, the capital's car population during the boom grew at an average rate of 12 percent.[13] Overall, between 1960 and 1993, automobile ownership in Bangkok grew roughly sixteen-fold, well outpacing most other Asian cities.[14]

What's more, Bangkok's roads are not laid out in the grid system favored by Americans steeped in the rationalist tradition. Instead, they follow a hierarchical pattern that eerily mimics patron-client relationships in traditional Thai society. Power in Thailand has historically been concentrated in the hands of a small elite, each of whom enjoys an extensive network of supporters providing allegiance in exchange for patronage. Similarly, Bangkok has a few broad avenues that randomly criss-cross one another; these are interlaced with dozens of small, twisting feeder roads and *soi*s (side streets). Perhaps the best analogy is to compare the roads to rivers. The *sois,* which no doubt were modeled after the pattern of canals that existed before, act as tributary streams, an appropriate system given the country's social and natural history.

Of course, many old cities with winding narrow roads have struggled to adjust to the automobile, particularly in Europe. In London, it's said that the average speed of cars moving during rush hour in the year 2000 (from 10 to 12 miles per hour) is equal to the average speed that horses and buggies moved about town back in the year 1900.[15] But the situation in Bangkok is exacerbated by the strength of the auto lobby. Thailand aspires to be the "Detroit of Southeast Asia," an auto production center for the entire region, and has successfully wooed foreign manufacturers with tax breaks and subsidies. Thailand is now the largest producer and consumer of vehicles in the region.

Thailand, in other words, is wedded to the automobile for the long term, and that means the industry has clout; this clout in turn can affect policy in subtle (and not so subtle) ways. One obscure but telling example is the success the industry had in lobbying the Thailand Industrial Standards Institute regarding vehicular emission standards. In 1992, their ability to delay the onset of stricter European-based standards for car exhaust even managed to frustrate one of the architects of Thailand's industrialization policy, Piyasvasti Amranand, the tough-talking director of the National Energy Policy Office (NEPO): "They [the committee members] are not balanced.

They only look after the interests of manufacturers. No one on the committee is out to fight for clean air."[16]

Meanwhile, the demand for cars is heavy,[17] stoked not only by the innate popularity of having your own set of wheels, but also because the infrastructure for alternative means of transport has simply been neglected. Although new expressways were going up throughout the 1990s, vital mass transit rail projects have suffered from a lack of funding and have been delayed for decades because of political squabbling. Construction on the first light rail line did not begin until 1995.

This situation points to the biggest problem facing Bangkok: a lack of proper planning. Even the most farsighted technocrats may have been unable to meet the demands of a mega-city swamped with migrants and booming with double-digit growth rates, but the authorities who oversaw Bangkok were woefully unprepared. Not only did they fail to provide the necessary transport infrastructure but, when the green light was finally given to three mass transit rail projects in the 1990s, each was supervised by different agencies with little coordination to make the systems link up. The massive Stonehenge-like cement towers lining the road to the airport—the relics of a road-and-rail infrastructure project, known as Hopewell, that was scrapped after it was only 15 percent complete—stand as testament to the hodgepodge nature of city planning in Bangkok. When Chatichai Choonhavan, a former Thai prime minister, was driving past them one day in 1997, he remarked, "Wow! These piles are just huge. What're they for?"

His son Kraisak responded, "Don't you remember, Dad! It's the project you approved."

"Oh! Really?" was Chatichai's reply.[18]

The city's zoning regulations, or lack thereof, are another telling example of Bangkok's management failures. The main legal planning instruments, the Land Subdivision Act of 1972 and the Building Control of 1979, were regularly manipulated to allow developers freedom to do as they pleased, the first law repeatedly being amended to *reduce* requirements for infrastructure provisions. Regulations have apparently been strengthened in recent years. But during the real estate boom of the late 1980s and early 1990s, huge condominiums and office towers sprang up on tiny back streets, and particularly along popular thoroughfares already groaning with traffic, making them almost impassable during rush hour. Accounts of how the new central business district in the Rama III area was planned in the 1990s are particularly instructive: Zoning rules were swiftly changed to allow for commercial development, and developers had clear precedence over city officials in drawing up land use plans.[19]

Meanwhile, Bangkok has only partially installed an automated traffic light system. Policemen sitting in little booths still switch the lights manually at most intersections. Suspicious minds believe this is so that VIPs can retain the privilege of having police motorcades whisk them through the snarls. Adding to the confusion are the seemingly continuous installation and repair projects carried out by the telephone, electricity, water, sewage, and city administrations: A lack of coordination results in a veritable tag team of contractors tearing up and filling in the roads.

Bangkok's traffic crisis actually serves as a distressing metaphor for a broader management crisis in Thailand, one that extends to every facet of the country's development. Think of road space as an "open-access" resource: Apart from a few privately built tollways in Bangkok, people can use the roads freely, the result being massive congestion. Other open-access resources such as fish, air, and water are treated in a similarly profligate manner, and the result is they all too quickly become depleted or polluted. Just as developers in Bangkok were allowed to build as they pleased, putting incredible traffic pressure on tiny streets, farmers living alongside rural canals take as much water as they can pump, and wastes are released in a wanton fashion into the air and streams. Environmentalists call this phenomenon the "tragedy of the commons."

Once the canals have run dry or the roads become clogged, the government's first impulse is always to try to increase supply by building more dams and expressways—a more popular policy than trying to reduce demand. There is reluctance to acknowledge that since resources are limited so are supply-side options. The result is drought and gridlock.

Corruption, lack of enforcement, and planning problems mean that most developing countries in practice follow a similarly haphazard approach to urban development and resource management. China, and other more authoritarian states, tries to restrict migration to the cities, but even there it's difficult to control. Chinese leaders often look to Singapore as a model. That city-state is amazingly well run thanks to its strong commitment to planning and its government's strict enforcement of zoning rules. It is clean to the point of sterility, and safe for the whole family, a kind of Asian Disneyland. Food vendors have been herded into hawker centers, and the red-light district has been cleaned up. Singapore's extensive subway system has allowed the establishment of a sensible traffic-management system in which drivers must pay a toll to gain access to the city center during rush hour. But Singapore has some built-in advantages over its boisterous neighbors: It does not have a countryside that sends in waves of poor migrants. And because the vast majority of its people live in condominiums bought in government-built high-rises, the government can rigorously control where development takes place.

Malaysia's example may be more intriguing, since it does not share those advantages. The country does suffer from corruption, but its civil service—its forest department, for instance—tends to be more professional than similar bureaucracies in most other Southeast Asian countries. Malaysia also has a federalized political structure that has helped spread out urban development in cities such as Kuala Lumpur, Johor Bahru, and Georgetown. These cities do have traffic and pollution problems, but nothing like those found in Bangkok, Jakarta, and Manila.

For all the hassles brought on by Bangkok's "arterial sclerosis," it has given the city a perverse identity. It was often said, for instance, that when one of the country's innumerable military coups occurred, you knew that it was a "good" coup or a harsh one by whether the tanks rolling down the streets stopped at the traffic lights. And as far back as 1975, when the Vietnamese army was triumphing in Cambodia, it was joked that the only thing preventing it from marching all the way to Singapore was Bangkok's gridlock.

Thais, as these quips suggest, are a famously good-natured people. They accept vicissitude with a shrug and a smile, or a knowing roll of their eyes. The country's favorite expression is said to be *mai pen rai,* which roughly means "never mind." True, the booming metropolis of Bangkok, a city ever ready to ignite your passions, is faster paced than the countryside. But unlike the residents of most capitals in Asia and the West, Bangkokians have managed to retain a relatively laid-back attitude, perhaps out of necessity. On some level, you must learn to laugh at the gridlock if you're to avoid crying about it, or flying into fits of road rage.

For visitors and foreign residents, Thailand's laid-back attitude can make life there alternately blissful and exasperating. On the one hand, the attitude has helped turn the country into a popular vacation destination, whether you're looking for *sanuk* (fun) or to be *sabai* (comfortable or relaxed, although those translations don't do the word justice). Thais are also generally forgiving when you commit a cultural gaffe or utter some unwittingly insulting remark. On the other hand, many Thais admit this sense of fatalism contributes to the country's problems because authorities are simply not held as accountable as they should be. There seems to be a general sentiment that Bangkok has always had terrible traffic, and always will.

GETTING THE LEAD OUT

Unfortunately, no one seems to have come up with any silver linings to the clouds of pollution spewed out by all that traffic. The air quality in Bangkok is so bad that a team of researchers from the University of Hawaii who came to study air pollution in the early 1990s actually fled the city and refused to return after realizing the significance of their initial measurements. The World Health Organization (WHO) reported in 1996 that Bangkok's air was fourteen times dirtier than the international standard, and simply breathing it was considered a health hazard to children not yet twelve years old.[20] You only need spend a few minutes walking on a main street or standing at a major intersection before you start to feel grimy, and a motorcycle ride through rush hour traffic can leave you nauseated. Even if you live on a secluded back street, you still have to travel along the main roads where the smog is thickest. And while those fortunate enough to travel in air-conditioned vehicles certainly feel more comfortable, studies have shown that they remain exposed to the smallest and most dangerous airborne particles. Under these conditions, even exercise can prove hazardous to your health because you end up inhaling greater quantities of noxious gases more deeply. Running for thirty minutes in a polluted environment is the equivalent of smoking a pack of cigarettes.[21] Air pollution, therefore, is much like the traffic: You just can't escape it.

There is an important difference between the two, however: Air pollution has a direct and dramatic effect on public health. Of all Bangkok's environmental problems, air pollution is probably the most dangerous. According to the World Bank, it is responsible for at least 200 to 400 premature deaths in the city per year,[22] and a U.S. Agency for International Development report estimated that it could cause as many as 1,400 deaths per year.[23] The toll is particularly high for those who work on the streets, for instance, traffic cops—half of whom suffer chronic respiratory problems—and roadside vendors.

Indeed, air pollution afflicts mega-cities around the world. In Jakarta, the premature death toll as a result of exposure to particulates is estimated at 4,000 per annum (and a further 1.5 million asthma attacks).[24] In China, thought to have Asia's worst air pollution, the World Bank estimated it caused 178,000 premature deaths and 1.7 million cases of chronic bronchitis among urban residents in 1995.[25] This public health crisis also exacts an economic toll on Asian cities: In China alone, the World Bank estimated the damage totaled $32 billion in 1995, or roughly 5 percent of the GDP (gross domestic product).[26]

Small particles of airborne dust, which sound annoying rather than terrifying, are actually considered the most damaging pollutant because they're carriers for disease-causing bacteria and fungi, and easily find their way into people's lungs. A 1996 study by Thailand's PCD found that airborne dust was responsible for the allergies and upper respiratory infections suffered by more than a million people in Bangkok.[27] Dust levels in some parts of the city were measured as being more than six times higher than the international safety standard. Roughly 40 percent of Bangkok's airborne dust comes from vehicular emissions, including the city's 3 million motorcycles, 90 percent of which use two-stroke engines; but efforts to phase them out in favor of cleaner-burning four-stroke engines have proved difficult to enforce. Construction work accounts for another 40 percent of Bangkok's dust. In 1995, during the building boom that took place before the financial crisis, there were at least 3,000 construction sites in Bangkok (and another 1,900 developers applying for new construction permits). The remaining 20 percent of the dust stemmed from the estimated 20,000 factories located in the metropolitan area. A particularly macabre source are the five hundred crematoria located in Bangkok, each of which burns from three to five bodies in coffins every day. Ashes to ashes, dust to dust, indeed.

Airborne lead is another insidious pollutant. Its effect on children is particularly nasty because it causes a decrease in intelligence, hearing loss, hyperactivity, and aggression. The World Bank found that in the early 1990s, the average blood lead levels of Bangkok's residents measured an incredible 40 micrograms per deciliter,[28] more than ten times the average level in the United States, and well above the levels found in other mega-cities such as Cairo and Mexico City. Thai children were losing an average of 3.5 points off their I.Q. tests as a result of breathing Bangkok's air in the early 1990s,[29] and similar figures were reported in Manila.[30] Indeed, the World Bank says that all urban children in the developing world who are not yet two years old are thought to have excess lead in their blood, and an estimated 15 to 18 million children may have suffered permanent damage as a result.[31] Grimly, in the early 1990s Bangkok and Manila hospitals were even finding high lead levels in the umbilical cords of newborn babies.[32] And lead contamination affects adults, too: It has caused more than 200,000 cases of hypertension in Bangkok and Jakarta annually.[33]

There is no mystery about the source of this airborne lead contamination: It comes from cars using leaded gasoline. Most people in the West have forgotten how important it was to switch to automobiles using catalytic converters and unleaded petrol, but in Thailand the conversion did not start until 1992 (and Vietnam didn't go through with the shift until 2001). Even then, there was surprisingly strong resistance: from refiners and gas station owners who found it inconvenient; from motorists who feared it would ruin their cars' engines; and from businessmen who claimed the license to produce catalytic converters had gone to a friend of former Prime Minister Chatichai. The government came up with an effective solution: It subsidized unleaded gasoline, making it slightly cheaper than the leaded version, thus persuading car owners to purchase it. As is true of so many other environmental reforms, fears that a technological or economic calamity would ensue were not borne out. The carping soon died down. It's now difficult to find a service station in Bangkok selling leaded petrol, and blood lead levels of Bangkokians have decreased significantly. By 1996, the average IQ of children in Bangkok had risen four points as a result of declining levels of airborne lead, the World Bank reported.[34]

So there is hope, and perhaps the biggest hope for solving both the traffic and air pollution problems is to provide more mass transit options to the public. Bangkok's first overhead rail system finally opened in December 1999, but so far it consists of only a couple of lines totaling 20 kilometers. For the vast majority of city residents, therefore, the only alternatives to the private car are other forms of motorized transport: taxis, shared taxis, motorcycle taxis, and the public buses.

MOVING CARS OR MOVING PEOPLE?

Taking a bus is a good way to people-watch, and that is usually how I got to work, commuting daily from Sukhumvit Road out to the *Nation*'s* office on a superhighway at the easternmost edge of Bangkok. It meant passing through some awful gridlock, including the dreaded Bang Na intersection, which could take an hour to get through during particularly bad times. But I was lucky in that I generally went to the office in the early afternoon and left late at night. Unless I had a morning appointment for an interview or a conference, therefore, I usually managed to miss Bangkok's extended version of rush hour. In addition to keeping me sane, it made taking the bus more feasible, and I could usually get to the office in less than an hour, not bad for Bangkok.

---

*All references to the *Nation* in this book are to the English-language newspaper in Thailand.

The city's bus system is actually quite impressive when you consider that it covers the vast metropolis fairly well and offers a wide range of services. At the cheap end of the scale are the careening green "baht buses" with their tiny seats and cramped aisles. A couple more baht will land you a spot on the rumbling *rot mae,* the sixty-seat mainstays of the Bangkok bus fleet. A step up in class is the air-conditioned *rot air,* a comfortable ride if you can get a seat, but it will set you back from 6 to 12 baht (roughly a quarter or two). Finally, there is the Microbus service, the Cadillac of Bangkok bus fleets, introduced in the 1990s. Although these buses are compact like the baht buses, that is all they have in common; for a little more than a dollar you're guaranteed a seat in climate-controlled comfort.

Each service not only caters to a different class of clientele, but also apparently has a different standard for driving skills. The baht-bus drivers place a premium on speed—and often seem to be on speed themselves. Their working-class passengers exhibit true Thai fatalism by entrusting their lives to the reckless maniacs at the wheel. The more expensive the service, the more responsible the drivers become. Best of all are the Microbus chauffeurs, who can actually be relied on to stop and pick up passengers trying to wave them down. The same cannot be said for the other drivers. There is nothing more frustrating than seeing your ride roar by with a shrug and a wave from the driver. Stories circulate about one would-be passenger who became so incensed after being spurned repeatedly that he waited until the same bus came back on its route and proceeded to lob a grenade on board in the direction of the driver.

The privately run baht buses do have one advantage over the other services, however: They actually make money for the operators. The state-run Bangkok Metropolitan Transit Authority (BMTA) has piled up huge debts in running the public fleets because of the government's demands that fares be subsidized and the agency's own chronic mismanagement. The clearest illustration of these twin problems can be seen in the murky clouds of black soot that typically spew from the tailpipes of Bangkok's buses. The situation is unfathomable, even by Bangkok standards. That a bus system supposed to help solve one set of urban ills should be allowed to exacerbate another is so illogical that at one point it was investigated.

The source of the problem turned out to be the maintenance subcontractors, whose primary goal is to keep buses on the road 90 percent of the time. Environmental performance is way down on their list of priorities. So when drivers—who surely face one of the most harrying jobs on Earth in trying to maneuver their clunky vehicles amidst rush-hour madness—ask for more horsepower to get them past the jams, mechanics oblige by altering the size of the fuel injectors, or adjusting the fuel pumps to increase the amount of gasoline flowing into the system. This provides more oomph, but not all the petrol is burned. Most of the residue is belched out to form those distinctive sidewalk-to-sidewalk plumes of smog. The remainder sticks to the exhaust pipe, causing back pressure to build up and ultimately decrease the buses' power. The operators no doubt realize they are wasting lots of fuel. But why should they care? The BMTA doesn't pay for the petrol anyway. The government provides it free of charge as a subsidy—a small but telling example of how subsidizing the use of resources can be disastrous for the environment.

Given the state of transportation in Bangkok, the key question is this: If the government is willing to subsidize the bus system, why was it so unwilling for so long to spend money on an urban rail system that would presumably be faster and cleaner? The short answer is that government policy seems aimed at moving vehicles rather than people. For decades, the authorities claimed it was simply too expensive to build a subway. The swampy conditions underlying the city, they added, would make it susceptible to flooding. So subsequent governments usually looked at building an elevated rail. Numerous systems were proposed over the years, and entire books could be written to explain why they foundered. Even when construction of the so-called Skytrain finally began just eight years ago, it was the Bangkok Metropolitan Administration, not the national government, that pushed it through.

More to the point, it did so without public funding. The BMA awarded a concession to build and operate the system to Tanayong, a major real estate developer. Backed by private financing, the firm hopes to recoup its costs by collecting fares and exploiting commercial opportunities within and around the stations. It is unlikely that Tanayong will be able to do so, however, and the system may eventually be taken over by the state. Meanwhile, a vigorous campaign against the Skytrain led by Chodchoy Sophonpanich, a Bangkok environmentalist famed for spearheading anti-litter campaigns, argued (controversially) that it would destroy neighborhoods and (correctly) that the project had never been subject to an environmental review. Chodchoy, who was widely criticized for her opposition, claims she just wanted the project to go underground.[35] However, only after work had begun did a new (national) government announce that all future mass transit rail lines in Bangkok would be subways, which would cost about twice the price of an overhead rail line. It remains unclear why the government finally caved in and decided to make its funds available, but Chodchoy's campaign undoubtedly had something to do with it.

The key difference between a bus system and a subway, therefore, is that politicians considered a subway an infrastructure project. That thinking leads to a double standard, because although the mechanisms for building infrastructure may vary—some are privatized, some are run by state enterprises, and still others are built by private firms and then given to the state—Thai governments have felt they should make money on such projects, or at least break even. That was especially so under the privatization policies that dominated the 1990s. No such demands are placed on the BMTA, which operates the bus system. Delaying matters even further is the tendency of corrupt Thai politicians to seek kickbacks from infrastructure projects.

The many factors behind Thailand's mass transit negligence include squabbling by politicians about the spoils of contracts; a general lack of foresight and enlightened leadership in the upper political echelons; and the short-lived nature of governments, a chronic obstacle to any environmentally beneficial policy because the benefits are likely to emerge only long after the leaders currently in power have gone. But once again, the lack of a sense of community must bear some responsibility. Bangkok shares a strong individualistic, libertarian ethos with other cities—Los Angeles, Houston, and Phoenix come to mind—that prefer to develop around the automobile. Sure, middle-class Bangkokians say they want a subway . . . but deep down most of them want *other* people to ride on it, which, they hope, will free the roads for themselves!

ROAD KILL

Through all this discussion about transportation alternatives in Bangkok, one means of moving around is inevitably overlooked: walking. With good reason, perhaps, for if Bangkok is an urban jungle, then pedestrians are at the bottom of the food chain. Zebra crossings in the city are so faded that they have become an endangered species, and they are treated with such disdain by Bangkok's drivers that pedestrians who actually dare to use them risk ending up as road kill. Meanwhile, city residents who brave the footpaths—where they exist—find they must share their sidewalks with roving vendors, whizzing motorcyclists, and the occasional parked truck. Merely avoiding the chopped-up paving stones and treacherous open manholes requires feats of agility that would test the Artful Dodger.

So, to coin an old phrase, there are just two kinds of pedestrians in Bangkok: the quick and the dead.

Considering all of Bangkok's traffic problems, you'd think the Bangkok Metropolitan Administration (BMA) would encourage people to walk as an alternative to using motorized transport, at least when taking short trips. But Santi Ruangwanit doesn't see things that way—and his thoughts count because when I interviewed him he was the director of the construction and maintenance division in the BMA's Public Works Department, responsible for maintaining the pavement: "Thais don't like to walk because it's too hot," he said. "In Western countries, it may be very cold, but at least walking warms you up."[36] With such an attitude—Santi hastens to add that he often walks—it is not surprising that Bangkok's footpaths more closely resemble obstacle courses. In the suburbs, there are few footpaths at all, even though they are supposed to be built alongside all new roads. According to Santi, they're often forgotten because "few people walk" in the suburbs.

Bangkok's pedestrian policy, or lack thereof, is "a matter of priorities," explains Thongchai Panswad, a professor of environmental engineering at Chulalongkorn University. "Building footpaths is too small a contract, so government officials say 'just let the contractors do it.' Roads are public property, too. Why is the government able to maintain them? Because officials tend to think more costly projects are more important."[37]

This brings up what is perhaps the most fundamental question about Bangkok's numerous ills: Why do people put up with it? When asked in polls and informal interviews, they generally tend to blame government for failing to manage the city properly, and failing to solve environmental problems in general—it's human nature, after all, to avoid responsibility for one's own contribution to the situation. But that still begs the question of why they tolerate this negligence. Thailand's political system has many flaws, but it *is* a democracy. The people elect their members of Parliament (M.P.s), and thus indirectly their prime minister, and in Bangkok they can elect their mayor. So why do they accept all the environmental abuse?

Bhichit Rattakul thinks he knows part of the answer, as he explained when describing why he set up his Anti–Air Pollution & Environmental Protection Foundation. "I started the drive after visiting a *mae kha* (vendor) in Huay Kwang. A bus started off, spilling out black smoke. I nearly choked, but the *mae kha* didn't feel anything. 'Don't get excited,' she said. 'Yesterday was like this, today and tomorrow will be the same.' That's the problem. People don't think it's abnormal. It's not their fault. Nobody has told them it doesn't have to be this way."[38] Bhichit is not just blowing smoke here. As a successful politician who served as an MP for eight years and also the scion of a famous political dynasty, he understands public sentiment. Bhichit also understands the technical issues involved because he's a trained scientist (he holds a Ph.D. in microbiology), and a former lecturer at Chulalongkorn University, considered the Harvard of Thailand.

But although Bhichit wears many hats, his most famous adornment is a surgical mask. "We try to tell people to wear masks. We distribute them to policemen, *tuk-tuk* drivers, bus drivers. We tell them it only offers partial protection—for dust, but not for gases. But it's the only way to let them know that this is an abnormal situation. It's a way of saying to the government, 'Shame on you, because citizens have to protect themselves.'" Bhichit's anti–air-pollution foundation has been active in other ways, too, putting up signs around the city explaining what hazardous substances are in the air, where they come from, and the government agency responsible for controlling them. In 1991–1992, it filed petitions urging many of the reforms—introduction of unleaded gasoline, mandatory installation of catalytic converters, tougher sulfur-content rules for diesel fuel—that were eventually passed by the Anand Panyarachun government.

Most remarkable of all, in 1996, Bhichit won an election to become Bangkok's governor (Bangkok is a province as well as a city, but his actual powers are probably less than those of a Western mayor). Debates have raged about how much he achieved while in office and how much of his green message was actually just public relations. But he had some innovative programs; he took on the powerful construction industry to create "dust control zones" that helped reduce air pollution, and helped to green the city with a street-side tree-planting campaign. His most symbolic victory came in facing down Bangkok's powerful golfing community to turn the State Railway Authority of Thailand golf course—the only one located near the city center—and create a badly needed public park. But as governor he did not have the power or money to launch sizable infrastructure projects, and the limit of his gubernatorial powers became evident when he was unable to force the police (who come under the authority of the Interior Ministry rather than the BMA) to install a computerized traffic light system.

Bhichit seemed so frustrated by the time his term was up in 2000 that he decided not to run again. Nevertheless, that an environmentalist such as Bhichit, running on an anti–air pollution platform, could be elected governor of Bangkok represented an important event in Thailand's political evolution. Before his victory in 1996, democratically elected officials in Thailand had ignored environmental concerns unless a crisis was involved. A ban on logging passed in 1988 was ordered by then Prime Minister Chatichai only after a landslide on a denuded hillside had killed more than a hundred villagers in a southern region. The Anand government of 1991–1922 passed a vital and comprehensive suite of legislation and regulations that included the establishment of permanent environmental agencies and the championing of lead-free petrol—but that was a short-lived government of technocrats installed by the leaders of a military coup. So the direct election of Bhichit marked the first time in Thailand's history that people had actually made their environmental concerns a decisive issue in the way they cast their ballots.

That's important because the main reason—apart from ignorance—usually proffered for why Bangkokians, and people throughout the developing world, have put up with a lousy environment is that they are willing to make immense sacrifices to improve their incomes. A common argument holds that protecting the environment (along with other social goods) is a luxury that people support only after securing their financial well-being. By working in sweatshops, traveling abroad as laborers, and selling themselves (or their daughters) as prostitutes, millions of Thais have shown they are willing to risk losing their families, their communities, their health, their very lives in their attempts improve their standard of living, or at least to send their brothers to school. Of course, many Thais choose not to make those sacrifices, or feel they are unnecessary. You'll find refugees from Bangkok all over the country, people who have fled the big city to seek a better quality of life in the greener, cleaner pastures of the countryside. But there is no doubt that Thais want development, however you choose to interpret that word. In movies and on television, they have seen the wealth of the world and all that it can bring, and they want their share of it.

We in the media often fail to realize our influence in such matters, whether overt or subliminal. As an environmental reporter, I am often reminded by critics of how many forests are cut down to make the newsprint on which my stories are published. But perhaps an even greater impact stems from our core business: advertising. My salary at the *Nation* was supported by full-page real estate ads promising luxurious new lifestyles in the latest housing estates.

I am reminded of my first trip to Laos in the spring of 1990, when I spent a blissful week in the sleepy village of Vang Vieng, a few hours north of the capital, Vientiane. The World Cup was in progress, so at night I would wander down to the tattered local saloon, which featured a cranky old black-and-white television that managed to get decent reception from Thailand. It was odd enough to sit in a wooden Lao shack amidst the rice fields and watch football being played on the other side of the world. But most surreal of all were the Thai whiskey commercials, full of handsome young men in shiny new cars and gorgeous young ladies bedecked in jewels and finery. Looking around at the local lads sucking down their Beer Lao, I wondered how could they not be titillated? And so it is no surprise that people in developing countries all over the world uproot themselves to seek a better living. Even if their village still has abundant resources and a close-knit community, they naturally take these conditions for granted, having never been without them.

A city such as Bangkok is certainly the cure for such naïveté, but don't forget it is also a tremendously exciting place to be. For all the problems rapid growth caused in Thailand during the boom decade of 1987–1997, it also inspired an amazing vitality, particularly in the Big Mango. There was a tremendous energy about the place that was almost palpable, an optimism you could virtually tap into. The vast increases in wealth spurred a materialist frenzy in Thailand and created an entire class of nouveaux riches. Whereas in 1985 the city had only two small department stores, ten years later it had sixty,[39] and it often seems the main leisure activity is to visit the latest new mega-mall and shop at the latest name brand stores. But the boom brought more than just material gains: It brought more schooling and better educational standards, improved access to medical care, an ability to travel and see the world. In short, along with wealth and excess, it brought opportunity and hope.

MOVING UP THE KUZNETS CURVE

Bangkok's development has been remarkable. Where once there was only swamp now stand great mountains of steel and cement. But instead of a garden city, it is a walled city, its homes separated by imposing fences; its neighborhoods fragmented by a transient, mobile citizenry; its commuters stalled in their own private passenger vehicles; its communities suspicious of official intentions; and its nascent civil society struggling to overcome all these barriers. The people of Bangkok must find a way, both literally and figuratively, to take down these walls. Because while the view is great from the mountaintops, we can't all live there.

So it's not surprising that the excesses of the boom also brought a demand for better urban and environmental management. The election of a green campaigner like Bhichit as its governor suggests that Bangkokians realize they don't have to sacrifice their quality of life for the sake of development. There are other indicators. A poll carried out in 1996 by Japan's

Institute for Developing Economies and a Thai market research firm revealed that 85 percent of young respondents in Bangkok considered environmental protection more important than economic development, and 85 percent of respondents in all age groups found an environmental tax acceptable.[40] Meanwhile, some Bangkokians have stood their ground and refused to sacrifice their homes for development. The two-hundred-year-old, predominantly Muslim community of Baan Khrua, for instance, has fought a tense tooth-and-nail battle for years with powerful business interests seeking to build an expressway interchange through their neighborhood.

Most Thais still look to the government to take action. To make further progress, Bangkokians will have to make a few sacrifices of their own to create a better environment, whether that means taking the train to work or paying to have their sewage treated. But electing a governor who sought to promote the civil society Bangkok so desperately needs could be seen as an important first step. Bangkok may have turned the corner; progress may not be even—the financial crisis didn't help, nor did the decision of Bhichit not to run for reelection—but the city's environment is gradually improving and, so long as the economy doesn't stall completely, should continue to do so.

Academics who study societies undergoing economic growth have noticed they tend to follow something called the "environmental Kuznets curve,"[41] which charts the recurring pattern of resource abuse that occurs in societies around the world. As countries undergo the initial stages of industrialization, environmental indicators—air quality, water quality, forest coverage—generally deteriorate. Then, a turning point is reached and many of these indicators gradually start to improve, although there are exceptions (carbon dioxide emissions, for instance, have continued to climb). Preindustrialization conditions may never be fully restored, but sometimes they can at least be approached.

The Kuznets curve is controversial with many environmentalists because of the prevailing assumption, particularly among leaders of industrializing countries, that societies *inevitably* follow this pattern and it is therefore okay to put off attempts at cleaning up.[42] Under this traditional development paradigm, the curve becomes a self-fulfilling prophecy—a result of the tendency to separate economic and environmental concerns, reflected most fundamentally in the way statistics are kept. GDP numbers account for the benefits of economic activity, but ignore environmental and public health costs. A car crash, for instance, generates lots of economic activity by sending people to hospitals and spurring them to buy new vehicles. Similarly, a polluting factory contributes its basic production to economic statistics, along with its clean-up activities and the medical spending required by victims of its pollution. But you wouldn't consider all that activity to be beneficial. As a result, the high GDP growth rates recorded by industrializing countries are misleading about the real progress being made in quality of life.

The Kuznets curve need not be an inevitable cycle. If governments accept the link between economic and environmental concerns, they can select from a range of market-based methods to ameliorate the two. China, for instance, has finally realized the cost of its massive air quality problem and is now developing a tradable emissions system for the highly polluted city of Taiyuan,[43] although this realization has only come after having sunk into the environmental abyss. And it's not assured that countries will come out of that abyss. Some mega-cities—Manila and Karachi come to mind—have failed to do so, and it's possible Bangkok could share that fate.

The holy grail of environmentalism is to avoid the curve altogether, or find a "shortcut" that would allow developing societies to become wealthy and green without going through the intermediate stage of massive resource degradation. Using "green GDP" accounting methods—which account for the costs of development as well as the benefits—would be an excellent start. Taxing waste and pollution rather than labor would also be a boon.

It's mostly too late for countries like China and Thailand, which have already exploited so many of their resources, to find a shortcut (although their neighbors could certainly benefit from it). The question for Thailand now is whether it can come up with the political will to resurrect its environment. Given the experience of industrialized countries, that is only likely to result from a long struggle on the part of communities and environmental groups—in other words, from a well-developed civil society using the courts, the media and public opinion to press for greener policies. And it's possible that Bhichit's election may have marked Bangkok's turning point.

One encouraging example is the authorities' determination to acquire more land for city parks, which are in desperately short supply. As of 1997, Bangkok had an average of only 0.59 square meters of park space per inhabitant, compared to the WHO standard for international cities of four square meters per person.[44] But several small parks have been opened in Bangkok in recent years, and there is potential to create larger ones. The extensive green surroundings near the Sirikit National Convention Center are slated to become a major new park centered on Lake Rajada—that is assuming the current occupants of the land, the Thailand Tobacco Monopoly, can ever be persuaded to leave (how symbolic is that conflict!). Even bigger is Bang Krachao, a huge tidal wetland lying just south of the city that is zoned for conservation and considered to be Bangkok's "green lungs."

Bangkok may even end up with some unique public parks that draw on its own traditions. Thailand is famous for its Buddhist conservation movement, hallmarked by the establishment of forest temples scattered in remote areas around the country. But some crusading monks have taken up the challenge of bringing this movement to the city. On Rama I Road, smack up against the mammoth World Trade Center and amidst the chaos of downtown Bangkok, sits Wat Pathum Wannaram, an urban forest temple whose abbot, with the help of Thailand's king, has managed to accrue several acres of land that is slowly being restored into woodlands. Stroll along the grounds and you can almost forget you're in a mega-city.

Public parks are important because they are the closest things to a modern commons. Benjasiri Park, opened in 1992 right on Sukhumvit Road, became my local haven. Built in cooperation with the neighboring Queen's Park Imperial Hotel, whose guests have direct access to the park, it's a model for joint public-private ventures in establishing open space. It has a nice blend of active and passive space, a swimming pool, a jogging track, exercise facilities, and a central pond that also acts as a storage reservoir during the rainy season. But it has some curious aspects as well, such as the loudspeakers that blare out music and BMA announcements. The park is actually built on the former site of the Meteorology Department, which had some huge and spectacular old trees on its grounds. But, amazingly, they were all cut down to make way for the landscaped design that the authorities were apparently determined to follow. The design included some peculiar sculptures but almost no shade. To rectify its mistake, the park authorities planted some scrawny saplings, but the park remains empty during the day when it bakes in the tropical sun.

During the late afternoon and early evening hours Thais call *tawn yen* (cool time), however, when people in the tropics love to come out on the street and socialize, the park is packed. It's a great place to watch friends, families, lovers, and would-be lovers stroll and lounge about, although if you're a woman, the roving packs of *jigo* (flirtatious young men who fancy themselves as gigolos) can be annoying. The ballplayers at the back of the park also offer a nice cross-section of Thai youth culture. The basketball players and skateboarders are trendy dudes who sport the latest American inner-city fashion. But the coolest guys are the lithe and limber *takraw* players. If you've never seen the sport, think of volleyball played by three men on a side, but instead of using their hands the players use their feet. A single point can include incredible feats of juggling, leaping leg blocks, and stunning bicycle kicks, the cat-like players somehow contorting themselves to land back on their feet. The neighboring volleyball court, meanwhile, is dominated by *katoey* (transvestites), who tend to punctuate their points with lively banter and the occasional effeminate shriek.[45]

Bangkok's parks—along with its markets, temples, and its historic Rattanakosin area—could form part of a pleasant urban future. What will the city look like twenty years from now? The Los Angeles model is clearly relevant, but another good comparison, believe it or not, could be Paris, where a similarly centralized state has made the central city a glittering showcase, but where much of the surrounding *banlieues* has been turned into a ring of gray and forbidding housing projects. The situation could be even grimmer for the outer rim of Bangkok and its surrounding suburbs—the so-called *parimonthon*—where the bulk of industrialization has taken place. Thai authorities have repeatedly urged these companies, and sometimes offered economic incentives, to migrate to other regions of the country to decentralize industrial development. Assuming they do depart, they will leave behind decrepit factories, toxic waste sites, and numerous social problems, much like those in the "rust belts" of the West.

Central Bangkok, however, could sparkle. Once completed, the city's mass transit rail systems will hopefully allow the authorities to get a grip on the traffic situation. Just as important is a decades-long project to build a city-wide sewage system (until the 1990s, just 2 percent of the city's inhabitants were connected to proper wastewater treatment facilities). Like so many infrastructure projects in Thailand, the sewage system has suffered delays and hiccups, but once completed, it should go a long way toward cleaning up the city's waterways. And it should provide economic benefits by raising the value of waterfront property. In the long term, Bangkok's city planners have their eyes on redeveloping Klong Toey, long the country's main industrial port and still famous for its sprawling slum. The residents there won't want to move, but shipping operations are increasingly migrating to the Eastern Seaboard. Redeveloping the area could make it a gateway to the enticing green space of Bang Krachao, just across the Chao Phraya River.

Finally, don't forget that although Bangkok no longer resembles Venice, it still has its canals. Once they have been cleaned up, imagine how pleasant it could be eating a scrumptious Thai meal at a waterside restaurant on a balmy tropical evening in the Big Mango.

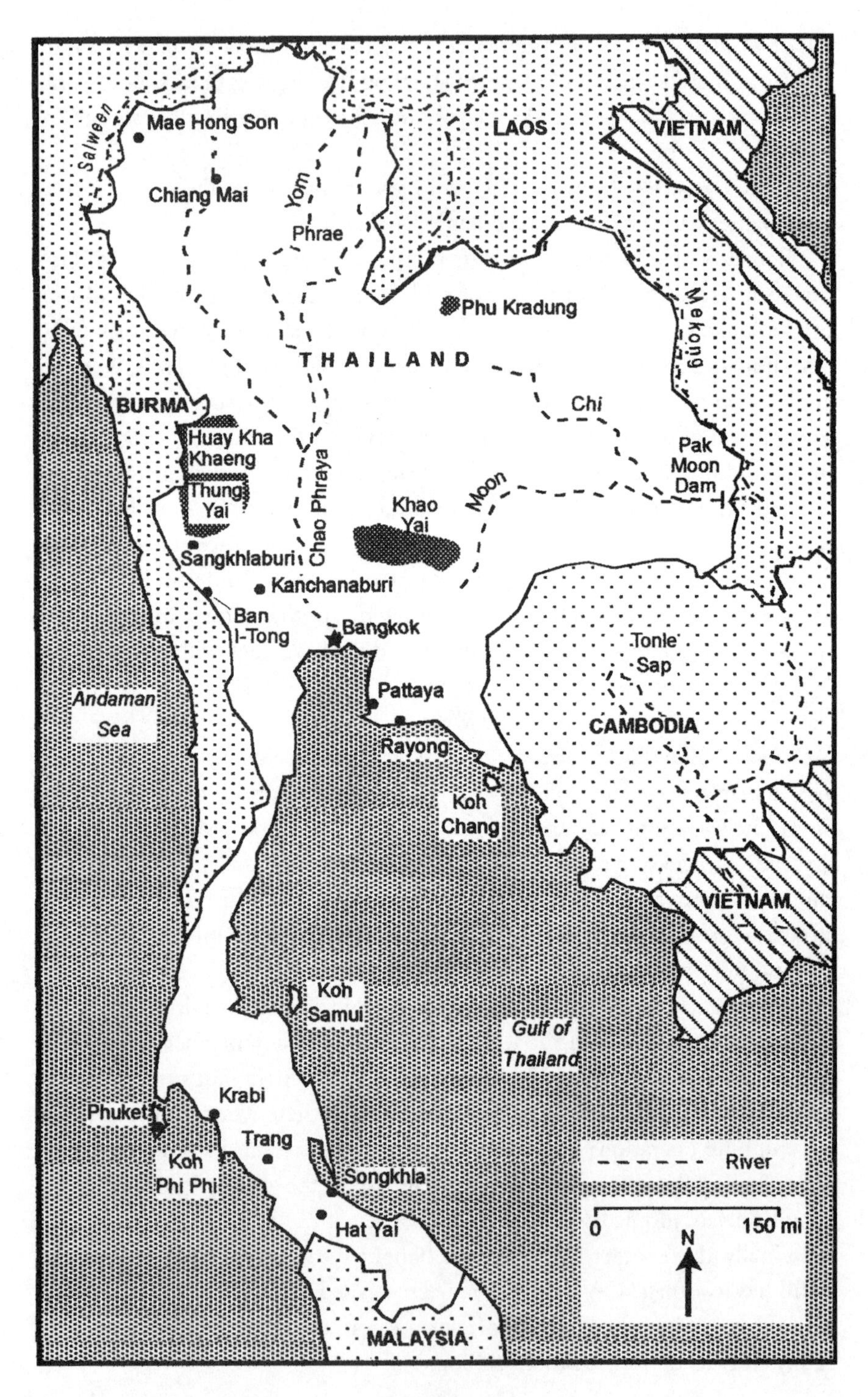

Salween
Mae Hong Son
LAOS
VIETNAM
Chiang Mai
Yom
Phrae
Phu Kradung
Mekong
THAILAND
Chi
BURMA
Huay Kha
Khaeng
Pak
Moon
Dam
Thung
Yai
Chao Phraya
Khao
Yai
Moon
Sangkhlaburi
Kanchanaburi
Ban
I-Tong
Bangkok
Tonle
Sap
Andaman
Sea
Pattaya
Rayong
CAMBODIA
Koh
Chang
VIETNAM
Koh
Samui
Gulf of
Thailand
Krabi
Phuket
Trang
River
Koh
Phi Phi
Songkhla
Hat Yai
0
150 mi
N
MALAYSIA

Thailand

2

TOURISM

MONEY CHANGERS IN THE MONASTERY

MELTING MOUNTAINS

On October 27, 1998, a man entered the compound of an eco-tourism company in Phuket, took out a gun, and began shooting the firm's operations manager, Panwong Hirunchai, hitting him in the leg, arm, and abdomen. Believing he had killed his victim, the assassin hopped onto the back of a motorcycle—one of those ubiquitous 100-cc Honda Dreams—and was driven away. No suspect in the crime has ever been arrested.

On April 22, 2000, the thirtieth anniversary of Earth Day, the U.S. television network ABC aired an interview of President Bill Clinton conducted by Leonardo DiCaprio, who questioned the chief executive about several environmental issues, in particular global warming. The main impact of the show, which received abysmal ratings, was to raise a furor among the news media because a movie star rather than a seasoned journalist was interviewing the president, thus blurring the line between news and entertainment.

What could these two events possibly have in common? As it turns out, both were galvanized by fierce environmental battles over one of the most

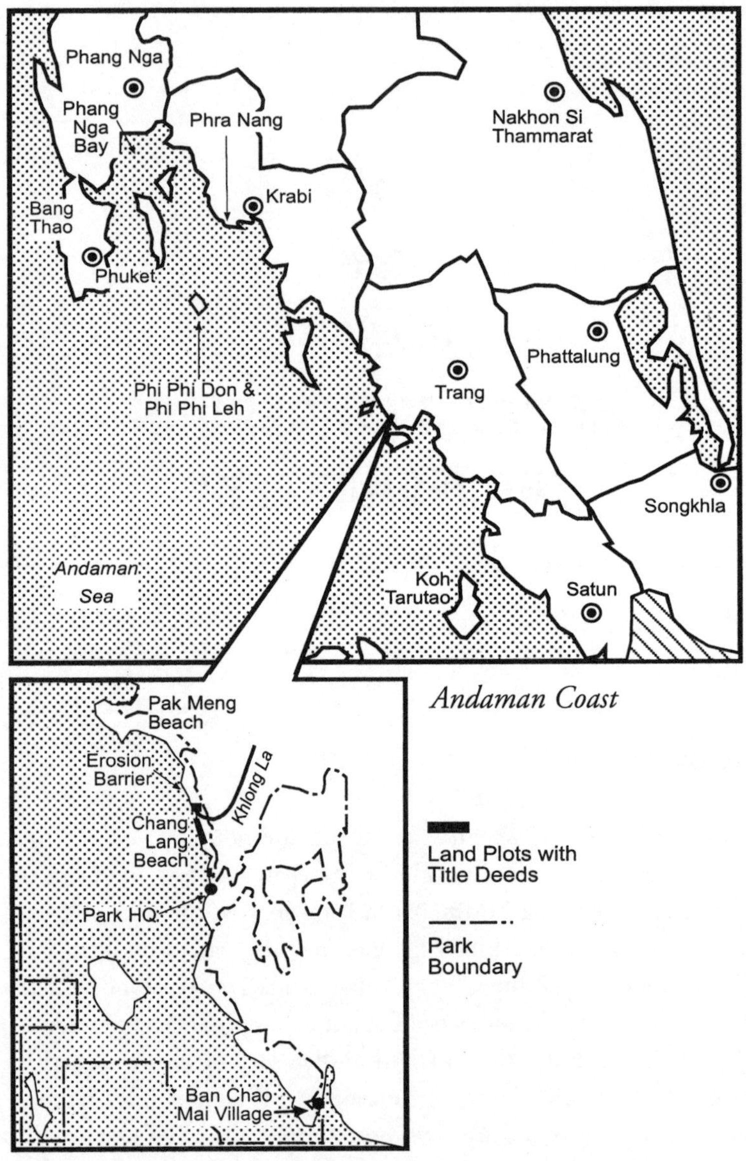

Andaman Coast

Haad Chao Mai National Park

remarkable landscapes on Earth: the soaring limestone towers of Thailand's fabled Andaman coast.

DiCaprio had been targeted by environmental activists protesting the filming of *The Beach,* a movie he starred in, on a stunningly dramatic island in one of Thailand's national parks. Upset at becoming the target of what he considered to be unfair charges of environmental negligence, DiCaprio vowed to increase his own green activism, and subsequently (among other activities) made the deal with ABC to collaborate on the Earth Day special.

Leo may have temporarily lost his good name, but Panwong nearly lost his life. The company he worked for, Sea Canoe Thailand, is a pioneering adventure travel operation that has sought to adopt many of the high-minded principles advocated by eco-tourism experts. But while guiding vacationing sea kayakers through the hidden tunnels and enchanting grottoes around Phang Nga Bay, the company eventually ran afoul of the area's notorious birds' nests concessionaires, who have used vigilante-like violence to control its territory for centuries.

Even if you've never been to Southeast Asia, you're probably familiar with the surreal landscape of limestone cliffs that stretches all the way from northern Malaysia up through Thailand and Laos into northern Vietnam, where an even larger cluster of the misshapen islands forms an area known as Ha Long Bay, and finally up into Guilin in southern China. The bizarre scenery was first made famous in old Chinese paintings, but has since appeared as a spectacular backdrop in numerous films, including *The Man With The Golden Gun* (Scaramanga's Phang Nga Bay lair in the movie is now known as James Bond Island), *Indochine,* which has scenes set in Ha Long Bay, and, most recently, *The Beach.* Many of these mountains are coated with a thick layer of jungle, which adds to their primeval appeal. Their steep slopes and tough terrain make them difficult to walk across, much less log, so they've become important islands of habitat for species such as macaques, gibbons, weasels, and hornbills.

The weird shapes of the limestone peaks, the remnants of ancient coral reefs packed together under tremendous pressure, arise from their calcium carbonate structure and are primarily sculpted by chemistry rather than erosion. The peaks are known as "karsts," a term used for rock formations that lose more material as a result of being dissolved than being worn away by the elements. So it's no wonder the cliffs of the Andaman look like lumps of dripping candle wax: They are virtually mountains that melt.[1]

The star attractions of the karsts are their caves. As water runs into the limestone's cracks, it eats away at the rock and widens the fissures causing them to intersect. They often harbor strange creatures, typically pale and blind, especially adapted to the pitch black conditions and found nowhere else: Thailand alone has more than 10 percent of the world's known species of cave fish, says the Royal Forest Department's (RFD) Dean Smart. But the caves' ecosystems are also incredibly fragile. If permanent lights are installed to guide tourists, they warm the air, kill off some of the cave's denizens, and increase the moisture in the air by sucking water out of the stalactites and stalagmites, which then start to dry out. Even breathing in a cave can affect its stability through the carbon dioxide that is exhaled.

Occasionally, as water eats away at the rock, a cave will grow so large that its roof collapses, creating a giant sinkhole known as a "doline." Exposed to the sun and open air, the *hong,* as it's called in Thai, is gradually colonized by all manner of plants and animals. In karsts along the coasts, the bottoms of these grottoes become hidden tidal pools, some of which are accessible from the sea only at certain times of day by passing through tunnels lined with razor-sharp oyster shells. Visitors to Koh Panak, for instance, must lie down in their kayaks to avoid lacerating their skulls on the way to visiting the island's *hong.* The tricky passage, however, is usually worth the trip: Entering one of these hidden lagoons, ringed by steep cliffs of foliage and buzzing with the hum of cicadas, is a truly Jurassic-like experience.

THE LIMESTONE COAST

The chance to explore this mysterious world is what drew John Gray to Thailand in 1989. A journalist, environmental activist, and kayak tour operator formerly based in Hawaii, Gray paddled up and down the Andaman coast, exploring the caves of Phang Nga Bay and points south. He soon decided to leave the United States—"I was preaching to the converted there," he says—and set up Sea Canoe Thailand. Taking tourists into the *hong*s on silent, gliding kayaks seemed to be the perfect way to put into practice many of the eco-tourism principles he and his colleagues had discussed. "I came to Asia to try and plant a seed of environmentalism," he explains with a sigh, "but I was in fantasy land."

Big and shaggy, Gray has inevitably been dubbed "The Caveman." But he seems more like a real-life version of Allie Fox, the protagonist of Paul Theroux's novel *The Mosquito Coast.* Consumed by his dream to bring progress to the tropics, he set out for foreign shores, only to be done in finally by gangsters, the realities of the developing world, and his own naïveté.

Phuket had already been developed into the Andaman coast's most famous resort island by the time Gray established his company there in 1990, and nearby Phi Phi Don Island—a stunning blend of limestone cliffs, sandy beaches, and coral-fringed bays—was already being ravaged by overdevelopment. But the southern reaches of Phang Nga Bay were still relatively undiscovered, and Sea Canoe sought to reduce the impact of its paddle tours by strictly limiting the number of visitors it took on its tours, a policy it still maintains. Customers are taught how to behave when entering the caves: no smoking, drinking, shouting, or collecting of souvenirs. Just as important, according to Gray, is Sea Canoe's effort to ensure that the local people benefit from its business—a crucial feature for any eco-tourism endeavor—and to that end Sea Canoe hires guides from local fishing villages, training and paying them well. Although Gray is the company's founder and remains its guiding light, the firm is actually owned by Thai partners and it hires guides from local fishing villages, training and paying them well. "If we are going to save the planet we have to share the knowledge," says Gray. "A lot of times, the Western world doesn't want to share its know-how for fear of losing its competitive advantage." At first, the formula seemed successful; Sea Canoe even won several prestigious eco-tourism awards.

But things turned sour when other firms started up their own sea kayaking tours. As many as eighteen companies currently run visitors out to the *hongs*, which now suffer from traffic jams during the high season. "It's more like a floating market than an eco-tour," worries Anupharp Tirarat of the Tourism Authority of Thailand (TAT).[2] Gray, never known for his diplomacy, insults his competitors and lambastes them for lacking the training and environmental awareness that his own firm tried to instill. "They're just in it for the money," he says with disdain, adding that Western inbound tour operators cater to a lowering of standards by seeking out operators who charge the lowest prices. Sea Canoe, which has kept its prices high, asked the authorities to impose a limit on the number of tourists who could enter the caves.

Essentially, Gray had come across the same problem that plagues eco-tourism everywhere: It needs to be rationed, otherwise it won't be "eco" for long. Access to natural destinations has to be restricted to keep them equally worth visiting for future generations. There are basically two ways to make such restrictions. You can either set high prices, in which case only the wealthy can afford to visit the site; or you can limit the number of visitors directly through the use of quotas, preferably set according to a site's "carrying capacity." But, as Sea Canoe quickly discovered, that's not a very popular option either, particularly in a country such as Thailand, which has a weak regulatory tradition. Sea Canoe's competitors accused the company of simply trying to gain a monopoly on the trade. The TAT urged the tour companies to form a self-regulating cartel. Efforts to cooperate foundered, however, and Sea Canoe's competitors blamed Gray for being arrogant and stubborn in his demands.

Matters took a turn for the worse in 1998 when a company called Phi Phi Cabana 1991, owned by the locally influential Kittithornkul family (the company is named after a hotel they run on Phi Phi Don) won the government bid to operate the birds' nest concession in Phang Nga Bay. Amongst the Chinese, birds' nest soup is considered a delicacy, and a bowl at some of Hong Kong's tonier restaurants fetches more than US$50. The main ingredient is actually the product of a swiftlet, a sparrow-like bird found throughout Southeast Asia that painstakingly glues its nest into shape by using its own saliva.[3] Thais have collected the nests for centuries and have done a roaring business. Today, one kilo of swiftlet nests can be worth 100,000 baht, roughly a year's income for the average Thai. But the local fishermen see virtually none of this money. Throughout the centuries, concession holders have guarded their claims jealously; indeed, they have the reputation for shooting interlopers first and asking no questions. In the 1990s alone, concessionaires killed more than a dozen Thai villagers.[4]

In 1998, the Phi Phi Cabana Company decided that the booming tourist trade on the limestone islands was disturbing the swiftlets and harming nest production, so it put up signs prohibiting entry to the caves. Since the islands in question are part of Phang Nga Bay National Park, the company's right to close the caves was questionable. But such legal niceties are often overridden in Thailand, particularly when powerful interests are at stake. At any rate, by that time as many as five hundred boats a day were entering some of the more popular *hong*s, and the TAT was also becoming concerned by growing complaints from visitors. A couple of low-end, high-volume tour companies catering mainly to rowdy Chinese tourists were undercutting the higher-end companies such as Sea Canoe. Eventually, the kayaking cartel was formed. The Phuket Paddle Club for Environmental Protection, as it came to be known, decided to limit cave entries to three hundred boats a day, and then negotiated with Phi Phi Cabana to pay it 100 baht for each boat entering the *hong*s. Even though Phi Phi Cabana had no legal right to levy a charge, explains Thiti Mokkapant, the director of the Paddle Club, "We were told by the TAT to sort the problems out between ourselves . . . reach a compromise or else we would have to drag it out in the courts. [So] we agreed for the sake of our businesses."[5]

Sea Canoe, however, refused to play ball, arguing that the money should go toward conservation rather than lining the pockets of local businessmen. Gray said he would be willing to pay 500 baht per head to the park service for the right to enter the caves, but would not yield to attempts at "mafia extortion." A couple of his lieutenants, including Panwong, were equally defiant. In return, Phi Phi Cabana refused to allow Sea Canoe boats to enter the caves it controlled. A game of cat-and-mouse ensued, and after one particularly heated incident between Panwong and the guards, Sea Canoe's staff began to receive death threats. A few days later, the unknown gunman entered the Sea Canoe compound and shot Panwong point blank. He was lucky to survive. Phi Phi Cabana chairman Dumrong Kitithornkul has denied involvement in the shooting: "We have had our differences in the past [with Sea Canoe], but we would never consider having anyone shot," he claims.[6]

For a while, Sea Canoe held out against paying the fees. But finally it gave in, "to save the jobs of our employees," explains Gray. Despite all the tumult, Gray's company has done well. It has opened branches in Krabi, farther down the Andaman coast; on Koh Samui in the Gulf of Thailand; in the Philippines; and at Vietnam's Ha Long Bay. In that sense, the analogy with Allie Fox isn't appropriate. But Gray seems scarred by all the fighting over Phang Nga Bay, and plans to spend as much time as possible outside Thailand. "Looking back on it, I don't know if we did the right thing in commercializing the caves," he concludes. "Eco-tourism rolls off the tongue easily. But quite honestly, there's very little around."

AN INDUSTRY WITHOUT SMOKESTACKS

On a pleasant December evening in 1992, a festive group of yachties wearing bright floral print shirts were celebrating the conclusion of the sixth annual King's Cup Regatta on the beach just below the Phuket Yacht Club Hotel at Nai Harn Bay. The event that year, sponsored by companies such as Yanmar Diesel Engines and Champagne Mumm, was supposedly being held in the name of "preserving the environment," but most of the sailors had simply gathered for some hard racing and harder partying. So imagine their surprise when up the beach came a multi-ethnic throng of demonstrators sporting protest signs written in Thai, English, Japanese, and even one in Danish in honor of the regatta's official guest, Denmark's Prince Henrik.

The protestors were participating in a seminar called the "People's Forum on the Impact of Tourism," which was coincidentally being held at the Phuket Teacher's College across the island, and they were arguably even more colorful than the sailors. They included such veterans of the Asian tourism wars as Roland Martins, an activist from Goa who had been arrested for throwing dead fish at German package tourists; Gen Morita, a

Japanese coordinator for GAGM, the Global Anti-Golf Movement; and Ing Kanjanavanit, Thailand's most outspoken tourism activist.

Bright and attractive enough to have done modeling work during her early days in the advertising industry, Ing has become an artist of all trades: journalist, author, painter, and filmmaker. Her work and insights are at times brilliant, and almost always shocking, because Ing is most of all a non-conformist. In fashion-conscious Thailand, she prefers to wear shorts, a T-shirt, and flip-flops. Although everyone else in Bangkok travels by car, Ing tools around on her old three-speed bicycle. And when it comes to her work, she is not satisfied unless she is rooting through the more hidebound conventions of Thai society and stirring up trouble.

Growing up in Bangkok in the 1970s, Ing was a self-described hippy who hung out with the alternative crowd at the newly built Siam Center, still a major crossroads for the city. She was sent to boarding school in England, but shortly after matriculating to an art college, she dropped out and headed to the Thai-Cambodian border, where she served as a volunteer in the refugee camps. Back in Bangkok, she eventually turned to activism and journalism, writing columns for the *Nation* and the magazine *Lalana,* particularly about women's issues. Her crusades against sex tourism eventually led her to look critically at tourism in general, and how it can lead to the exploitation of local people and the ruin of formerly pristine wilderness. "It was an issue that appealed to me because it's not black-and-white," she explains. "Most people think of it as harmless, but it causes tremendous damage." Ing's writings were eventually turned into a book, *Khanglang Postcard (Behind the Postcard).* But Ing says she tends to think more in images, so she eventually turned to film making. Her first documentary, about the impacts of tourism on Phuket, was called *Thailand for Sale,* and it focused on a land dispute surrounding the yacht club.

The yachties at the regatta had never seen Ing's film, of course, and the reaction to the protest was predictably surly in some cases, and quizzical in others: What could the protestors possibly have against a harmless sailboat race? What the yachties didn't know was that part of the hotel where they were celebrating had been built on top of a public footpath (which had been turned into a road). Thailand's Juridical Council had ordered that a section of the hotel be demolished to provide unimpeded access to the community that lived beyond the resort. But the company that then owned the yacht club, First Pacific Land, a subsidiary of a Hong Kong–based real estate firm called First Pacific Davies, vowed to appeal and tie the matter up in court.[7] The tactic must have worked, because the hotel remains where it was.

The concerns of tourism activists extend to far more than just the alleged transgressions of one hotel. The fact is, the importance of tourism to Southeast Asian economies and the impact it has on the region's societies are both usually overlooked. In 2001, tourism was Thailand's largest foreign-exchange earner; more than 10 million foreign arrivals spent around 300 billion baht (an amount equal to approximately 6 percent of the country's GNP) during their trips to the kingdom—a huge jump from the 3.4 million visitors and 50 billion baht in tourism earnings that Thailand received in 1987.[8]

That is generally considered the year Southeast Asia's economic boom began in earnest. The 1985 Plaza Accord on currencies had sent the value of the yen soaring against the dollar (to which the baht was then tied), and Japanese investors targeted the region as a low-cost manufacturing hub. Meanwhile, the TAT's landmark Visit Thailand Year campaign gave the country a huge promotional boost, resulting in enormous increases in tourism arrivals and earnings the following years. Since then, virtually every other country in the region has set up a similar campaign,[9] with varying degrees of success. But there is no denying the success that tourism as a whole has had in the region. In 2000, ASEAN (The Association of Southeast Asian Nations) countries attracted nearly 38 million visitors and yielded an estimated earnings of around $30 billion.[10] And as it is in Thailand, in most countries tourism, tourism is one of the largest earners of foreign exchange.

To top it off, tourism is typically seen as a less environmentally destructive trade than the low-wage manufacturing and resource extraction industries that so many Southeast Asian countries have relied upon in their bid for rapid industrialization. Boosters call it "an industry without smoke stacks," and argue that it could become a force for conservation because tourists pay to visit beautiful and pristine nature spots such as Phang Nga Bay.

Thus the concept of eco-tourism. For this vague and much debated concept to stand a chance of working, however, several conditions need to be met. A significant portion of the proceeds from these activities must go to the locals, who then have an incentive to help with conservation; otherwise, they will simply look on tourism as another outside force seeking to exploit the resources they need for their own livelihoods. Also, tourists must become responsible consumers when traveling, just as they are when buying food and furniture. That could mean researching a trip to ensure they book with a responsible tour operator, or to make sure a hotel they plan to stay in cleans up its wastewater. It might mean accepting quotas or higher fees in parks to ensure they are properly managed. But tourists have to demand that the "eco" part goes into tourism. Attempts are now being made to set up an eco-tourism certification system, based on a set of social and environmental standards, so that tourists can choose their destinations with more care. That should help provide an incentive for the ultimate goal: to manage tourism in conservation areas sustainably—no easy feat, as the kayak tours showed, especially when regulators are weak and venal.

As it stands, virtually every major tourist destination in Thailand has been ravaged by unplanned overdevelopment, particularly the beach resorts. One by one, the Thai strain of "tourism rot" has spread from Pattaya to Phuket, Koh Samui, Phi Phi, Hua Hin, Krabi, and it is currently ravaging Koh Chang. The pattern is by now all too familiar. First, the backpackers and more adventurous tourists seek cheap, out-of-the-way places where they can relax amid quiet, beautiful surroundings. Gradually, a village becomes "discovered" by the guidebooks, travel writers, and travel agents, and tourism becomes the area's main livelihood. An influx of workers is needed to support the new trade, and the destination grows rapidly. Local people who have not benefited from the boom—either because they don't own land or possess the necessary skills—are marginalized. The coral reef offshore quickly dies and becomes an impediment to docking boats; so a channel must be blasted through it. As the destination becomes more popular, high-rise hotels and condominiums begin to crop up; they in turn cast shadows across the beach and spoil the view. Then the package tourists arrive, and so do the go-go bars, the karaoke clubs, the massage parlors, and the mafia.

By now, the area has become a full-blown city, but it has grown too fast for the authorities to handle. They have supplied plenty of water—typically, by building a dam inland—and electricity, but have neglected to build a wastewater treatment plant to handle all the new sewage, which pours into the sea and begins to pollute the beach. Local fishermen, if they want to continue their traditional trade, have to go elsewhere to practice it. Far away from the beach, slums and shantytowns spring up to house the migrants who have come looking for jobs. Meanwhile, the resort area's narrow streets can no longer handle all the new traffic. They become choked with cars and air and noise pollution. Standing amid the bustle and fumes, you can be forgiven for thinking you never left Bangkok.

Part of the blame for this disturbing trend must rest with the Tourism Authority of Thailand, which has promoted growth but overlooked the impact of the surge in arrivals and failed to prepare the country to handle them. In the early years of the tourism boom, this shortsightedness led the state agency to ignore environmental concerns and the shouts of sex-tour protestors who warned about the danger of AIDS. In more recent times, during its Amazing Thailand Year campaign, the TAT's obsession with increasing the number of tourists led to an influx of cheap package tours from China, thereby ensuring the agency's preset targets for inbound arrivals were met. It's a typical example of a state that is still operating under a "five-year plan" mentality.

Critics argue that Thailand should seek out a tourism strategy that maximizes revenues, not the number of tourists. Thailand could never have imitated Bhutan, which has actively sought to minimize the cultural and environmental impact of tourism by keeping a strict limit on foreign tourists and making them pay through the nose (it's not even clear whether Bhutan can maintain such a model in an increasingly globalized world), but it certainly had an opportunity to cater to the higher end of the eco-tourism market, thus elevating income while putting less strain on Thailand's environment and infrastructure. With a bit more strategic planning and local consultation, developing countries could offer a range of tourism options that would allow some places to undergo mainstream development while preserving others through strict regulations—they could create "mini-Bhutans," if you like.

The other major criticism of the TAT is that it is far more concerned with the image of Thailand than the reality. The TAT is a promotional agency that sells an idealized version of the country to the world—it's the way the United States would be represented if Disney were put in charge of America's public relations. Thais often laugh at TAT ads that portray them as a nation of fruit carvers and flower arrangers, but they can also take offense if they think the agency is trying to pimp the country to foreigners. To be fair, the agency has in recent years become more outspoken about the need to protect the environment; and it has made efforts to foster eco-tourism because it realizes that tourists don't want to visit polluted beaches and degraded parks. But TAT does not have any real regulatory powers. That may be a good thing, since government agencies habitually struggle when they are supposed to serve both as promoters and regulators. Unfortunately, the local and national authorities that do enjoy such powers have been slow to act; a pity, since a little foresight and planning could have helped prevent the pollution, the overcrowding, and the seediness from taking over Thai resorts, and perhaps helped these areas to grow sustainably.

But say this for tourism: It offers a vivid demonstration, one that even Thai officials couldn't ignore, of how economic advancement requires environmental protection. For tourists do shy away from polluted areas. So when the Thai government passed a series of sweeping environmental reforms in the early 1990s and gave the Ministry of Science, Technology, and Environment special powers to manage certain areas as "pollution control zones," it's no coincidence that the first such areas to be declared, Pattaya and Phuket, were centers for tourism (Hua Hin was later added to the list). The local authorities were subsequently ordered to build wastewater treatment plants, and in Phuket a limit on high-rise development was put in place. Originally, financing came from the Environmental Fund, which obtains money from a tax on gasoline. Later, a tax on hotel rooms was imposed. So perhaps one benefit of tourism is that it provides governments with a clear incentive to organize environmental cleanups.

Ing sees things differently. To her, Thailand's declaration of its pollution control zones merely demonstrated once again how the government cares more about the well-being of foreign tourists than of its own citizens.[11] Besides, she points out, revenue figures are a misleading indicator of how much the country benefits from tourism since many of the creature comforts demanded by foreign visitors have to be imported. Ing cites a study carried out by Thailand's National Institute for Development Administration, which found that 56 percent of the foreign exchange earned from tourism during Visit Thailand Year went to pay for such imports.[12]

The key point that Ing and other tourism activists make is that although the industry may not require smoke stacks, it does require extensive infrastructure: not just sewage treatment plants, but roads, airports, dams, water, and electricity. Most of all, it requires real estate development—hotels, condominiums, and the land on which to build them. In *Khanglang Postcard,* Ing documents how well-connected developers helped create the promotional blitz of Visit Thailand Year and then benefited from the resulting boom. She writes,

> What people did not realize [when the boom began] was that tourism and the real estate/construction industry are really one and the same. Land speculation became a national pastime, permeating every beautiful village, however remote. Land prices skyrocketed. Villagers sold agriculturally productive land to speculators. Practically overnight, fertile land became construction sites. The plague kept spreading, corruption got out of control.

> National parks and forest reserves were encroached upon by golf courses and resorts.[13]

Finally, she adds, the tourism boom led to a property bubble that burst in the 1990s and eventually helped cause the financial crisis.

GOLF WARS AMID "GREEN DESERTS"

The best examples of how land speculation in Asia can devolve into a life-and-death issue are the fierce battles waged over golf course development. They've led to murder in the Philippines, where three villagers opposing the construction of a golf resort at the beachfront community of Hacienda Looc were killed.[14] They've led to riots in northern Vietnam, where a demonstration by villagers in Tho Da hamlet against a Korean-backed project was violently broken up by the police, leading to the accidental death of a female protestor.[15] Some of the bitterest fighting has taken place on Java, in Indonesia. In the village of Rancamaya, golf resort developers displaced 1,000 families; and in Cijayanti, more than 300 families were forced to move. The Cijayanti conflict turned violent when some villagers tried to re-occupy their land, beating up security guards and burning down a construction camp in the process.[16]

These are just a few of the often-bloody conflicts that have erupted over golf courses and which seem to emerge from the grass roots in the most unlikely places—the protest in Vietnam, just a few kilometers from Hanoi, being a good example. And by no means are such conflicts limited to Asia. In Mexico in 1996, opposition by residents from the town of Tepoztlan to a golf course designed by Jack Nicklaus and slated to be built inside Tepozteco National Park turned violent when a man was killed and three opponents of the project were accused of the crime. A group of Tepoztlan villagers subsequently traveled by bus to meet with then-President Ernesto Zedillo in defense of the "golf war political prisoners," but were ambushed by heavily armed policemen; one elderly man died and eighteen others were injured.[17] There has been plenty of opposition to specific golf courses in the West as well, particularly in Hawaii, where, for instance, tenant farmers were evicted[18] to make way for the Japanese-owned royal Hawaiian Country Club in Maunawili Valley.[19] Elsewhere in the United States, opponents of suburban sprawl have decried the conversion of former farmland into real estate projects that are often built around sprawling golf courses.

Nevertheless, the fiercest opposition to golf seems to be in Asia, where a loose network of grass roots anti-golf activists from all over the region has coalesced to form the Global Anti-Golf Movement (GAGM). What is it about this seemingly innocuous game that makes it so violently controversial? After all, one of the things golfers like about the sport is the chance to relax amidst green and pleasant surroundings. What could be so harmful about grass, trees, lakes, and sand bunkers?

Well, golf courses may be green, but there is almost nothing natural about them in the tropics. Critics call them "green deserts," and if you've ever seen one being built you'll understand why. They require moving a massive amount of earth, an average of 2.5 million cubic meters in Thailand, and as much as 11 million cubic meters for courses in Japan.[20] The course is built atop a thick layer of fine sand, underlain by an extensive network of drainage and irrigation pipes. The sand, usually trucked in from far away, undergoes strict quality control to make sure rainfall quickly drains away. The special Bermuda grass, meanwhile, is imported to meet international standards.

A golf course, in other words, is actually a monoculture made up of delicate, non-indigenous grass species kept alive through the extensive use of irrigation, fertilizers, and pesticides. Since golf courses need to look green year round, even during the dry season, and are purposely built to drain quickly, they require a huge amount of water: an average of 3,000 cubic meters per day, according to Anita Pleunarom, an anti-golf crusader in Thailand, or enough to supply a village of about 2,000 households.[21] GAGM also claims the average Asian golf course needs about 1,500 kilograms of chemicals a year—from seven to nine times the amount needed for farming—much of which leaches into the groundwater and ends up polluting the surrounding water supply. (Thai farmers, too, are notorious for overusing pesticides, a practice that affects not only the environment but also their own health and that of consumers.) In Thailand, workers and caddies often complain of dizziness, nausea, and rashes after spending time on certain courses. The golfers, warns GAGM's Morita, are "playing in poison."

The game simply wasn't designed with the tropics in mind. As originally conceived in Scotland, golf was a pastoral game where the fairways and bunkers were natural parts of the landscape. There is nothing wrong with importing sports, of course, but golf course designers' slavish devotion to Western standards has increased the burden on tropical ecosystems. For instance, planting a local species of grass (or allowing fairways to go brown during the dry season) would reduce the need for water and chemical inputs. But designers say that golfers won't like that because it doesn't provide them with a "good lie." Many owners, meanwhile, want to build "championship courses" and fulfill their dreams of getting on the pro circuit. But other sports, such as tennis, play matches on different surfaces and seem none the worse for it, so why can't golf also add a little diversity to its tournaments?

Another option might be to add more trees, even if it would make a course more challenging. Currently, there is an element of absurdity to golf in the tropics: Developers first go around cutting trees down to create fairways and greens; then the players (or more likely their caddies) carry umbrellas to protect them from the scorching tropical sun. Tiger Woods found out just how brutal the conditions are when he came to Thailand for a much celebrated visit in 1997: He was forced to bow out on his first day on the links when he was overcome by the heat.[22]

Nevertheless, golf has become increasingly popular with the Asian middle and upper classes. Many play for the love of the game and the chance to exercise amid green surroundings, but supporters and opponents agree that a major reason for golf's growing popularity is the status it conveys. Pichien Amnatworaprasert, the managing director of Bangkok Golf Service Co. Ltd., explains why: "Golf allows people to relax and forget their troubles, but people are also attracted by the exclusive nature of the sport. It's good for business. You can meet clients and partners, and expand your society. A membership card is . . . like a credit card, but it's more than that because you must have more than money to get one; you must have connections." Morita, a handsome musician and passionate critic of the game, sees things in a more negative light: "Golf has become a social status symbol, especially in Japan. You go out on a huge course and you can feel like you own a vast area. You have a caddy to serve you and you feel like a king. . . . Conformity is the most important thing. If your business partner plays, you feel obliged to play also to keep the business, even if you don't want to."

There is also disagreement about the economic benefits of golf courses. Supporters say they not only attract a good class of tourists but also provide jobs for people in rural areas. Critics say they employ only thirty or forty people as guards, greenskeepers, and clubhouse staff, and the number can drop as automated sprinklers and mowing machines are brought in. Additionally, from three hundred to six hundred caddies are paid on demand. In Thailand, caddies are invariably female and preferably attractive. "Prostitution has become a common part of the Thai golf course business," claims Pleunarom. "In Japan, the sex trade and golfing are the same business." There is, in other words, a class element to the conflict about golf that makes it a highly emotional issue in the developing world, similar to the way fox hunting has become so controversial in Great Britain. In the United States, the exclusive nature of golf and country clubs has meant that ethnic minorities or women aren't always welcome. In Asia, they've become settings for crooked deals. In Thailand, for instance, it's said that many of the country's military coups were plotted on its golf courses, which are sometimes built as a result of crooked deals themselves.

This is a crucial point: In the final analysis, the social pretensions of golfers matter little; what matters is that golf has such powerful supporters and patrons that developers are often able to steal the land and water they need through corrupt and shady deals. At Hacienda Looc, according to filmmakers Jen Schradie and Matt Devries, the Filipino government allegedly sold land farmed for generations by local peasants for less than it was worth to a developer attempting to build a four-course golf and tourist resort. Their documentary, *The Golf War,* focuses on how the situation has turned into a political conflict between the villagers—who have been offered support by the rebel New People's Army—and local and national politicians backed by the military. Similarly, the conflict at Tho Da in northern Vietnam escalated when farmers learned that the land they had long tilled—which, like all land in Vietnam, is owned by the state—had been mysteriously given to a Korean firm for the construction of a golf course. In Thailand, says Pleunarom, a favorite tactic of developers is to target a piece of land, buy up parcels around the edges, and then restrict access for those farmers who live within this boundary until they are forced to sell. A visit by Nicklaus to inspect the Golden Valley golf course he designed created a stir in Thailand when it turned out that the course had encroached on Khao Yai, Thailand's flagship national park, and had blasted into the mountainside in the process.[23] Nicklaus later claimed he knew nothing about the incident.

Developers' schemes to acquire water are often equally dubious. In Thailand, golf courses are supposed to obtain water from their own reservoirs, but resorts cluster suspiciously around public reservoirs and waterways. In *Green Menace,* a documentary about golf produced by Ing K and Brian Bennett, a tanker truck from one Thai golf course is shown illegally taking water from a public reservoir on the Eastern Seaboard. A neighboring course has a hose and pump connected to the same reservoir. A blasé official from the Royal Irrigation Department (RID), meanwhile, simply insists that such theft doesn't exist. The film also contains an enlightening interview with Suradej Vongsiniang, a water resource engineer who quit working for golf courses after witnessing all the problems they create. "I saw outrageous exploitation," he says. "[One] golf course usurped a water source that was used by three to four villages of over 1,000 people. . . . The villagers suffer, but can do nothing. The golf course owners are influential people with everything in their power, including high government officials." When asked how golf courses steal water, Suradej replies, "From what I've seen at many sites, they are not subtle at all. They just take what they want. In some places . . . they build concrete channels directly from the irrigation canal, or even lay pipes from the reservoir itself. Some golf courses near big rivers dump rocks (and sand) into the river to make the water level rise so it would flow into their golf course. And what of villagers downstream? They used to travel on the river; they used to fish from it. It's their livelihood. So what to do now? Become a caddy?"

Golf critics like to say that such abuses can't be tolerated "for the sake of a mere game." But this is misleading, because as practiced in Asia, golf is far more than just a game: It is an exercise in land speculation driven by real estate bubbles. Currently, around two-thirds of the world's roughly 25,000 golf courses can be found in the United States,[24] but a growing number are being built in Asia. Japan now has more than 2,000 courses.[25] However, increasing opposition to new courses there, combined with the high price of land, has led investors to seek out greener pastures in the Asia-Pacific region. The small state of Hawaii now has more than 70 courses, according to GAGM, more than half owned by Japanese interests. Singapore also has one of the highest "golf densities" in the world. Thailand has only 150 or so courses, but even the standard-bearers of the golf industry admit that its boom of the early 1990s left it overdeveloped: "Thailand was just unbelievable," Nicklaus told *Asian Golfer* magazine. "We signed 11 golf courses in Thailand in just about ten months. There's always a danger about going too quick any place, but what are you going to do about it?"[26]

The unique feature about Asian golf resorts that drove the speculative boom is displayed in a chart prepared by Kuji Tsutomu, a mild mannered anti-golf activist from Japan, that records the price of Japanese course membership fees—it fluctuates erratically enough to give the most hard-bitten securities dealer a peptic ulcer. "Golf is not a sport, it's a money-making game for both developers and members, a form of speculation," insists Morita. Many investors actually purchase memberships at a reduced price before a course is even built, hoping to sell at a hefty profit once construction is completed—rather like those who get privileged shares for a firm's IPO. Club memberships are then bought and sold like shares on the stock exchange.

Since the price of memberships tends to follow the value of real estate, it's an excellent way for foreigners to invest in land, since under normal conditions they are barred from owning any in Thailand.[27]

Thais have also played this speculative game with gusto. According to Pichien, the golf entrepreneur who set up a membership exchange, during the property boom under the Chatichai administration, the price of golf club memberships went up almost every day. In an extreme example, he recalls, the cost of membership at Pinehurst Country Club increased ten-fold in a year. In Japan, too, says Tsutomu, "membership fees fluctuate with the bubble economy." The average price peaked at 39.24 million yen ($320,000) in February 1990. Then the bubble burst. Between January 1991 and July 1992, eighty-eight golf-related businesses went bankrupt, racking up losses of 1,390 billion yen (roughly $11 billion), and exposing various scandals and pyramid schemes.[28] Alarmed at the prospect of suffering from its own property bubble, China temporarily banned the construction of new golf courses in 1993.[29]

The anti-golf crowd has succeeded in foiling the construction of many new courses. In Malaysia, protestors managed to halt the construction of a resort atop Penang Hill.[30] The Japanese network, meanwhile, has created the Standing Tree Trust. If they find a landowner unwilling to sell to developers, the trust "buys" the trees on his property, claiming ownership above the land so that the developer can't buy him out. Asked how many golf courses he has halted, Tsutomu pauses for a moment, then says with a quiet smile, "The anti-golf organization I head in Ibaraki Prefecture has stopped fifty-seven golf courses. I am now number one on the developers' black list. I am very honored to be in this position." Altogether, he and his colleagues estimate they have blocked more than three hundred golf courses in Japan, "but for every one we've stopped, another has been built abroad by Japanese interests." Indeed, the biggest obstacle to the spread of golf courses has been the property slump that has gripped Japan and most of Southeast Asia.

Land is the key to the conflict about golf in Asia. Most of the other issues can probably be resolved, at least in the more democratic countries. Now that Thailand is facing chronic water shortages, developers increasingly accept the need to arrange for their own water supplies, either by building reservoirs or practicing greater conservation. Greenskeepers may also turn to integrated pest management techniques to reduce the use of chemical inputs (although some will probably always be necessary given the demand for spotless greens and the use of foreign grasses). But Asia is so densely populated, and land so precious, that golf course development is likely to remain a visceral issue. That golf in most of Asia is seen as an imported game, one turning farmland and forests into immaculately groomed private parks catering to foreigners, adds a whiff of nativism to the debate and makes it even more emotional. As a result, the conflict about golf has become symbolic of tourism in general: On the surface it seems harmless, even beneficial to the environment, but underneath are some truly troubling issues.

THE BATTLE OVER NATIONAL PARKS

Soon after arriving in Thailand in 1990, I received a symbolic glimpse of the grip that tourism and commerce have on Thai society. I'd initially settled into a Bangkok guesthouse not far from Khao Sarn Road, a string of guesthouses, travel agents, restaurants, shops, and curbside stalls that serves as a nexus for backpackers. I regularly took a shortcut through a monastery called Wat Chana Songkhram, and there on the grounds could usually be found a mobile currency exchange van belonging to Bangkok Bank. And I had thought that the expression "money changers in the temple" was only a biblical reference!

My first trip outside the city, to a blissful island off the eastern coast called Koh Chang, eventually provided a more heartfelt lesson: Virtually everyone who has lived in Thailand has seen a favorite nature spot change for the worse over the years. Access to Koh Chang was limited at the time—one boat a day going to the island from the eastern port of Laem Ngob—the best sign that a place hasn't yet been overdeveloped. I headed to a small beach on the southwestern coast called Kai Bae. There was no road and no pier, so you had to disembark with your backpack onto a ridiculously tiny wooden dinghy and be ferried to shore, if you didn't capsize first. The lodgings were basic, just a few wooden huts strung out along a coastal trail. But the scenery was fantastic. To the north, you could walk along a meandering trail up to a spectacular waterfall. Off the coast were some scattered forest-covered islets bobbing out of the sea. The coral reef nearby was degraded but still colorful. And I vividly remember standing in the warm, shallow ocean at sunset watching a full moon rise over the mountainous interior.

Best of all, a short hike to the south lay a beach that was not only beautiful but also empty. Reaching it required only a twenty-minute walk across a forested headland. As the trail descended sharply, a long golden arc of sand suddenly stretched before you. Once down by the water, you felt heroically isolated, as if you'd discovered something kept secret from the world. Looking out at the sea and sky, you could observe every shade of blue from aquamarine to azure. Floating in the water, you could look back at a jungle-covered mountain framed by cobalt-tinged clouds.

I went back to Koh Chang frequently during the following years, and each time I would see startling changes. It was like watching the video of a construction site being played at fast forward. Bungalows and resorts sprang up everywhere. Some were tasteful wooden structures built a discrete distance from one another; others were concrete monstrosities all crowded together. The pleasant little trail to the waterfall was replaced with a road. A little port was built at Kai Bae allowing boats to dock there, but it was not used much because the area was connected by road to the island's main pier, where car ferries now dock throughout the day. Kai Bae even has a little shopping strip now, with bars, mini-marts, travel agents, souvenir shops, and Internet cafes. Like so much of coastal Thailand, Kai Bae has turned into a clone of Khao Sarn Road.

With each trip I took, Kai Bae became a little less pleasant. The coral reef quickly died off. It became impossible to rent a bungalow on weekends during the high season. Even the magical sunset view of the mountains was ruined by the ugly pier now jutting into the water. But I kept returning because somehow or other the empty beach—I never even learned its name—survived. Until the last trip, that is, in 2001. This time, I saw some ominous warnings. A map of the island had been published and it showed some beaches to the south of Kai Bae. The trail was still there, but now it was paralleled by a coastal road. As I hiked down from the headland with a sinking heart, I soon realized that the beach was no longer empty. A couple of resorts had been built and bungalows now lined the shore. It was still a lovely setting. The development was not excessive. If I had never been there before, I would have thought it a pleasant place to stay. But I *had* been there before, I had experienced it when it was empty and wild, and that makes all the difference. I was spoiled, and although I tried to enjoy the afternoon, I couldn't help feeling depressed, especially when I learned what the place had been named: Lonely Beach.

In Thailand, the development of such destinations as Koh Chang is inevitable. Once it becomes known, once it becomes named, once a road is connected to it, it's going to be built up. The discovery by backpackers generates the spark, and the demand by tourists for modern conveniences feeds the flames, but two other factors fuel the fire of progress. The first is the need for businesses to grow. Investors want to increase their profits, which they can do by catering to more people or offering better services, or both. The second is the demand by local villagers for development. They quite naturally want cars and paved roads, they want piers where they can dock their boats, they want piped water and air conditioning and television and electricity. You can't begrudge them that; these are things we generally want, too—except for the times when we go off to some blissful tropical isle for a few days to get away from it all. There are still a few places left in Thailand where you can do that, but they are becoming ever harder to find.

Eventually, tourists who are looking for some semblance of solitude in a wild and natural place will have only two options. They can either pay through the nose to stay at a private resort that has its own wilderness reserve. Or they can stay at a national park. And it seems that Koh Chang will be a stark example of this dichotomy because the Thaksin administration is promoting the park as Thailand's next five-star destination; a frenzy of land speculation and howls of protests from visitors already priced out of Phuket has ensued.[31]

Despite its rapid development, Koh Chang is part of a park. It is just one of many Thai islands and beaches—including Ko Phi Phi and Phra Nang in Krabi, and Koh Samet in Rayong—that have been overwhelmed by the growth of tourism despite being protected areas. How is such development possible in areas supposedly dedicated to conservation? The short answer is that there is a huge, but necessary, loophole in the national park law. People who have been living on land or who can prove they have used the land before it was declared a park can claim ownership of it. So although most of Koh Chang's interior is considered parkland, much of the coast where people have long lived and worked is considered private. As we shall see, developers all over the country have connived with corrupt officials to take advantage of this loophole and gain ownership of valuable public land.

First, however, there is the huge question of how to manage Thailand's parks both for tourism and conservation. The National Parks Act explicitly states that they should serve both purposes (wildlife sanctuaries, on the other hand, were not set up for tourism; they are reserved for research, educational, and conservation purposes). But in practice, reconciling the two goals is no easy feat. The actions of just a few tourists can have a serious impact on endangered ecosystems and species, and during long holiday weekends, some of the more popular parks such as Khao Yai, Doi Suthep, and Phu Kradung may receive as many as 10,000 visitors. Most of the more serious impacts of mass tourism in these areas—such as the disturbance of wildlife mating patterns—are too subtle for the average tourist to notice. But obvious signs of degradation are everywhere: Graffiti and litter are common; noisy parties scare off wildlife, as do the increasingly popular "caravans" of off-road vehicles,[32] and tourists' carelessness is a major cause of forest fires.

The RFD's National Parks Division (NPD) has struggled to come to grips with these problems, despite opportunities to learn lessons from abroad. Several years ago, for instance, a team of RFD officials went to South Africa to visit Kruger National Park, considered a model for eco-tourism. Visitors must make reservations to enter the park, and the rest camps inside are rudimentary—basically bungalows and barbecues. Visitors can stay at more luxurious accommodations in the area surrounding the park, much of which has been turned into private game preserves, sometimes by farmers who want to live in their ancestral homes but otherwise wouldn't have the means to do so. Tourists to the private facilities may have more liberties—they can go on night safaris, perhaps, or hunt—than in the government-run part, but they also pay more for such privileges. These private parks act as a kind of buffer zone: Wildlife is able to roam farther afield since development tapers off in the areas closest to the park. And because it's a market-based solution, people living around the preserve have greater incentive to protect it, leaving the park and its wild inhabitants less vulnerable to corruption and poaching.

Of course, South Africa had certain advantages over Thailand: Kruger is huge and was created a century ago, when local people were unable to object. The Makuleke people of Limpopo province, for instance, had seen their land forcibly taken away and added to the park. Following the end of apartheid, they fought to have their land returned, and eventually gained it back in 1998. They now jointly manage the land with the South African National Parks agency in an innovative attempt to merge conservation with community development.

Frequent calls have been made for Thailand's parks division to establish carrying capacities for its protected areas; namely, figure out how many visitors a park can take at any given time, and then limit the number of visitors accordingly. That was in effect what occurred with the sea kayak tours at Phang Nga Bay, although it was not done in a scientific manner. But it's not only tour companies who dislike such limits. They are also unpopular with the public. Even modest proposals to raise entrance fees—which for Thais are usually less than a dollar (foreigners pay more)—generally result in howls of protest. The NPD therefore still relies on government allocations to run the parks, which are invariably undermanned and underequipped. The agency does get some revenues from its tourism facilities, but with the exception of the more popular parks, where it has some pleasant bungalows, most of its operations are of low standard. Lodging often consists of dingy and overpriced dormitories. The food offered is usually quite basic. And the service is indifferent because park employees receive the same standard low wage no matter how happy the guests are. The RFD is after all a forestry agency; it has never gotten the hang of running a service industry. In addition, chronic corruption and political interference has made it difficult simply to manage the forests. As a result, most park visitors either camp in tents or stay in resorts outside park boundaries.

Tourism investors have long been frustrated by seeing the immense potential of nature tourism in the parks go unexploited as a result of the RFD's questionable management. Predictably, they and their allies in the TAT have made several attempts to gain control of tourism in protected areas. In 1993, Savit Bhotiwihok, a Cabinet minister who also owns the Rayong Resort not far from Koh Samet, sought to give control of certain areas within the parks to the TAT, which would then turn them into "tourism estates."[33] In 1997, the TAT put forward a plan to take the National Parks Division out of the RFD and make it a state enterprise that would promote commercial tourism in the parks.[34] The RFD, recognizing that the parks are a potential gold mine, naturally opposed these plans, which were quickly shot down after an overwhelming public outcry. The TAT inspires a great deal of mistrust, not only because it has neglected environmental issues in the country's resort areas but also because it mismanaged resort facilities in Khao Yai—the country's first and highest-profile national park—which at one point included a golf course smack in the middle of the park. So environmental groups, despite regularly criticizing the RFD for its deficiencies in forestry management, are also the first to jump to its defense when the TAT tries to hoard in because they judge the RFD to be the lesser evil.

Public groups have also been able to fend off quite a few development projects inside the parks.[35] A good example is the long-running attempt by investors in the northeastern province of Loei to build a cable car up to the summit of Phu Kradung. A striking table-shaped mountain that juts up from the Isaan plain, Phu Kradung was named Thailand's second national park in 1962. It is actually many worlds in one. Walking up the mountain, you first pass through a bamboo forest—dry, brown, and leafless during the hot season. Farther up, you become enveloped in the cool shade of a lush evergreen forest. Then, at the summit, you round a bend and are suddenly struck with an entirely different vista as the sun glares down on Phu

Kradung's broad plateau, an expanse of grassy savanna dotted with groves of gnarled pine. It's like stepping out of Asia into Africa. You almost expect a herd of giraffe to wander by. These varied ecosystems help make the mountain unique: It is home to four species of plants found nowhere else in the world and seventeen species of threatened birds. And Phu Kradung (the name means Bell Mountain) has also become a cultural resource because it's a rite of passage for thousands of Thai college students to make the trek up the mountain together. "Young Thais generally stay with their families until they graduate," explains Krongkaew Chaiyapa, a professor at Loei's teacher's college who helped establish a park support group. "[So this is often] the first chance they have to go away from home with their peers."[36]

The park's popularity has attracted the attention of investors who argue that a cable car would make it easier for people to visit the summit and bring more tourists, and revenues, to the province. Environmentalists counter that too many visitors already strain the park's resources, particularly around New Year's, and are clearly impacting both the wildlife as well as the experience of the visitors themselves. The cable car would become an eyesore, they say, and would perhaps lead to the construction of resorts on the summit. Opposition has also come from nearby farmers, who double as porters for trekkers hiking up the mountain. It can be grueling work, but it fits in well with their farming activities since the park is closed during the rainy season, when they are busy growing rice. But if the cable car comes, they fear they will lose their supplementary income. Hae, a local guide, offers perhaps the simplest reason to avoid a cable car: "People should just walk up the mountain," he says, "it's *sanuk* [fun]."

STRANGERS IN THEIR OWN LAND

Hae is right, and not just about Phu Kradung. Traveling just about anywhere in Thailand is fun. Thais have practically made a religion out of *sanuk sanan.* They are quick to laugh, even at themselves. Blissful relaxation, the feeling of being *sabai,* is the other deity in Thais' emotional pantheon. It's a feeling that's hard to comprehend until you've been lulled to sleep in a hammock on an idyllic, tropical beach, having just gorged on a splendid feast of Thai seafood (Thais are relentless eaters while on vacation, yet amazingly manage to stay rail-thin). The best times of all come in the evening, when you can relax with your friends on the beach, sip a little Mekong whiskey to loosen your vocal chords for when someone breaks out a guitar, or just stare at the stars and the phosphorescent sea. It's so simple, so *sanuk,* and so *sabai.*

You can imagine how pleasant life used to be for people on Phuket, an island of rolling hills and splendid beaches, rich in all kinds of resources. "When I was in my early teens, my friends and I would go to the beach and have a little picnic. We would make some *gai yang* [barbecued chicken], swim and just talk," says Lek, a youth leader from Bang Thao Bay. He speaks with a wistfulness not unlike that of tourists who have seen their vacation spot ruined by development, but for Lek it's his home that has suffered. From 20,000 visitors in 1976, the number of tourists visiting Phuket had soared to 2 million by 1994, and 3 million by 1999.[37] Although the increases have helped raise incomes on the island and have provided gainful employment for thousands of workers from elsewhere in Thailand, there have been undeniable costs, as well.

Lek didn't want his real name used because he and his fellow villagers were fighting with Thai Wah, a subsidiary of Singapore-based Wah Chang International, which has built a series of huge resorts—including the Sheraton Grande Laguna, Dusit Laguna, Pacific Island Club, and Banyan Tree Golf Course—at Bang Thao Bay, on Phuket's west coast. The locals made many complaints about the company's behavior, accusing it of blocking their drainage canals, for instance, and saying that hotel security guards had mistreated them. Such tension between big hotels and local residents is unfortunately all too common in Thailand, and it is one reason that Phuket's residents seem to have mixed feelings about the spread of large-scale development on their island.

But perhaps the most poignant complaint heard on Phuket is that people like Lek now often have difficulty going to their own beaches. This was a major factor behind the opposition to the North Phuket resort proposed at Nai Yang National Park, the only undeveloped beach left on the island. Actually, by law all beaches in Thailand are supposed to be public property, but the big resorts like to offer their guests private seaside enclaves and keep local hawkers away. Many have therefore found clever ways to block public access, much as the Phuket Yacht Club did and for which it was reprimanded. The cove at Phuket's Meridien Hotel, for example, has effectively been privatized because the resort controls the land that leads to it. At Bang Thao Bay, the locals complained that they could no longer use the public roads that led to the beach. When I traveled there, one road had a sign saying "public," but then immediately behind it another said the road was private property and that people should keep out. It's no wonder many Phuket residents feel like strangers in their own land.

Thai Wah's representatives responded that Laguna Development, a firm that it only partially owns and does not control, erected the sign in question. They also claimed that following a confrontation with the villagers, three roads to the beach had been opened. The conflict turned into a typical cat-and-mouse game between the various groups. For journalists and other outsiders, it's often hard to figure out what exactly is going on in these arguments, since the statements from each side often have nothing to do with one another: The villagers will complain about how they've been wronged, but the company will claim that the mob of protestors is merely a front for a secretive business competitor.

The Thai Wah situation was particularly frustrating because I had hoped to interview the president of Wah Chang International, Ho Kwon-Ping, a former student radical and journalist who had been a colleague of the *Nation*'s founder, Suthichai Yoon. Given his background, he must have had an interesting perspective on the situation. But in the end, the firm's only comments came from functionaries who seemed unfamiliar with the details of the dispute. I did receive a personal letter from Ho, a bitter and impassioned defense of his company's actions that claimed the firm was the victim of attempted extortion and a biased press. But he would not allow the contents of his letter to be quoted, and so they became just empty words.

Most frustrating of all, we never received a comment on the main feature of our report: an exposé into how the land at Bang Thao Bay had become private in the first place.[38] Land rights investigations became a specialty of ours at the *Nation*'s environment desk because the government has been so reluctant to address how the system is regularly abused. Thailand has virtually no property tax, a major reason for the country's income disparities. And even compared to its neighboring countries—where the colonial powers systematically surveyed and taxed land, and also kept good property ownership records—Thailand's land rights system is confusing and subject to manipulation. So although developers everywhere have tricky ways of gaining land, their methods seem especially murky in Thailand.

By digging up the actual title deeds to controversial land claims and investigating how investors acquired them, we were able to offer reports based on facts rather than on the charges and countercharges of various stakeholders. Because these land rights documents are kept jealously private in Thailand, uncovering them took a lot of sleuthing. Sometimes, we were able to obtain records presented in court during trials; we even managed to meet with government investigators who were upset because findings had been covered up. Often, we simply had to plead with sympathetic officials.

Our reward came in acquiring some major scoops. My colleague Kamol Sukin revealed how the Nimmanahaeminda family, Chiang Mai's wealthiest clan, had managed to obtain title to a seventeen-rai plot of land in Doi Suthep-Pui National Park and then lease it out to Juldis, a major development company, to build a resort there.[39] Another colleague, Nantiya Tangwisutijit, reported on a disputed piece of beachfront land at Khao Sam Roi Yot National Park that is claimed by the Diamond Group.[40] Ornithologists fear that the development of the Riviera Beach resort threatens several endangered species because the site is a world famous bird sanctuary.

Pennapa Hongthong wrote about a land scandal on Koh Samet in Chumphon province (a different island from the more famous Koh Samet in Rayong province). The local villagers had long tried to obtain title to land they had worked on for generations, but were rebuffed by Land Department officials who claimed that regulations did not allow land ownership on islands.[41] Unbeknownst to the villagers, the regulations were modified in 1994; by the time they found out, much of the land was in the hands of some well-connected outside investors, including the family of Suchart Tancharoen, who just so happened to be the deputy interior minister in charge of the Land Department. Klomjit Chandrapanya and I worked on a dispute involving land owned by a company called Si Chang Thong Terminal and slated to become a controversial 10-billion-baht industrial port in Chonburi province.[42] We discovered that all the original proof-of-land-use documents, on which the title deeds are based, were reported missing by the owners and the Land Department—a suspiciously common occurrence in land scandals. In each of the above cases, the landowners and developers denied any wrongdoing. None was ever punished or even brought to trial.

As these examples suggest, Thailand is plagued by land scandals all over the country. One of the more famous of these in recent years was the discovery that the family of Prasad Wessabutr, a high-ranking Interior Ministry official, had built three houses on land at Kanchanaburi's Sri Nakharin National Park that had supposedly been set aside for villagers displaced by the construction of a dam. Such scandals tend to be particularly common in popular tourist areas, not just at beaches and parks, but also in the northern highland provinces of Chiang Mai and Chiang Rai. For brevity's sake, I will stick to examining examples in another of the country's most popular and enchanting regions: the Andaman coast.

MAGIC COCONUTS AND FLYING DEEDS

For this alternative tour of southwestern Thailand, let's first return to Bang Thao Bay in Phuket, where Klomjit and I, with the help of another journalist who must remain anonymous, uncovered our first big land scandal. In 1953, Khoon Visesnukulkij, a local businessman known by his Chinese name, Eng Kee, received a government concession to mine for tin on 456 rai* of land at Bang Thao. (Phuket has extensive tin deposits that, along with its rubber plantations, helped make the province relatively wealthy even before large-scale tourism development began there.) After he died in 1960, his heirs continued to mine the land and eventually gained title to 1,000 rai of land on the site. It remains unclear why the government granted them ownership, since the Mining Act of 1940 states that if a concession is applied to public land, it must be returned to the public when the concession is completed. Meanwhile, the Land Act of 1954 states that title deeds will be granted only to those who possess documentary proof of land use. But according to the records we obtained, Visesnukulkij Company never requested such documents for the land it claimed at Bang Thao. We tried to contact Suchin Udomsap, the heir who runs the firm, for our report, but he was unavailable for comment. He assured a friend, however, that all the documentation for the land was correct.

In 1984, Visesnukulkij Company sold its land holdings at Bang Thao, which by then totaled 1,500 rai, to Thai Wah (the Singapore-based firm later acquired another 1,500 rai or so of land along the bay). By all accounts, the land was in pretty poor shape when Thai Wah bought it, having been degraded by the mining operation. Thai Wah set about cleaning it up, not only making the site suitable for tourism but also a tidy profit for the firm. Visesnukulkij Company reportedly sold the first bit of land to Thai Wah for 40,000 baht per rai. Nine years later, Thai Wah was selling its land for more than 4 million baht per rai.

In these situations, attempts to return the land to the public domain prove difficult. The current owners invariably claim that they merely purchased the land from the previous owners and have since invested huge sums of money in it. So do you then go after the original owners? Or the officials who approved the awarding of title deeds? Doing so would require not only strong evidence but also courage and determination, which are distinctly

---

*One acre equals about 2.5 rai of land.

lacking amongst Thai provincial authorities. "I do not have the authority to investigate the origin of land title deeds because that would concern civil law," said Phuket's then-governor, Yuwat Vuthimedhi. "We can't just follow hearsay that the land rights were granted falsely. If we investigate private projects, it would delay their projects and banks will not lend money to companies. Companies could sue us." It was to become a familiar refrain, one we would hear frequently in Phuket's neighbor to the south: Krabi.

The caves and coves of Krabi have long been famous as the lair of pirates who set out to rob passing merchant ships. Today, the situation has been reversed: Despite being protected as national parks, the beaches and cliffs of the Krabi shore have been invaded by land encroachers. But as we've seen with the unfortunate manager at Sea Canoe, the threat of violence is not merely confined to history.

The most famous tourist destination in the province is Phi Phi Don Island, a narrow sandy isthmus connecting two landmasses ringed by towering limestone cliffs. Sarah MacLean, a long-time ex-pat who first visited the island twenty-five years ago, remembers it as a jewel: clean beaches bordering two bays of crystal clear water, each home to vibrant coral reefs. But upon returning to the island fourteen years later, she was shocked by the changes crude development had wrought. Today, the isthmus is chock-a-block with resorts and bungalows, including several high-rise hotels such as the Phi Phi Cabana. The beach at Ao Lo Dalam is filthy, and the cove is filled with boats. The water is still clear, but the coral has died, choked by the run-off of pollutants and silt. And parts of the island have turned into what can only be described as a slum: mounds of rotting garbage, streams of stinking sewage, and tumbledown shacks sitting amidst piles of wreckage. Everyone on Phi Phi Don complains about the situation, but little is done. The island has an incinerator and a wastewater treatment plant, but neither is in use, and no one is sure why. Some say faulty technology was employed, others that there is not enough electricity. Some even claim there isn't enough garbage, and yet boatloads of rubbish are sent to the mainland every day for disposal.

How could so much of Phi Phi Don and other nearby islands become private land when they are supposed to be part of Nopparat Thara Beach–Phi Phi Islands National Park? It was Ing who first highlighted this issue after investigating the land claims of wealthy investors around the park and uncovering encroachment techniques that take advantage of the legal loophole allowing people to claim previously used territory. Wealthy investors usually can't claim they had once worked or lived in these remote regions. But local villagers are willing to make such claims, spurious or otherwise, since they often use public or vacant lands to graze livestock, to grow and tend plantations, or as fishing bases. On their own, the villagers' attempts to claim the land usually go nowhere. But with sufficient financial backing to grease the wheels of officialdom, they can claim their title deeds, which are then quickly sold off at a pittance to their wealthy backers.

One of the cleverest techniques used in southern Thailand is known as the "magic coconuts" scheme. Coconut trees growing elsewhere are transported to a desired location where they are turned into the "instant plantations" that form the basis for claiming ownership. Alternatively, young trees may be planted in rows beneath the forest canopy, which is then culled with herbicides as the coconuts grow taller. Sail around Phi Phi Don Island and you will see lots of coconut plantations growing up magically out of the jungle. Even a palm plantation has sprouted near park headquarters on the mainland.

RFD investigators suspect this tactic may have been used on Koh Poda, an island in the park about an hour's boat ride from the mainland that has been turned into a resort by Chuan Phukaoluan, a former president of the Krabi Chamber of Commerce.[43] The title deeds for the land were issued in 1983, the same year the national park was declared, based on a land-use document that supposedly stated the land was used as a coconut plantation back in 1955. But a special task force set up by the Prime Minister's Office found that, as so often happens, the land-use papers were reported lost in 1983 by both the alleged owners and the Land Department, which then issued a new set. Aerial and ground surveys of the island in 1957–1958 reported no signs of a coconut plantation, and a plant analysis on the trees made in 1987 showed that the trees were not even twenty years old. Chuan, meanwhile, defended the legality of his resort: "If someone bought land from someone else and that person gained it illegally, is [the purchaser] wrong? It is a major mistake on the part of the authorities who drew up the national park boundaries. Krabi is a hundred years old. The national park boundaries were just announced in 1983. People have been living on the land forever." Nevertheless, in 1990, the special task force instructed the Interior Ministry (which oversees the Land Department and runs the provincial administration) to revoke the title deeds. But the governor of Krabi refused, and the ministry's permanent secretary supported him, claiming the deeds were merely "inaccurate." In 1993, the RFD vowed to take the case to court, but apparently the local prosecutor simply refused to file suit, and the matter has been in limbo ever since.[44]

That same year, another dispute between investors and the park erupted into prominence when we at the *Nation* uncovered evidence that the land claims for the 500-million-baht Dusit Rayavadee Resort at Phra Nang Bay were partially based on documents which had, once again, gone missing.[45] Phra Nang is actually a peninsula on the mainland cut off by a curtain of limestone cliffs, making it accessible only by boat. Bordered on three sides by spectacular beaches and punctuated by yet more karst towers, the area's beauty rivals that of Phi Phi. Development is also going the way of Phi Phi, however, and the peninsula has become packed with resorts and the usual Khao Sarn–like businesses. Many of the land claims on the peninsula are not disputed, but a section near Phra Nang Bay, where a cave is home to a revered local shrine, was thought to be forest reserve until 1987–1988, when an architect named ML Chainimit Navarat and another businessman claimed ownership of the land, according to Ing. Chainimit first built scores of bungalows on the land, but then decided to go upscale. He bought as much land as he could and sold it to the Premier Group. Together, they built Dusit Rayavadee, an array of high-priced, low-rise luxury bungalows. Now, the only way for the public to reach Phra Nang Beach from the rest of the peninsula is to travel along a narrow walkway designed to keep the resorts' well-heeled guests secluded from prying eyes, and that easement was won only after a campaign overcame the resort's initial objections.

Meanwhile, there remain problems with the resort's land-rights documents, including discrepancies in the sizes and shapes of the land plots to which they refer. Premier's CEO, Vichien Phongsathorn, says that all the documents were investigated when it bought the land in 1990–1991, and were proved authentic. Chainimit also defends his actions:

> I was shocked at the way the bungalows were managed. Tourism, sold stupidly, can be destructive, just like on Koh Samet or Koh Phi Phi. We have sold *farang* all of our national resources, very cheap. . . . There was no garbage dumpsite [at Phra Nang]. I was not making enough to conserve the environment. My theory is that natural resources should be maximized. Why not bring it to the highest value and sell it at a reasonable price? But I didn't want to do it with *farang*. We've lost many good places to *farang* in Phuket. So I turned to [Thai-owned] Premier. . . . Now I sell the property at the highest price. Okay, it is a haven for the rich. But does

> Thailand have a right to have a haven for the rich? [Natural resources] are not only for the poor.[46]

Chainimit is bitterly resentful of journalists, particularly Ing, whom he verbally threatened in front of a couple of *Nation* reporters: "[Other powerful people's] land has smaller-sized coconut trees than my land. You put my name on the thieves' list. The real thieves just smile, they are not touched. . . . You have the power of the pen. Be careful. Or the pen could stab other people. If anything comes up, I'm coming straight for you." The RFD's legal department investigated whether to take action concerning the case, but apparently decided that either the evidence wasn't strong enough or the project's backers were too powerful to take on, for all has been quiet since.

Given the advanced state of development in Phuket and Krabi, and the obvious difficulties in revoking land deeds in these situations, it seemed worthwhile to see whether there were other similar examples that could be exposed before development got a foothold. I eventually came across just such a place at Haad Chaomai National Park on the Andaman coast in Trang province, south of Krabi. The water is not as clear as it is at Phi Phi and Phra Nang, and the scenery is not as striking, but this area also has some magnificent karst cliffs. What's more, it is still largely pristine and its long beaches are mostly empty, apart from the rubbish that washes up from the sea.

When I first visited the park in 1995, there was little development apart from a few small restaurants along Pak Meng Beach, but it was clearly being primed for a tourism boom.[47] The TAT had provided Bt27 million for a pair of projects: a pier from which boats could ferry tourists out to the islands, and a mile-long road that opened up the area around Yong Ling Mountain—a cluster of enchanting limestone towers adorned with riotous jungle, hidden caves, and sparkling pools of tranquil water. The Yong Ling project seemingly ignored the principles of eco-tourism: There was no study of the area's carrying capacity; and there are no educational materials at the newly built tourist service center, apart from a poster on coral reefs, which are not even found there. "I agree with having tourism in national parks, but not the way they are planning to develop it at Haad Chaomai," said one RFD official stationed at the park.

The worst project of all was taking place at the southern end of Pak Meng Beach, where the Public Works Department built a so-called "erosion barrier" with 24.5 million baht in provincial funds. The big cement wall is supposed to shore up a road that was built too close to the shore, between the beach and the mangrove forests that form a much more effective defense against erosion. Both Trang's then-governor, Yongyuth Wichaidit, and the RFD's marine parks chief, Nopadol Briksvan, considered it a waste of public money; and sure enough, the barrier's base is already being eaten away by the pounding surf. But there may have been an ulterior motive for building it. Its broad walkway—a perfect place for vendors to set up shop and for families to go on an evening stroll—and broad staircases leading down to the beach make it an ideal promenade.

Perhaps most suggestively, it is only a short distance away from Chang Lang Beach, where investors have staked claims to private land marked off by long stretches of ugly barbed wire fence. If the park ever takes off as a tourism destination, this is probably where the hotels and shops will be built. During the 1980s, ten title deeds covering 180 rai of beachfront land were handed out on the strength of some land-use papers that allegedly proved villagers had used the area before it was declared a park. But an RFD investigation noted many discrepancies in the sizes and shapes of land plots between these proof-of-land-use documents and the title deeds that were subsequently handed out. For instance, papers allegedly proving previous use for 30 rai of land were turned into deeds that covered a total of 78 rai. In another example, ownership for a plot of land at the mouth of a stream was awarded even though, according to the RFD report, the area in question was under the sea at the time the land-use papers said it was being used. Meanwhile, plots described as "grasslands" on paper actually host dense forest.

Privately, investigators believe this is a typical case of *chanot bin,* or flying deeds, a common ploy used by encroachers whereby papers for one plot of land are used to gain ownership of another, more valuable, plot. Following its own investigation, Thailand's Counter Corruption Commission (CCC) agreed, at least in part. In 1992, it concluded that four of the original ten title deeds were "wrongfully issued." Although the title deeds were initially awarded to local villagers, suspicions were heightened when the deeds were quickly sold to big-time investors. Pichet Phanvichartkul, the Democrat member of Parliament for Krabi, bought 5 rai of land just ten days after it became private property. Salil Tohtubtiang, a wealthy businessman whose brother Surin owns Trang's Thumrin Hotel and serves as president of the province's chamber of commerce, bought 8 rai of land approximately three months after the deeds came out. Meanwhile, Prakij Rattamanee, a former Democrat M.P. from Trang, was awarded ownership of 10 rai of land, and then bought another 2 rai. He has since built a hotel, the Chang Lang Resort, and a real estate project there. Another landowner is Wech Kiman, whose wife worked in the provincial planning office when it designed the road construction projects inside the park. Pichet, Salil, and Wech all claim they have done nothing wrong (Prakij would no doubt say the same, only he has refused to speak to me). In fact, they say, they are victims in this dispute because they have purchased private land that now appears to be of dubious origin.

But since the CCC report was released, nothing has happened. Yongyuth, the former governor, said he didn't have enough proof to move ahead, and the RFD has not filed suit to revoke the questionable deeds. So far, most of the land remains undeveloped. Surin, the province's biggest businessman, once planned to build a resort complex there, but he now says it won't happen. "We won't build it because it is virgin land. If we build facilities to stay overnight and too many people come, it will damage the environment, just like at Phuket or Samui. So we will promote day trips to the park instead." At the moment, tourists come, but as campers and day-trippers. But the Amari hotel chain is reportedly planning to open a new resort on Chang Lang Beach.[48] Haad Chaomai National Park tenuously hangs on as one of the last surviving examples of unspoiled coastal wilderness in Thailand. But for how much longer?

BEAUTY AND THE BEACH

"Did you see the picture in today's newspaper of the French actress who will play in *The Beach,*" a senior parks official asked me wistfully in October 1998. "She's beautiful, huh? And Leo, the *phra ek* [hero], is very handsome, too." Yes, Thailand's forestry officers had stars in their eyes, and perhaps it clouded their judgment. For they apparently did not expect the bitter protests that resulted when they allowed a 20th Century Fox producer to alter part of a national park to film the movie *The Beach* at Maya Bay on Koh Phi Phi Leh.

A smaller cousin of Phi Phi Don, the island is equally spectacular but uninhabited, and thus in much better shape. The RFD chief, Plodprasop Suraswadi, acknowledged that he signed off on the project largely for economic reasons. "Thailand is broke, and we need the money the film will bring in," he said. "I can well believe there will be some damage done to Phi Phi Leh, if only by accident. But it is only one small island, and the film will benefit the country as a whole by promoting it as a tourist destination. Just like with James Bond Island [in Phang Nga Bay], foreign tourists will want to see where *The Beach* was made."

The filmmakers had searched far and wide for the ideal setting for their movie—about a young American backpacker who finds a hidden beach where a group of travelers have established a seemingly idyllic commune secluded from the rest of the world—and they found it at Phi Phi Leh. Except for one thing: The beach at Maya Bay did not have any coconut trees. How could you have a movie about a tropical paradise without any coconuts? And so the filmmakers decided the native vegetation at Maya Bay would have to be removed and replaced with more than a hundred coconut trees. Since willfully disturbing a protected area is illegal under the National Parks Act, producer Andrew Macdonald, presumably obeying orders from director Danny Boyle, sought permission from the RFD. And after paying 4 million baht in fees for use of the island, they got it. Plodprasop even suggested the newly planted palms should stay once the movie was completed, as tourists would want to view the bay "just the way it was filmed."

It was all so ironic. Here was a unique place sought out for its natural splendor, and yet it needed to be tarted up to satisfy Hollywood-inspired fantasies about tropical islands—and with transported ("magic") coconut trees, no less, the very instrument that had become a tool of encroachers all around the park. Was this a matter of Hollywood's celebrating Thailand, as the movie's supporters claimed, or Thailand's celebrating Hollywood?

Thai environmentalists had no doubt; the authorities had sold the country's resources to foreigners, and, along with local tour firms, they worried that Phi Phi Leh would be damaged by the filmmakers' activities. Removing the native plants and bulldozing the sand dunes, they argued, would cause the fragile beach to erode. The sand running off into the bay would then kill the coral. "I'm definitely concerned that it will damage the coral," said Pongsak Mukda, a divemaster on Phi Phi Don. "Coral can't run away, so any sediment that covers it will cause it to die off. It may take a few years to happen, but it's a real threat."[49]

A raucous protest campaign commenced, led by who else but Ing. The issue was a natural for her in more ways than one. She had always been entranced by the beauty of Phi Phi Leh, and considered it to be one of the last outposts of wilderness in the country. Meanwhile, Thailand's National Film Board, which had approved the filming of *The Beach,* had forbidden the release of Ing's first feature film, *My Teacher Eats Biscuits,* because of its piercingly satirical look at religion and the Thai clergy.

Under Ing's guidance, the media-savvy campaign began with a protest featuring performers wearing Leo masks decorated with fangs dripping blood in front of 20th Century Fox's office in Bangkok. Eventually, they carried out a sit-in at Maya Bay. DiCaprio complained that he was treated unfairly, but turning the issue into a cause célèbre worked because it drew media attention from around the world. Even local politicians and businessmen got into the act by complaining they had not been consulted; eventually they filed a lawsuit to halt the activities.

The RFD and the filmmakers reacted defensively. Plodprasop vowed that his agency would defend the island's environment vigorously, and that no permanent harm would be done. A committee was set up, if belatedly, to study the impacts, and it gave the project a green light. Meanwhile, DiCaprio came out and said the filmmakers had shown meticulous care in looking after the island's environment. The producers announced they would plant only sixty coconut trees and that they would be removed immediately after shooting was completed. The native vegetation was put in a nursery and would eventually be replanted on the beach. The filmmakers also put up an environmental bond for 5 million baht in the event of damage done to the island. They also claimed they had improved Phi Phi Leh's environment by picking up all the garbage that had washed up from the sea (a welcome contribution, no doubt, but also a pretty weak one: They had to clean up the beach to make the movie, and once they left, more detritus would simply wash ashore).

The filmmakers were essentially tangled up in a dispute between Thailand's greens and the forestry department. They had received permission from the authorities to film on the island, and they assumed that would be the end of it. Yet here they were in a strange country besieged by enraged Third World environmentalists and the press. They were even being criticized by some of the hoteliers who had done so much damage to Phi Phi Don. Nor was the message from the protestors always clear. Some critics complained that because they had paid off the RFD, the filmmakers could do as they wished inside the park, even while local housewives were detained for collecting snails to take home for dinner. Others argued that the filmmakers should have been forced to pay a lot more than 4 million baht for using the park since they had given DiCaprio a reported $20 million for appearing in the movie.

On the other hand, the filmmakers at times also displayed a surprising lack of sensitivity. DiCaprio ultimately complained that the protests were just "a big waste of time."[50] Alex Garland, the author who had written the novel on which the movie was based, dismissed the whole affair as "crazy."[51]

And the producer Macdonald seemed to gloat when the sit-in demonstrators were finally hustled off the beach and filming began. "It was a very exciting day," he said. "These ten wimpy Greens from Bangkok facing off against sixty to a hundred of these tough fisherman types. There weren't machetes flashing, but it was a bit Jimmy Hoffa."[52] Most important, the filmmakers could have ended the controversy by scrapping the planting of the coconut trees, which were necessary only for a brief scene or two. Perhaps they could have used computer effects instead, or mixed in shots of other locations (after all, they found a way to film a ganja plantation on the mainland without growing real marijuana plants). But they ruled out such options. Either they didn't agree with the environmentalists' objections or else decided that fulfilling Boyle's vision of filming real coconut fronds was more important. In the end, the director got his wish.

Many films have been made in Thailand without controversy, but *The Beach* was different because it entailed altering a national park. To environmentalists and many of the locals, purposefully changing the beach at Maya Bay was a form of sacrilege. There is a legend about the place. The story goes that a man and his son named Maya were once exploring the island, perhaps looking for birds' nests, when the young boy was trapped in a cave by a landslide. Try as he might, the distraught father could not dislodge the boulders. He attempted to keep the boy alive by passing food and water in through cracks in the rocks, but eventually realized it was no use. Rather than have his son waste away, the grief-stricken father finally stuck a rifle in through the rocks and killed the boy. So there is a sad sanctity to Phi Phi Leh, and the thought that its name might be changed to, say, "Leo Island" as a marketing ploy seemed to many like a desecration. Would the reaction be so different in the United States if a foreign movie crew had sought to alter Yellowstone or Yosemite, even temporarily? To those who revere nature unadorned, it would be as if the money changers had entered the temple.

For Ing in particular, the battle over *The Beach* was an intensely emotional affair. Having spent a decade fighting against encroachers in the national park, having reported on how virgin forests are cut down to make way for "magic coconuts," and having seen her own low-budget film banned by Thailand's Film Board, here she was battling against a Hollywood movie that was seeking to plant coconut trees on the park's one remaining undeveloped island. After a while, she found it difficult even to discuss Phi Phi Leh and *The Beach* without breaking into tears. "I have repeatedly laid down my life for Nopparat Thara Beach–Phi Phi Islands National Park, but I would never be allowed to shoot there," she wrote. "The Film Board would never pass my screenplay about the Phi Phi Islands, a story of corruption and land scandals which would not be portraying the park as some tropical fantasy but as its beleaguered self."[53]

The filming of *The Beach* on Phi Phi Leh has also created a worrying precedent. In 2002, the U.S. television network CBS received permission to shoot its hit show *Survivor* on Tarutao Island, in another national park farther south, despite renewed objections from environmentalists. The show's participants were able to gather food in the park, which is normally forbidden by Thai law.[54]

The Beach affair resulted in a disappointing revelation for me, as well. I saw the controversy as a way to focus on the broader issue of managing tourism in conservation areas, and highlight the troubles facing this park in particular. The movie's backers claimed the movie would help promote tourism in the area, but its critics argued that was the last thing Phi Phi needed. "We know that the film will bring us more tourists, and that this will put still further strain on our environment, which is already at the breaking point," explained Surat Jepkhok, the head of Phi Phi Don's villagers' committee, who led a petition drive expressing concern about the filming of *The Beach* on Phi Phi Leh. "If we don't get a handle on the situation soon, it will be our ruin." So it seemed an ideal time to talk about the need for carrying capacities and to look at how media promotion contributed to tourism's problems. The tourism blight ravaging Thailand has a very predictable life cycle: Promotion leads to encroachment and development, which is followed by pollution and decay. All too often, people notice only the final stages, when it is usually too late to prevent the environment from being despoiled.

By the time the controversy about *The Beach* broke out, although I was still writing an occasional column for the *Nation,* my main job was working as a reporter, scriptwriter, and part-time host on Thai television for a weekly feature program about the environment. We got the go-ahead from our sponsor, the government's Department of Environmental Quality Promotion (DEQP), to produce a show looking at the issue. It proved a hectic task. Our own filming was delayed by a fierce storm, but we persevered and managed to arrange interviews with people from all sides of the issue (although the producer Macdonald declined to meet with us), including the Koh Poda resort owner Chuan. We lugged camera and sound equipment up mountains, onto boats, and out to Phra Nang, Koh Poda, Phi Phi Don, and finally Phi Phi Leh.

Things turned bleak, however, when we returned to Bangkok. Our government sponsors decided at the last minute to scrap the program. They were vague about why, but the reason was obvious: It was simply too sensitive an issue for them. Having experienced so much freedom at the *Nation,* I found it something of a shock, not to mention a bitter disappointment (although we did manage to sneak in much of the footage on another episode we ran about coastal resources). But it confirmed what I'd always heard: Although the press in Thailand is quite free, television is far from it.

In fact, everything and everyone who became involved with *The Beach* seemed to end up the worse for it. The filmmakers' reputations were tarnished by the controversy. So was DiCaprio's, who vowed that "this year I'm really going to do some big things as far as the environment is concerned. I'm going to become a lot more active."[55] Bitterly disappointed that the filming had gone ahead, Ing withdrew from public view out of a sense of hopelessness for her country. After the coconut trees were removed from Maya Bay, the attempt to replant the native vegetation failed. The plants died, and a storm during the following monsoon season wiped a good portion of the beach out to sea. The RFD posted a sign warning visitors not to tread on the damaged dunes. And finally, the movie bombed. The scene with the coconut trees passes so quickly you can barely notice it: A surreal anticlimax to a real-life drama.

3

DAMS
THE PRICE OF POWER

MYSTERIES OF THE MEKONG

The explorers seeking the source of the Mekong River underwent severe hardships. Their quest had taken them to the remote highlands of eastern Tibet, a region sparsely inhabited by the nomadic Khamba people, whom the authorities had never been able to bring under control. The fifty-eight-year-old leader of the expedition, Michel Peissel, a veteran of two dozen journeys across the region, called it "the very inner heartland of the highest and most inhospitable part of the Central Asian highlands."[1] By day, the explorers braved thunderous hailstorms. By night, they were attacked by wolves. Several of their horses disappeared, and then so did one of their guides. Finally, high up in the Rup-sa Pass, nearly five kilometers above sea level, they found a "totally unspectacular . . . marshy field from which water was oozing." This was it: the origin of the mighty Mekong, the tenth biggest and twelfth longest river on Earth.

A typical tale of Victorian-era exploration, you might think—except that it occurred in 1994. That's an indication of just how much remains unknown about the river. Although the Mekong basin encompasses 800,000 square kilometers and is home to 55 million people, war followed by political and economic quarantine rendered much of the area off-limits to outsiders for decades.

From its wellspring near the Tibetan plateau, the river picks up speed and volume, rushing through steep gorges in China's Yunnan province during the first stretch of its 4,200-kilometer journey. The Chinese know it as the *Lancang Jiang* (the Turbulent River), but it may not stay that way for long, as they've already begun a series of massive hydro-engineering projects that together will rival the output of the much better publicized Three Gorges Dam. Heading south, the Mekong becomes less turbulent as it weaves through a rich cultural fabric of hill tribes and valley dwellers who rely on fish from the river and its tributaries as their main source of protein. Here the Mekong serves as a political boundary for Laos, separating it from Burma in the Golden Triangle, and then forming most of its border with Thailand. In southern Laos, the Mekong pours over the Khone Falls, a stretch of rapids so wide you can't see from one side to the other.

The river then flows south into Cambodia, where it forms part of the ecological marvel that is the Tonle Sap Great Lake. Angkor Wat may be the soul of Cambodia, but this immense body of fresh water is the country's beating heart. Each year, the Tonle Sap pulses to a natural rhythm as it becomes swollen by the monsoon rains and expands across Cambodia's central plains, only to shrink back to one quarter the size during the dry season. Life in Cambodia moves at the same annual cadence. With the rains, the Mekong rises, spilling water into the Great Lake through the Tonle Sap River. Usually the lake's main artery, the river actually reverses itself, becoming a vein through which the lake is fed with seeds and fish eggs. These are then spread across the flood plain, a fertile feeding ground for spawning fish that provides a copious bounty of protein for the people of Cambodia.[2] And the Mekong keeps on giving. As it enters Vietnam and nears the sea, it splits into what Vietnamese call the "Nine-Tail Dragon" and supplies a huge delta with enough water and nutrients to make it the country's richest rice-growing region.

In modern times, the mystery of the Mekong is portrayed by war movies such as *Apocalypse Now* and its scenes of a gunboat moving up jungle-cloaked waterways. But before Peissel's discovery, European explorers had been trying to locate the source of the Mekong since 1866, when the first French expedition set off from Saigon[3] to seek a trade route that would unlock the riches of inland southern China. Moving upriver, within a month, they came to the Sambor rapids in Cambodia, then the massive barrier of Khone Falls—less dramatic than Niagara, but larger in volume during the rainy season. The explorers' hearts must have sunk. But they portered their way upriver and persevered for two more years, making it all the way to Yunnan.

The Mekong never became a major trade route for Europeans. Another stretch of rapids north of Thailand made it unnavigable on its upper reaches. "The river is a paradox," says historian Milton Osborne. "It's very big, but in historical terms it has not brought people together because of its physical obstacles. It was a barrier people had to cross." The French later built a railway around Khone Falls, but it still took six weeks to travel from Saigon to the new Lao capital at Vientiane, the same time it took to travel all the way from Marseilles to Saigon. "It's a wonderful example of people's hopes flying in the face of reality," Osborne concludes.

That may also be a parable for modern times. After World War II, there were dreams of building a cascade of dams on the Mekong that would stretch from Cambodia to the Golden Triangle, lighting up the region just as the taming of the Columbia River had powered the American Northwest. In 1955, a United Nations commission established the Mekong Secretariat to promote cooperation among the riparian states and joint development of the river's resources. But financing the dams was always a challenge. And once the cold war heated up in the region, and open warfare broke out, financing became impossible. Only the Nam Ngum project, built with Western aid in 1974 on a Lao tributary to the Mekong, went ahead. Most of the 240 megawatts (MW) it produces is exported to Thailand, since that is far more power than Laos uses.

Hopes rose anew as peace settled over the region beginning around 1990.[4] The Thai prime minister, Chatichai Choonhavan, made a famous call to turn Indochina's "battlefields into marketplaces." Even Burma began to open up, cautiously. Trade grew among the riparian states. In a burst of optimism, regional institutions such as ASEAN (The Association of Southeast Asian Nations) and the Asian Development Bank promoted major joint development plans within the "Greater Mekong Subregion" (Vietnam, Laos, Cambodia, Thailand, Burma, and Yunnan). Road projects linking the countries are now underway. Rail and telecom projects have been dreamed up. Tourism networks have formed. A regional power grid is envisioned through which energy will flow from the less-developed countries to the industrialized centers of Thailand and Vietnam. Laos, with its tiny population and tremendous hydropower potential—it contributes 40 percent of the flow to the Mekong—is seen as becoming "the battery of Southeast Asia."

Thailand, increasingly faced with droughts caused by rising consumption and its profligate use of water, also began eyeing the flow of its border rivers—the Mekong and the even more mysterious Salween to the west. But because tapping international waterways is diplomatically tricky, the government has instead set about damming their Thai tributaries. It found the rules under the Mekong Secretariat restrictive, so it pulled out and renegotiated with Cambodia, Laos, and Vietnam to establish a new Mekong River Commission (MRC). Agreements were drawn up to govern water use in the basin, but whether they are strong or clear enough to prevent future conflicts is yet another mystery. The cascade idea was also reborn, but by now, even scaled-down versions of the dams would involve displacing tens of thousands of people. Within Thailand, opposition to such projects, even on its borders, would be immense. So most of the attention in the lower Mekong region is now on the tributaries (although mainstream projects outside of Thailand, for instance, at Stung Treng in Cambodia, may some day get a green light).

That is of some relief to the river's defenders. But damming the tributaries, as Thailand's Pak Moon project has done, can also be enormously disruptive to the basin's ecology because it blocks fish migration routes. Even species that spend most of their lives on the mainstream often travel up tributaries during the rainy season to forage or lay their eggs in flooded forests. "You can't just look at the river itself," warns Witoon Permpongsacharoen, a leading anti-dam crusader in Thailand who now heads the regional NGO named Terra. "You have to look at the whole basin. There are more than a thousand fish species in it. It may be the second most diverse riverine ecosystem in the world," a kind of Asian Amazon. Walter Rainboth, a biologist from the University of Wisconsin and an expert on Mekong fisheries, estimates there are about 1,200 species in the basin. "Right now, the total number is anybody's guess, and the more effort we put into looking, the higher the number will be," says Rainboth.[5]

Surveys on human ecology have also underscored just how dependent people are on this river of abundance. The MRC's Jorgen Jensen reports that fishermen in the basin catch about 1 million tons of fish every year from the Mekong and its tributaries, production that is worth some $700 or $800 million.[6] But since most of it is then eaten rather than sold, its value goes unrecorded in GDP accounts.

Another fact that has come to light since the region opened up is the continuing survival of a freshwater species of dolphin in the Mekong. The Irrawaddy dolphin was once found along many Asian rivers from China to India, but has disappeared from most due to pollution, and is now highly endangered. Its days may be numbered in the Mekong, as well. The dolphins once ranged all the way from the delta to southern Laos, but Ian Baird, a Canadian fisheries expert, estimates that only about a hundred still cling to life, mostly in northeastern Cambodia and southern Laos.[7] Lao villagers consider them friendly creatures who help guide fish into their nets, keep boats from tipping over, and even rescue people from drowning. Some revere the animals, believing they are human spirits reincarnated. But the dolphins are threatened by new fishing practices. Some are caught in the increasingly popular gill nets and drown; others are killed by blast fishing, a common technique in Cambodia made possible by all the munitions left over from its civil war.

The river's most famous creature, however, is probably the giant catfish, known as *plaa beuk* in Thai. This huge herbivore, the biggest freshwater catfish in the world, can grow as long as three meters and weigh up to 300 kg. From mid-April to June every year, fishermen in Laos and northern Thailand wait ceremoniously in their boats to catch the giant beasts as they make their long journey up the Mekong. Once snared in the nets, the catfish engage in titanic struggles.

No one is quite sure where the catfish are going—they are thought to have spawning grounds in the upper reaches of the river, somewhere in China—but it's evident that fewer and fewer are making the trip. Thai fishermen report the catch peaked at sixty-nine as recently as 1991; but by 2000, they managed to net just two, and in 2001 they didn't catch any.[8] Nor is anyone sure why they seem to be almost extinct in the wild—the Fisheries Department has set up a captive breeding program to try to sustain the species[9]—but overfishing could be the culprit. Giant catfish are so prized for their meat and oil that one specimen can fetch 80,000 baht. Attempts to persuade the fishermen to halt their catch have so far been futile, in part because it would require cooperation from Lao and Thai villagers. China may also be responsible for the demise of the fish. In 1994, it blasted away the rapids, a popular spawning ground for fish, along its portion of the river to make it more navigable for boats. It is now paying Burma and Laos to do the same so that the long-fabled "river road" can finally be opened up to large-scale trade, and Thailand is keen on the project as well.[10]

China has also blocked fish migration in the mainstream by building hydroelectric dams on its section of the river. It has already completed the first project on the Mekong itself, the 1,500-MW Manwan project, is working on two others of similar size at Dachaoshan[11] and Jinghong, and has started construction on a fourth at Xiaowan. The 4,200 MW slated to be produced would make it the third largest hydroelectric project in the world (after China's own 18,200-MW Three Gorges and 5,400-MW Longtan projects). Altogether, Beijing plans to build a cascade of eight to fourteen major dams on its stretch of the Mekong that together should end up generating more than 20,000 MW.[12]

China controls less than 20 percent of the Mekong's total flow, but its downstream neighbors are naturally anxious about the impacts all these projects will have on the river. Beijing (along with Burma) has pointedly declined to join the MRC; as the upstream state, it can basically do as it pleases. "China acts like it doesn't need to care about countries downstream," says Witoon. "It has to recognize that the Mekong isn't just theirs." An ominous sign of China's power came in 1995 when the four member countries of the MRC decided to commemorate their treaty by taking a boat across the Mekong from Thailand to Laos; they found the water was too shallow because China was busy filling the Manwan reservoir upstream.[13] Vietnam has particular reason to worry. It continues an ancient rivalry with China—the two countries engaged in a fierce border skirmish in 1989—and its other most important waterway, the Red River in the north, also has its upper reaches in China. "Downstream countries will suffer, but every country wants to build its own dams, so that makes it difficult for governments to fight against upstream ones," explains Witoon.

What will these projects do to the giant catfish and hundreds of other less well-known species? How would a dam at Stung Treng affect the Tonle Sap Great Lake, whose fisheries provide Cambodians with 60 percent of their protein? There are so many questions to which no one seems to have the answers. With its awesome size, its amazingly diverse peoples, its abundance and variety of life, the Mekong is one of the last great rivers to retain its primitive character. But it is the Mekong's fate that is now the biggest mystery.

THEY MAKE THE EARTH MOVE

From the 1950s to the 1970s, dam building exploded in the developing world. A total of 45,000 large dams[14] have now been built, and their impact has been enormous. Perhaps most astoundingly, their combined reservoirs have shifted so much weight they've allegedly changed the Earth's rotation.[15] But that's just the beginning. Dams trigger earthquakes[16] and create inland seas. They destroy communities and help create cities. They inundate forests and irrigate fields. They flood ancient archeological sites and power modern industry. They block migrating fish, but their reservoirs can be stocked with fry. Even their effect on climate change is ambiguous. Hydroelectric dams are an alternative to burning fossil fuels, but they also release methane, a major greenhouse gas, from vegetation that rots within the reservoirs.

There's no mystery why governments and energy authorities like dams. If they are built in suitable locations, they provide cheap electricity. Just as important, the gates can be opened to provide power when it's needed most. If no lakes or aquifers are handy, big reservoirs are also needed to provide cities and factories with water, particularly in a monsoon climate that lacks rainfall half the year, explains Prakob Wirojanagud, a water resources expert at Thailand's Khon Kaen University. Dams provide other vital services: storing water for irrigation, controlling floods during the rainy season, pushing back salt-water intrusion from the sea. They even become tourist destinations. But in reality, warns Prakob, "There's no such thing as a multipurpose dam. That's only on paper. Either it's for hydropower or irrigation or flood control."

Examined closely, big dams are generally built primarily to produce electricity. That is what pays the bills (unlike with electricity, consumers typically pay little or nothing for their water supply) and provides financial backers the return on their investment. So when push comes to shove and drought forces dam operators in Thailand to decide between releasing water for irrigation or for electricity, they almost invariably choose the latter.

Further examination reveals a distressing reason for why hydro-electricity can be cheap. As the World Commission on Dams (WCD)[17] concluded in its landmark report released in 2000, the social and environmental costs of these mammoth projects are generally ignored; it's the people living around the dam site who end up paying so dearly for them. Dams, in other words, transfer resources from the countryside to the city, and often from small-scale agriculture to industry (or to agro-industry). They have become a giant symbol of industrialization, the technological apotheosis of tension between city and countryside. And it's difficult to compromise over a dam. Yes, you can tinker with its height, or make it a run-of-the-river project with a smaller reservoir—but in the end either you build it or you don't.

For all these reasons, dams are the most contentious environmental issue in Southeast Asia, and perhaps the entire developing world. Because dams have huge impacts on rural villagers and the countryside around them, they perfectly illustrate how social and environmental concerns can merge to unite broad-based green movements whose factions often disagree on other issues. They also demonstrate how environmental issues aren't merely "fringe" politics but instead are extraordinarily revealing about societal values.

It is usually the impacts on local communities—mostly forgotten in the West, which built its big dams generations ago—that make dams the source of so much conflict. These impacts are two-fold. Not only do reservoirs displace rural villages but they also destroy the means of villagers' livelihoods by flooding fertile riverbanks and damaging fisheries. That is particularly significant in Southeast Asia, where agrarian societies, and civilizations, have flourished along river valleys. The Thai word for river—*mae nam* (mother of waters)—is indicative of the special nurturing role rivers have for Southeast Asian cultures.[18]

Culturally and historically, argues Witoon, "rivers are not borders" in Southeast Asia. That stands in contrast to Osborne's characterization of the Mekong as a barrier. Traditionally, trade and contact have always flowed along and across rivers, the Mekong included (with the exception of particularly obstructive areas such as Khone Falls). People living on opposite sides share cultures, languages, even relatives, although today they often live in separate countries. The population of northeastern Thailand, a region known as Isaan, is predominantly Lao, and most residents there speak Thai only as a second language. In Cambodia, which historically encompassed what is now southern Vietnam, including the entire Mekong delta, the lower river basin essentially serves as a giant circulatory system. Even for modern Laos, the Mekong is not so much a barrier as a backbone.

Dams also have an important political dimension, explains the cerebral Witoon: "They are part of the centralization process. Large projects change power. Someone gains, and someone loses." Traditionally, particularly in northern Thailand, farmers would gather together to build weirs, known as *muang* or *faay,* on their local waterway, and then agree about how to share the water. But when the government came in and started building dams, power shifted to the state agencies that controlled them, particularly the Electricity Generating Authority of Thailand (EGAT) and the Royal Irrigation Department (RID). Villagers not only lost control over their most important resource but also lost a main reason for the community to join together and cooperate. It's a classic example of how governments use technology and large-scale development to appropriate resources, and the authority to manage them, from local groups, in the process destroying traditional forms of civil society.

Controlling resources, particularly water resources, has always been the key to power in Southeast Asia. "The ruler, as 'lord of land and water,' and as mediating agent with the natural forces, was perceived to be in ultimate control of the environment," writes Jeya Kathirithamby-Wells of the University of Malaya. "Contingent upon the role of the ruler as *dharmmaraja,* in the Buddhist mainland, and *klipatullah* in maritime Southeast Asia, was the successful and just management of material resources."[19] As proof, you need only visit Angkor, its temples surrounded by vast moats and reservoirs; this was a civilization that thrived on taming the waters of the Tonle Sap region for irrigation.

To this day, water remains a powerful symbol in Southeast Asia, and a symbol of power—a link embodied by His Majesty the King of Thailand, Bhumibhol Adulyadej, ninth in the line of the Chakri dynasty. One of his chief aides, Sumet Tantivejkul, head of the Royal Projects, says the king is dedicated to being a provider of water for his people: "His Majesty realizes that most of the people are farmers, and that's why the most important [of his development] programs concern water."[20] The king is a strong advocate for building dams—the country's biggest are named after members of the royal family; the biggest of all is the Bhumibhol Dam. And if you look on the back of the new thousand-baht notes, you'll find a picture of the king and queen; the king is holding a hydrographic chart, and a dam is in the background.

Nevertheless, Thailand was startled in 1993 when, during his annual birthday address, the king called for the construction of three new dams—the Pasak, the Nakhon Nayok, and the Pak Phanang—to be finished within six years, in time for the celebration of the king's sixth cycle (his seventy-second birthday). At the time, public support for dams was clearly ebbing, weakened as it had been by a violent clash earlier in the year over the Pak Moon project. The dam on the Pasak would be quite large, and the Khlong Tha Dan reservoir in Nakhon Nayok, although smaller, would encroach on 2,000 rai of forest in Khao Yai National Park. For the king to make a public call for such controversial projects was unprecedented. "This is very serious. In my whole life I have never seen him do anything like this. Everybody was shocked. But His Majesty sensed great danger if nothing is done," explained Sumet, who was in charge of studying the two dams.[21]

I asked Sumet about the environmental impacts of dams, mentioning as an example that the Bhumibhol Dam had flooded the last known tree of the species *Damrong chia,* named after a famous Thai prince. "That's just one side of the story," he said. "The main factor of the environment is humans, not trees. You must accept the old saying that God created nature for human use. 'Ecology must go along with the life of the people,' His Majesty has said. The problem is how to use it correctly and in a sustainable way." Sumet talked about how the forests used to help regulate water flow, and he agreed that reforestation is important. "Once the forest grows [back], the need for dams can disappear naturally," he said, but added that in the meantime, dams were necessary.

I mentioned my theory to Sumet that dams are so controversial because they transfer resources from the countryside to the city. The dam on the Chiew Larn River in southern Thailand, for instance, produces electricity; but because the reservoir's underlying geology has polluted the water, villagers downstream can't use it, and many of them don't even have access to electricity. Isn't that like stealing from farmers to give to the city? I asked. "Let's suppose you want to construct a dam just for agricultural purposes. [Should we] just let the water flow away? If we put in some turbines, then we have electricity. Do we need electricity or not? Stealing or not stealing. Do you have air conditioning in your office? Should we go for nuclear power? Because anything you do, the consumption of electricity still goes up. Stealing or not stealing, we cannot separate [urban people] from villagers. We must consider the nation as a whole."

After the king made his call for the dams to be built, Sumet said that the time for public hearings was over. The project designs had already been altered to reduce their impact, he claimed, and now it was time to begin building. "I must accept that [dam-building] agencies in the past did not care enough for public opinion. But this time . . . people in the area will be the first to receive the benefits."

Pramote Maiklad, another close aide to the king, became head of the RID, and he has often expressed what might be considered the conventional view of dams, particularly among engineers and officials: that fresh water running into the sea is "wasted." Building reservoirs can help siphon off the surplus water during the rainy season to power the cities and industries that are the region's future. But NGO activist Witoon contends that a river needn't have dams on it to be considered "developed," and he says the media "spreads the assumption that being pro-dam is pro-development, and anti-dam anti-development." Tens of millions of villagers already use the Mekong's resources for farming, fishing, hunting, transportation, and other economic goods and services; and, like the Mekong's fish production, these resources aren't recorded in GDP figures. But the ecosystem is already being used, is already productive on a subsistence level. And many of these goods and services, Witoon points out, are placed in jeopardy if dams are built.

THE FATE OF THE DAMMED

Boonmee Khamruang was seventeen years old when his village in northeastern Thailand disappeared under water. His father was supposed to receive 1,800 baht as compensation for the loss of his house and farmland, but never saw the money. The family was offered a place to live in a resettlement zone, but they turned it down, claiming the land was useless. His elder brother took up arms with the Communists—"at that time," says Boonmee, "it was the only way to fight state abuse"—and was eventually killed. "We first settled in a degraded forest, but then we were evicted again to make way for the state's eucalyptus plantations," Boonmee continues. "Without land, we must leave our families to work as laborers in big cities. Many villagers ended up as garbage collectors in Bangkok."[22]

In many ways, this is a typical life story for the victims of dam displacement. But the Sirindhorn Dam that inundated Boonmee's village was completed more than three decades ago. By the time he told his tale to the *Bangkok Post*'s crusading columnist Sanitsuda Ekachai in August 2000, Boonmee was fifty years old and a father of four. He was participating in a mass protest in front of Government House in Bangkok, one of many that has been organized by a group called the Assembly of the Poor to demand redress for past wrongs committed by the state against rural villagers. At the time the dam was built, Boonmee and his fellow farmers believed it was impossible to oppose the government. "It was a dark time under military dictatorship. Anyone who protested was labeled a Communist and risked persecution. So no one dared," he explains.

Sanitsuda points out that Thailand's "so-called democracy" hasn't been kind to the protestors, either. Consider the chronology of their campaign. After years of demonstrations and talks, a committee to assess their claims was finally established in 1996 under the Banharn administration. The committee concluded that most of the 2,526 uprooted families had received poor farmland and only partial compensation. The affected villagers were told they'd each be given a 15-rai plot of farmland in addition to long-term, low-interest loans. But because the farmland wasn't available, in 1997 the committee agreed to establish a special fund that would allow the displaced to buy land themselves. But then a new government came in, led by the Democrats and Prime Minister Chuan Leekpai, and announced there would be no compensation for old dam claims. In 2000, another protest led to another committee and another recommendation for recompense, followed by another government refusal. And so it goes, on and on.

Although there's some public sympathy for the protestors, many city people are fed up with the continuous demands, and suspicious that the displaced are just trying to welch off the state. The villagers aren't likely to back down now, however. "My community dates back centuries, with local legends and a history of struggle against injustice. We're just carrying on our ancestors' spirit," Boonmee asserts. "I don't want my children and grandchildren to live a homeless life like I have." He has little to lose by continuing to fight; but by exercising his democratic rights, he has a new future to gain.

These rights represent a situation quite different from that of thirty years ago in Thailand, or today in neighboring countries, where people who protest are subject to government or military crackdowns. Thailand's democracy—imperfect as it is—has been a boon to those wronged by the state. They can for the most part express themselves freely, and if they are persistent enough or catch a sympathetic ear among those in power, they may eventually get their way. That said, military repression in Thailand has in a sense been replaced by mafia rule, a kind of private-sector tyranny. It is still distressingly common for rural activists in Thailand to be murdered by the influential groups whose projects they oppose: Jun Boonkoonpot was shot for fighting against a dam in Chaiyaphum province; Thong-in Kaewwattha was gunned down on his front doorstep in Rayong for leading the opposition to a toxic waste treatment center in a nearby watershed area; Prawien Boonnak was executed mafia-style for speaking out against rock quarries in Loei; and unfortunately so on.[23] The problem may even be worsening: In the first six months of 2001, four local environmental activists—Jurin Ratchapol, Narin Podaeng, Pitak Tonewuth, and Suwat Wongpiyasathit—were killed.[24]

Nevertheless, each protest about old dams makes new ones harder to build—if only because everyone simply becomes fed up with the continuing arguments—which is why full-fledged democracies have a tough time building big dams. By wincing down Thailand's dam-building era, protestors helped Thailand come full circle, because it was the anti-dam movement that re-sparked Thailand's broader democracy movement. Several years after the Sirindhorn project was completed came the democratic uprisings of 1973 and 1976; these were followed by an authoritarian period as the military fought Communists in the jungles. As the insurgency ebbed in the

1980s, more peaceful forms of protest became tolerated, and the first widespread expression of this freedom came in opposition to the proposed Nam Choan Dam.

The brainchild of EGAT, the 580-MW Nam Choan project was slated to become one of the largest hydro-electric dams in the country. But its two-hundred-square-kilometer reservoir would have flooded a major portion of the Thung Yai Naresuan Wildlife Sanctuary in Kanchanaburi, and split the remainder in two. A broad coalition came together against Nam Choan that included local groups, students from Bangkok, academics, forestry officials, and NGOs such as Wildlife Fund Thailand (WFT), and the Project for Ecological Recovery (PER), led by Witoon, who hails from Kanchanaburi. Environmental issues were a new phenomenon in Thailand at the time, and the English-language press was instrumental in raising awareness about them: Ann Danaiya Usher of the *Nation* and Normita Thongtham of the *Bangkok Post* wrote groundbreaking columns that looked critically at the project. The scrapping of the dam in 1988 was a major victory for the fledgling movement, and perhaps remains its finest hour.

The reasons for building the dam remained, however. As Thailand's economic boom took off in the late 1980s, energy demand skyrocketed, growing at roughly 10 percent a year. Unlike its petroleum-rich neighbors—Malaysia, Indonesia, Burma, Brunei, and potentially Vietnam, all of which have significant reserves—Thailand enjoys relatively few fuel resources. It does have a rich vein of lignite (a dirty form of coal with a high sulfur content), which EGAT burns at the massive Mae Moh power plant complex in the north, but that has created serious pollution problems. Thailand also taps some cleaner burning natural gas deposits in the gulf; but because they are contaminated by mercury, they create yet more environmental problems (see Chapter 8).

Its only other domestic energy resource (apart from wind and solar) is hydropower, so EGAT is keen to develop that wherever it can. The state electricity authority is one of Thailand's most profitable state enterprises, and, frankly, one that is better run than most others. But it is dominated by engineers and has a reputation for arrogance in its dealings with the public. "[That] is EGAT's own fault," admonished Ammar Siamwalla, a distinguished economist who headed the Thailand Development Research Institute (TDRI). "Its terrible record of compensation has made it lose all its credibility with the public. Improper compensation is haunting it to this day."[25] But EGAT did not take its defeat in Kanchanaburi lightly.

Enter the Pak Moon Dam. In 1989, the same Chatichai administration that had scrapped Nam Choan approved a new project at the mouth (*pak* in Thai) of the Moon River in the northeastern province of Ubon Ratchathani. This remote corner of the country, located not far from where Thailand, Laos, and Cambodia all meet, was to become ground zero in the war over dams. On one side, EGAT revved up its promotional campaign, led by its combative public relations director Subhin Panyamag. Thailand desperately needed the 136 MW that Pak Moon would produce, EGAT argued, because it would power poverty-stricken Isaan's attempt to become a center of industry for the emerging economies of Indochina. Besides, the dam had been turned into a run-of-the-river project to reduce its reservoir size. The government assured affected people that they would be duly compensated; this time, a significant chunk of the budget would go toward helping the villagers adjust. There would even be a fish ladder, just like those built for salmon in the Pacific Northwest, to help fish migrate upriver.

The environmental movement was equally determined to follow up its victory at Nam Choan by defeating EGAT at Pak Moon. It, too, offered an impressive array of arguments. Surveys carried out by the state agency predicted that the dam would affect roughly 250 families (about 1,000 people). But opponents claimed the actual number would be four to five times larger. The discrepancy was partially explained by the environmentalists' arguments that the dam would block the movement of up to 150 species of fish that regularly migrate between the Moon and the Mekong, and therefore wreck the livelihoods of local fishing communities. In addition, it was feared that creating the reservoir would lead to an outbreak of schistosomiasis, a disease caused by a parasite residing in freshwater snails that favor stagnant water.

In August 1991, a report prepared by the U.S. Agency for International Development recommended against building the dam because of "probably irreversible . . . environmental and social impacts." The following month, the World Bank decided to delay handing out a 550 baht million ($22 million) loan to the Thai government for the project. It was an unprecedented move, marking perhaps the first time bank directors had bowed to pressure over dam issues from international environmental groups, which were working closely with Thai NGOs.[26]

But Thailand was under a military dictatorship at the time (established following a coup in early 1991), which almost certainly weakened opposition to the dam. The Cabinet headed by Prime Minister Anand Panyarachun was made up of reformers, but they were also hardcore industrialists and economists, and they reacted strongly to the World Bank's waffling. At the annual World Bank/International Monetary Fund (IMF) meeting in October, which was held in Bangkok that year, Finance Minister Suthee Singhasaneh accused certain World Bank directors and member countries of "tyranny"; he added that further indecision on its part would "truly endanger" the bank as a collaborative institution. The president of the World Bank, Lewis Preston, quickly caved in, saying he expected approval of the loan "within weeks." Sure enough, on December 10, 1991, the bank's executive directors voted to fund the Pak Moon Dam, despite opposition from U.S., German, and Australian directors.[27]

That was not the end of the controversy. The next year, EGAT suddenly began blasting the rapids that formed the heart of Kaeng Tana National Park downstream of the project site, because, it said, the work was necessary to create an outflow channel for the dam. "Blasting is just normal engineering work," said EGAT's deputy director general, Sawarng Jampa. "It is a technical matter and quite complicated. So we did not tell the public."[28] Environmentalists protested that damage to the rapids would harm the area's fisheries, and was illegal under the National Parks Act. Phairote Suwannakorn, a former director of the Royal Forest Department (RFD), refuted EGAT's claim that it had received permission for the blasting. Then came another blow: A massive pollution spill from a factory upriver killed tens of thousands of fish in the Nam Phong-Chee-Moon river system. But a call by the Fisheries Department to delay the dam's construction—thus allowing fish from the Mekong to migrate up the Moon and repopulate Isaan's rivers—went unheeded.

The Pak Moon controversy finally boiled over the following year. Beginning on February 27, in a last-ditch bid to stop the project (which by then was already half-way complete), or at least to negotiate better compensation, a group of local villagers and students staged a sit-in at the dam site and brought construction to a halt. There they stayed for a week, hoping that the new democratically elected government led by Chuan would be more sympathetic to their plight. But Savit Bhotivihok, the Cabinet officer who oversaw EGAT, declined to get involved. Negotiations with the local authorities dragged on and tensions mounted.

Finally, on the night of March 6, the dispute turned violent. Trouble began when a group of "dam supporters"—either a hired mob, or a group of construction workers angry because they were losing their pay, or perhaps both—gathered near the protestors. There are conflicting accounts of what happened next. Suphon Sriphon, a *Nation* radio reporter covering the story, says that starting at around 11:00 P.M., the pro-dam crowd started hurling rocks at the anti-dam protestors, and the police were caught in the middle. Some of the dam's opponents sought to attack the stone throwers and ran into the police. "The police didn't hit the pro-dam demonstrators with batons at all. Instead, they mixed themselves with this group and beat the anti-dam protestors. EGAT wanted [them] out, and it may have been behind the dam supporters," Suphon said.[29]

Thongcharoen Srihadhamma, a fifty-year-old local leader of the protestors, said he didn't know who started the fighting. He came out of a meeting with EGAT to find his fellow villagers up in arms, shouting at the dam supporters: "Stop the dam! We don't want it!" He tried to calm them down, but suddenly the two sides were throwing stones and firing slingshots at each other. The clash left about forty people injured. Thongcharoen and other protesters leaders were arrested. According to Police Major Wichian Pulsap, commander of the local police force, "The demonstration had to be dispersed because it obstructed the work of the construction company which was losing a million baht a day due to the work suspension." Construction continued and the project was completed the following year.

Enough time has now passed to ask this question: Who was right about the dam? Well, here are the conclusions of the World Commission on Dams (WCD), which in 2000 completed the first independent analysis of Pak Moon's costs and benefits. Although the project was originally supposed to produce 136 MW of power, the WCD determined that in fact it produces roughly 21 MW—that's less than half the energy consumed by the World Trade Center, a giant shopping mall in Bangkok.[30] Meanwhile, the cost of the project rose from an original estimate of 3.9 billion baht to 6.6 billion baht, a 68 percent increase. Compensation and resettlement costs also increased from a predicted 232 million baht to 1.1 billion baht, or about 17 percent of the total budget. The number of families displaced by the dam turned out to be 1,700, almost seven times the original number. An additional 6,202 households have been compensated for the decline in fisheries—but only for the three-year construction period; no compensation has been granted for a permanent loss of fisheries, even though fish catches have dropped by 80 percent, and of the 265 fish species recorded in the river before the dam was built, only 96 remain. "If all the benefits and costs were adequately assessed, it is unlikely that the Moon project would be [built today]," the study claimed.[31]

DAM-AGE AND DEMOCRACY AT THE MOUTH OF THE MOON

Those numbers are pretty damning, but to understand the full impacts of the Pak Moon project, you need to go see it for yourself. One of my last trips to the area occurred in 1998 when I was hired, along with British journalist David Nicholson-Lord, by the Indochina Media Memorial Foundation to help train a group of journalists from Vietnam, Cambodia, Laos, Burma, and Thailand in environmental journalism. Over several weeks, we taught lessons in basic journalism skills and led discussions on environmental topics, mixed in with talks by invited experts and field trips to visit the site and interview the villagers. It's a great program; it offers journalists a chance not only to learn about the issues but also to find out about each other, and it's fair to say we trainers learned almost as much from the experience as the trainees.

We visited the fish ladder by the side of the dam and watched as a few tiny fry struggled to leap up the steps. "I guess Thai fish don't jump," one of the locals said with a bitter laugh.[32] We went out in the pre-dawn hours and sat with the villagers as their small boats plied for fish under the huge crest of the dam. Pulling out the fingerlings caught in their nets, they told us how much better life was before the dam. "Before, I could easily make 35,000 baht from simply working during the fishing season," said Chan Roobsong, a forty-eight-year-old fisherman who has survived a bout of malaria and a landmine accident. "Now it takes one or two weeks to earn enough for two liters of fuel." Chanphen Chaiyasatra does not even bother to fish anymore. She has taken up weaving, dairy farming, and odd construction jobs, but the 8,500 baht she now makes pales in comparison to the 20,000 baht she easily used to earn from fishing. "We planned to expand the house for my children, and send them to study in Bangkok for a good education. But with an income this low, there's no way," she said.

"Life was easy," says Thongcharoen, the fisherman turned protest leader who was arrested in the 1993 clash. "I never thought about what my kids would do. I always assumed they could become fishermen, too. Now, two are in school, one is a soldier and the other is a security guard in Bangkok. They didn't want to go, but they didn't have a choice." Nicholson-Lord, my fellow trainer, summed up the fishing community's situation nicely in a piece he wrote following the trip:

> The extraordinarily rich fisheries of the Mekong and its tributaries [had] not only fed the fisherman and their families. They provided a generous

> surplus to sell at market in exchange for other foodstuffs and products the river could less easily provide. And the lifestyle that went with this surplus was, in many respects, a comfortable one—a matter of setting nets and then, a few hours later, emptying them. No 12-hour grind in the office for the fishermen of the Mekong. No three hours in a Bangkok traffic jam either. Early mornings, maybe—but lots of clean air and quiet, shady riverbanks. Some people pay good money to holiday in such places. And, somewhat poignantly, they were clear about what they had lost. Their old way of life, they remarked, was one of ease and leisure. No wonder they wanted it back.[33]

Nobody knows how many villagers from Khong Chiam have drifted off to the city to look for work. One indication came on Visakha Bucha Day, the Buddha's birthday, when we attended a warm and colorful ceremony at the local *wat,* marked by a candlelight procession. But only a few hundred people attended. "There used to be over 3,000 people who would come to the ceremony," says Thongcharoen. "But the community is collapsing, along with our traditions."

Thongcharoen represents what's being *created* at Pak Moon: a broad-based protest movement, a key element of civil society that has become a political force and someday could conceivably turn into a more formal political entity. He arrives for a meeting in a covered pickup truck, a sticker posted across the windshield: "Assembly of the Poor." As he steps out, the first thing you notice about him is the broad-brimmed leather hat tilted rakishly atop his weather-beaten features. With a goatee gone gray and a colorful *pakamah* (scarf) draped around his neck, he looks as if he should be herding cattle. But he's got charisma, and with his slow and easy voice, he clearly knows how to play the crowd. In his own way, he's a born politician. Asked to recount the events of March 1993, he says he was sad to see villagers fighting each other. According to Thongcharoen, the formal politicians—the parliamentary candidates—say all the right things during election time, but then they disappear once they are elected. "Anyway, they just use money to buy votes," he adds. So, as head of his village's "protest committee," he has doggedly kept up the informal fight against the dam and for more compensation.

The Pak Moon struggle has served as a crucible for the NGO movement, and Thongcharoen is now one of the leaders of the Assembly of the Poor. With great pride, he showed me a statue erected on a boulder not far from the dam. It's a provocative sight: three protestors holding each other up, one with his fist boldly raised. It was purposely placed in a far-off corner of Isaan, he says, rather than in Bangkok with most of the other monuments. But the statue is emblematic of the movement; will it ultimately be seen as just a curiosity, remote from the corridors of power, or will it have some sustained political meaning?

In 1997, Thongcharoen and 10,000 other villages from all over the country rallied for ninety-nine days in front of Government House in Bangkok. In what has become something of an annual event, they aired a long list of grievances that included land disputes, government development projects, slum conditions, and workers' health and safety problems. What makes such an inchoate group come together under a banner that celebrates their poverty? Anger, defiance, a bitter sense of righteousness, perhaps. The Assembly of the Poor was pieced together from pockets of resentment by leaders such as Bamrung Kayotha, a former pig farmer who set up the Northeast Federation of Small-Scale Farmers, and Vanida Tantiwitthayapitak, a city-bred daughter of the middle class who cut her teeth in the fight against Pak Moon.

After talking to Bamrung, you get the feeling that the name "Assembly of the Poor" was chosen for its irony. That no doubt was the sentiment underlying the "Miss Poor" competition held as the protestors whiled away the months in front of Government House. A deliciously earthy satire of the glamorous beauty contests attended by high society, it featured wrinkled old crones competing on a jury-rigged stage to see who had the best hard-luck story.

No one is sure what to call groups such as the Assembly of the Poor. The organization has been referred to as a social movement, a protest movement, and a people's movement. I like to think of it as an "environmental democracy" movement because it combines issues about resources and social justice. But there's no doubting what it really is: the most organized authentically left-wing movement in Thailand. And similar movements have appeared wherever developing countries allow such freedom of expression. A famous example is the Chipko movement, which arose in India in the 1970s. Made up largely of village women who fought against loggers by hugging trees[34] and practicing nonviolent resistance, "it was not a conservation movement, but rather was organized to assert local rights over who can cut trees," explains Sunita Narain, deputy director of India's Center for Science and the Environment. "The environment [in the global South] is not a luxury, but basically a matter of survival. It's different from the North, where the environmental movement is distinct from the development movement."

The main theme for these groups is a call for greater "public participation" and "decentralization," euphemisms for more local influence over decision-making, particularly over the use of resources. But these terms, often like the groups themselves, are so amorphous that no one really knows what they mean. Thailand's Assembly of the Poor seems to have a much better idea of what and whom it's against than what it is for. It fights most of its battles with the state agencies—EGAT, the RFD, the RID, and so on—who currently control the country's resources. Politically, the group has not affixed itself to any mainstream political party or leader, although populist politicians such as Chavalit Yongchaiyudh and Thaksin Shinawatra have flirted with it. But its members and supporters seem uniformly antagonistic to the Democrats of Chuan Leekpai.

With its strongholds in the affluent southern part of the country and in the cities, the Democrats are the closest thing Thailand has to a liberal party. Established more than fifty years ago, the party is positively venerable by the standards of Thai political parties. In fact, some consider the Democrats the only real party Thailand has. All the others seem to come and go, each created by and for a particular leader who, before each election, courts machine politicians in a bid to gain the most seats, as if playing a high-stakes game of musical chairs. The Democrats, on the other hand, have some polish. They tend to be favored by technocrats, at home and abroad. And in Chuan they have a leader with a clean image. But they also have a reputation not only for favoring the interests of urban Chinese businessmen but also for scorning the rural villager. Their lack of the common touch in the countryside has probably kept them from consolidating power as liberal parties elsewhere in Asia have done.

Could the Assembly of the Poor become a political party? Perhaps, if this were the West. But left-wing parties have not been particularly successful in Asia, and in Thailand they are nonexistent. Whereas environmental concerns among city dwellers in Bangkok coalesced into the election of a governor campaigning on a green platform, rural discontents haven't been converted into a formal political presence (another indication of the huge gap between city and countryside). Only a few of the Assembly's leaders and fellow travelers have dipped their toe into electoral politics, and then only on a local level. Most fear becoming tainted by mass politics and its corrupt ways. But the assembly is an effective pressure group that expresses its opinions on a growing range of issues—from community forestry to free trade to the patenting of life forms, and every year it returns to Bangkok to continue the fight. It has made some real gains by settling many of the disputes it has brought to the table, although as the Sirindhorn Dam situation shows, progress is unsteady because of the continual shifting of governments.

In 2001, Thongcharoen and the Assembly of the Poor achieved another breakthrough when the Thaksin government ordered EGAT to open the sluice gates on the Pak Moon Dam and let the river flow normally for four months. EGAT finally complied in June. Elated villagers soon reported the return of from forty to fifty fish species that had been missing from the river since the dam was built.[35] The government subsequently ordered that the gates stay open for an entire year while studies are carried out to see whether the river's ecology can be revived. There is even talk, incredible as it may seem, of decommissioning the dam.

DAM DOMINOES

The Pak Moon saga has given big dams such a bad reputation in Thailand that it has become difficult for the government to approve other projects. Only the king's intervention managed to get the Pasak and Nakhon Nayok dams passed. The RID has repeatedly tried to gain approval for a major dam on the Yom River at Kaeng Sua Ten (Rapids of the Dancing Tiger), even though the reservoir would displace three sizable villages and flood Thailand's richest remaining teak forest.[36] An analysis carried out by the TDRI showed the project did not make sense economically.[37] But former Environment Minister Yingphan Manasikan, a member of Parliament from a downstream province that would have benefited from the project, supported the dam. Following severe floods in 2002, the dam even seems to have the support of the newly appointed forestry chief, Dhammarong Prakorbboon.[38] What has really prevented the Kaeng Sua Ten Dam's construction, ever since it was first proposed decades ago, is that the local villagers who face resettlement have remained steadfastly united against it.

It appears that Thailand simply doesn't have the room to build any more big dams. It's not just that the most promising sites have already been taken, but the central government no longer has the political space to impose its favored projects upon compliant peasants. In Thailand, the fate of the dammed is no longer just to suffer in silence; it is to help create a participatory democracy that acts as a check on the abuse and appropriation of resources by the state.

EGAT has reluctantly accepted that it can no longer build big dams in Thailand. Instead, it has looked abroad, and its hunger for power has caused dam projects to spring up all over the region—a new version of the domino effect. Thailand has signed a Memoranda of Understanding to buy energy from Burma, Yunnan, and especially Laos, which seems eager to fulfill its destiny as the battery of Asia. Since 1993, Vientiane has signed preliminary contracts for the construction of twenty-three dams.[39] Along with the earlier Nam Ngum project, three other large dams have been built on Mekong tributaries, and construction on the controversial Nam Theun 2 project may start soon. The country's largest project ever, its 450-square-kilometer reservoir, would displace more than 4,000 people and flood a quarter of the Nakai-Nam Theun plateau, an ecological "hot spot" and home to seventeen globally threatened species, including two newly discovered mammals—the goat-like saola and the deer-like giant muntjac. A million cubic meters of old-growth trees in the reservoir area have been logged by the BPKP (Bolisat Pattana Khet Pudoi), a military-backed timber firm. The forest around the reservoir is to become a biodiversity conservation area.

Besides the environmental and social issues, there are also questions about the economic viability of the project, whose foreign backers—Electricité de France, Australia's Transfield, and Merrill Lynch Phatra Securities—will own and operate the dam for thirty years. It will cost $1.2 billion to build, roughly equivalent to Laos's annual GDP. As for the electricity Nam Theun 2 generates, apart from about 80 MW that will supply the local market, the remaining 1,080 MW will be sold to one customer: Thailand. EGAT will thus hold a virtual monopsony over the power it produces.

The region's financial crisis seems to have short-circuited the Lao government's hopes of becoming the regional powerhouse, at least temporarily. It had previously announced plans to export $350 million worth of electricity within a decade, but electricity exports in 2001 are projected to total only $30 million, and several dam projects have been put on hold.[40] As it stands, Laos's 4.5 million people receive little direct benefit from all this hydropower development; average electricity consumption is only 55 kilowatt-hours per person, compared to 1,296 kilowatt-hours per capita in Thailand. And yet, Laos exports three-quarters of its electricity to Thailand,[41] the argument being that the foreign exchange can be put to good use. But there is danger in exporting energy (or virtually any kind of resource) if the government is corrupt and unaccountable. As Nigeria has discovered with its oil exports, kleptocratic officials all too often fritter away the profits while the locals get stuck with the impacts.

Now Burma is exploring the possibility of building a dam on the Salween River that would export electricity to Thailand. From its origins in Tibet, the Salween flows for roughly 2,400 kilometers. First it runs parallel with the Mekong through China, then it veers west when it flows into Burma; and it barely skims Thailand's border before it reaches the sea at Moulmein. But almost nothing is known about the river or its ecology in Burma[42] because the areas it flows through have been closed to outsiders for so long. Environmentalists have scoured libraries looking for research reports on the river's ecosystem, but could find only two pages from a book published in 1945 on Burma's fisheries. People who live alongside the river tell some intriguing tales. Ti So, a Karenni hunter and fisherman I met in Mae Hong Son, described a denizen of the Salween—called a *na mye* (horse fish) because of its horse-like beak—that sounded like a freshwater shark: "It is a huge fish. It can weigh as much as 200 kg. Gray and white in color, it has no bones except the backbone and in the fins. It is very strong and has very sharp teeth. The *na mye* has been known to attack people and rip off their limbs."[43]

Plans for the Salween Dam are still in their early stages, but a "technically promising" site is being studied at Ta Sang, a river crossing 80 kilometers north of the Thai border in Burma's Shan state. The region is rife with ethnic insurgents and drug warlords, and Burma's ruling military junta has forcibly relocated up to 300,000 people there, according to human rights groups. Japan's state-owned Electric Power and Development Corp is carrying out the feasibility study for the project, along with consultants from a well-connected Thai firm named MDX. The engineering firm Ital-Thai and Japan's Marubeni are also said to be involved in the project, according to Salween watcher Steve Thompson.[44] The generals in Rangoon probably view a dam on the river as a way to neutralize the Shan military opposition by cutting off support from Thailand, much as a gas pipeline did with another insurgent group, the Mon (See Chapter 7).

Vietnam has had its own adventures in building dams. It is planning to build fifteen new projects by 2010, including the Yali Falls Dam in the Mekong basin. Five are already under construction, with assistance from Russia, the Ukraine, the Czech Republic, and Japan. But the country's biggest projects are on the Da (Black) River, northern Vietnam's second largest river system. Depending on its eventual size, the Son La Dam could displace anywhere from 16,000 to 200,000 people.[45] Many Vietnamese members of Parliament expressed reservations about the project's safety and the need to resettle so many people, a situation reminiscent of the approval of China's Three Gorges project by a divided National Assembly in Beijing.

Questionable dam projects seem to appear in every developing country. The Bakun Dam, in the Malaysian state of Sarawak, has generated familiar complaints—shoddy resettlement and compensation schemes, the flooding of tropical forests, and contracts and concessions going to well-connected firms. In Indonesia, four villagers opposing a dam planned for the island of Madura were gunned down. In India, the battle over the Sardar Sarovar project has turned into an epic struggle. The World Commission on Dams was created because dams now inevitably lead to conflicts. Its final report, while acknowledging that "dams have made an important contribution to development," cautioned that "in too many cases an unacceptable and often unnecessary price has been paid by displaced communities and by the natural environment."

There's another important trend going on here: The more democratic a country is, the more difficult it is to build big dams. Building a dam despite vehement local opposition isn't necessarily anti-democratic; that might be setting the bar of democracy too high. "No democracy in the world can have 100 percent agreement," says Sumet, the king's aide, "but we have to get the [support of the] majority." Democratic countries that have a lot of sparsely populated land, or an uncommonly clean government enjoying public trust, or a good compensation program that is widely accepted, may be able to do it without much opposition. But those assets are rare. And in mature democracies, respect for the rights of a minority is usually strong enough to make something as serious as forced resettlement unacceptable, even if it's for the benefit of the majority.

Because building large dams does not sit well with democracy, an intriguing question arises: If dams are needed for industrialization, does that mean you have to sacrifice democracy to industrialize? Perhaps. Consider all the wrenching changes an agrarian society typically goes through to build an industrial economy: the redirection of resources, the pollution, the migrations of people, the breaking up of communities, the atrophying of traditions. Then consider that the results are unsure, that industrialization may cause the gap between rich and poor to widen, that many people may not grow much wealthier from industrialization, or at least much happier. Would a majority of the populace willingly vote for all that if they knew what was in store for them? Is that pot of gold—the opportunity to gain greater wealth, better education, more mobility—at the end of the Kuznets curve really tempting enough to make them willingly suffer all those changes? Or are they being hoodwinked by an elite that knows it alone will benefit?

It's hard to say, but you get the sense that many of the elite don't want to risk finding out. During the 1990s, a huge debate ensued over "Asian values." Asian leaders such as Lee Kuan Yew of Singapore and Mahathir Mohammed of Malaysia insisted that a society's need for development and stability must take priority over the individual rights as championed by the West. And they were by no means isolated in their views. They were saying only what many other well-off and well-connected Asians were thinking: Too much democracy would make it too difficult to achieve their personal vision of progress.

EPILOGUE: NEGA-WATTS INSTEAD OF MEGAWATTS

"People want their showers hot and their beers cold," asserts Amory Lovins. "They don't care how they get that way." That's the key to energy efficiency. And Lovins, founder of the Rocky Mountain Institute, is one of the main proselytizers for what he calls the "Nega-watt Revolution." Its fundamental principle is that people and firms aren't interested in consuming energy for itself, but rather the services it provides. So if people use efficient appliances and design efficient buildings, they can save energy and save money; and the utility can save the cost of generating more electricity. No conservation is involved. Nobody is making a sacrifice. People can still have their air conditioning and their computers.

The catch is that people first must fork out a little extra money to buy efficient equipment or well-designed homes. But they can be helped along by promotional programs that provide consumers with rebates or discounts for purchasing, say, an energy-efficient refrigerator. Funding such demand-side management (DSM) programs—Lovins encourages utilities to think of them as "Nega-watt power plants"—can be far cheaper than building the more conventional kind of power plant, and far easier on the environment, too. So energy efficiency can be a green alternative to dams.

Lovins visited Bangkok in 1992 and claimed that although the demand for energy services was growing faster in Thailand than in developed countries such as the United States, the greater inefficiencies found in Thailand mean that "Thailand could meet 100 percent of its growth in demand for energy services by promoting efficiency."[46] The technocrats at Thailand's National Energy Policy Office (NEPO) were eager to try it. With few energy resources and soaring demand, energy efficiency had the potential not only to reduce the financial[47] and environmental burdens of rapid growth but also to make Thailand less dependent on energy imports. The Global Environmental Facility (GEF) and Japan's Overseas Economic Cooperation Fund were ready to provide loans. The International Institute for Energy Conservation (IIEC), a Washington-based NGO, had opened its first Asian office in Bangkok to assist EGAT in carrying out the program. So in late 1993, Thailand started Asia's first ever DSM program (outside Japan).

But EGAT was skeptical. Largely staffed by engineers who see their job as building power plants, the state enterprise has a conservative culture that equates development with energy consumption. They see the utility as an electricity wholesaler; transforming it into an energy services provider requires marketing skills that the engineers typically don't possess.[48] So despite the success of DSM abroad, EGAT's 1992 power development plan claimed it is a "greater risk to invest in the DSM programs than to invest in new power plants." As a result, its initial goal of saving 238 MW was quite modest. A larger budget could have yielded ten times the savings, according to the IIEC, which assisted EGAT in drawing up the efficiency plan.[49]

As it was, the program churned out far more nega-watts than predicted. By 2000, it had saved 755 MW, more than the amount produced by Thailand's largest dam, at a cost of only $189 million.[50] Consumers have become aware of the potential savings thanks to marketing programs, and local appliance makers have been assisted in manufacturing more energy efficient products. One far-seeing developer, Rangsit Udomphol, hired a Singapore-based design consulting firm named Supersymmetry Services to make his new Olympia Thai office tower resource-efficient; not only did he save energy costs but also recycled water from the air conditioning system. "Engineers rip people off" by overequipping buildings, says Lee Eng Lock, the consulting firm's technical director. "They thrive on overkill."[51] A major reason why DSM has worked in Thailand—and why many other developing countries will find it difficult to emulate[52]—is that energy prices in Thailand aren't subsidized. The price of oil is floated, and consumers generally pay the full costs of producing and distributing electricity (except in industrial estates, where electricity is sometimes subsidized).

But despite its success, the future of DSM in Thailand is far from assured. The second phase of the program may not meet its projected target of 600 nega-watts because of a lack of funding. EGAT reportedly still considers it to be a "non-core business" and plans to spin off its DSM unit in a few years as deregulation of the energy market takes hold.[53]

Meanwhile, Thailand has done little to use its water more efficiently. The same principles that apply to energy can generally be applied to other resources, and the former prime minister, Anand Panyarachun, has been a frequent public advocate of "eco-efficiency," urging Thai businesses to examine whether they can save money and protect the environment by making better use of their inputs. But most of the country's water supply—an estimated two-thirds—is still wasted. Around one-third of Bangkok's water supply is lost due to a leaky distribution system, and the waste is even greater in the countryside, which uses from 70 to 90 percent of the country's supply. The Agriculture Ministry estimates that only 30 to 40 percent of irrigation water available is put to use. As a result, Thailand consumes about 600 cubic meters of fresh water per person, compared to 453 cubic meters per person in Korea and 128 cubic meters per person in Singapore, which both have much higher GDPs per capita.[54] Thailand's inefficiency is caused in part because more than thirty agencies scattered through seven ministries all have some say in water management.[55]

But the major underlying reason for this inefficiency is that, although energy is scarce in Thailand, water is not, and much of the country's valley-dwelling, rice-growing culture centers on the belief that water is abundant. Thais have never been asked, nor have they been willing, to pay the full price for its supply. City-dwellers benefit from subsidies, as does industry, and farmers receive irrigation free of charge. Since many are poor and rely on this subsidy to maintain their livelihoods, NGOs of the environmental democracy movement oppose attempts to charge irrigation fees. Here, environmental concerns conflict with farmers' (short-term) interests, and most NGOs have sided with the latter. In the long run, however, all the waste has exacerbated the yearly droughts, causing the government to pursue supply-side projects (that is, dams) which are then opposed by the same NGOs. It all seems counterproductive. Farmers are subsidized in many countries; surely there are better ways to do it than by giving them free irrigation.[56]

The issue came to a head in 1999 when farmers' groups alleged that the Thai government had agreed to begin charging irrigation fees to secure a $600 million agricultural sector loan from the Asian Development Bank.[57] The environmental democracy movement protested vociferously, and irrigation water remains free of charge. But it's hard to believe that the NGOs by themselves have the clout to block such policies. Word has it they have a most ironic ally on the issue, the best ally anyone could have in Thailand.

4

LOGGING

GUNS, TREES, AND REFUGEES

WHY YOU SHOULDN'T CLIMB DEAD TREES WHEN LOOKING FOR LOGS

"Hey, uncle, can you tell me where the sawmill is?"

The man I was asking, an elderly Thai farmer living in a small village in the remote northwestern province of Mae Hong Son, was not really my uncle, of course. But in Thailand, even when addressing strangers, it is polite to refer to them in familial terms: *pi* (elder sibling) for someone who's slightly older than you; *nong* (younger sibling) for someone slightly younger, or *loong* (uncle) for a man who's much older. It's a delightful tradition, and like so many in Thailand, it conveys a sense of warmth that takes you back to simpler, friendlier times.

Pursuing a news story in Thailand often required adopting a simple approach. Particularly when you're stuck for information while investigating environmental issues, it usually pays to go back to the grass roots. Talking to local villagers is almost invariably the best way to find out what's really happening on the ground; they may not know the big picture as well as, say, a government official, but they tend to be more honest and willing to talk.

I had already spent nearly a week in Mae Hong Son during the hot summer month of June 1995. I was investigating a tip that hardwood logs were being illegally imported into the province from just across the border in

Burma.* Not only was the trade contributing to deforestation in Burma but the smuggling traffic was also reportedly interfering with the flow of refugees fleeing the war-torn country.

It was a sensitive subject in the province; but after talking with various Burmese and Thai sources in town, I had managed to confirm the existence of the illicit logging trade, and I had also been told where the timber was allegedly crossing the border. I'd even discovered the name of the local logging kingpin. Then I'd spent several days wandering around Thai government offices, particularly the Royal Forest Department (RFD), where I badgered local officials until they grudgingly made some of their files and records available. According to the documents, there were not supposed to be imports of timber in this area—certainly no customs duties had been paid on them—nor were there supposed to be any active sawmills. I had even met with the governor of the province, who had assured me that no legal cross-border trade took place in the area.

Now I had to prove that the logs were indeed being smuggled. Pictures of timber coming across the border and run through a Thai sawmill would give me a major scoop. I had been to the border checkpoint at the sleepy village of Ban Nai Soi several times, but there were no logs or rogues in sight, just some empty logging trucks. So I'd decided to look for the sawmill first. Tooling around on the highway north of Mae Hong Son town on my rented moped, I'd finally heard the telltale whine of a buzz saw near the small village of Ban Kung Mai Sak.

I could hear the sawmill, but couldn't see it. Thus my question to the old farmer. "Just go that way . . ." he grinned in reply, pointing across some paddy fields toward a grove of trees in the distance. If he was curious about why this *farang* was speaking to him in Thai, or why I was interested in the sawmill, he was too polite to say so.

After crossing a patchwork of rice fields, walking on the raised edges as if traversing the lines of a maze, I came to a small stream. Thankfully, although the rains had already begun in this corner of Thailand, they were still only intermittent. Holding my camera high, I was able to wade gingerly into the current and clamber ashore on the far side. The mill was close now, just on the other side of some thick brush, but I still couldn't see it. A dead tree with low-hanging branches seemed to offer a good vantage point, so I began to climb. About fifteen feet off the ground, I was able to spy over the sur-

*Burma is now officially known as Myanmar, but most people still call it Burma.

rounding bushes into the mill yard, where several trucks and a pile of timber were heaped against the mill itself. A few workers were there, but they were absorbed in their work. So I started taking pictures, one arm wrapped around the trunk for support.

SNAP! I was suddenly in midair, plummeting. I landed in a heap on the ground with a solid thump, and I lay there for a moment, dazed, trying to figure out where I was; but after gingerly testing each of my limbs, I decided I was unharmed. My camera seemed to be okay, too. I still couldn't figure out what had happened, but when I saw a branch lying next to me on the ground, I realized it must have been the one I was standing on. It had snapped cleanly off under my weight. Now my arm began tingling. I looked down and saw a swathe of skin was missing from my right forearm, the one I'd been using to clutch the tree trunk. The flesh must have been stripped off as I fell. Already, little red and white beads of fluid had started to ooze from the raw flesh. Strangely, it didn't hurt at all.

Gathering myself, I counted my blessings. None of the workers seemed to have heard the fall; the mill was too noisy for that. But I still needed more pictures. What if the one I'd taken didn't come out? The entire trip could be wasted. But I was stuck with the same problem. Even if I could make my way through the thicket of trees, I wouldn't be able to see much. I had to climb the dead tree again. So, feeling faintly ridiculous, up I went.

A swarm of ants had been disturbed, and now they started to crawl across my hands. I tried to ignore them as I clambered onto another branch, which seemed sturdier. This time, I managed to take several pictures of the sawmill, and then started to make my way back down. My arm was starting to throb now. I looked at it once again when I reached the ground, and this time discovered a new horror. Dozens of ants were swarming across my arm, and they seemed to be inspecting my wound.

In a panic, I rushed to the stream and stuck my arm into the waters up to my elbow. A searing bolt of pain jolted up my arm as the cool current washed away ants, blood, pus, and any thoughts that might have been attempting to cross my mind.

THE ROOTS OF DEFORESTATION

The forests of Southeast Asia have been decimated over the last few decades. Usually, when people in the West think about tropical forests, they think first about the Amazon. And with good reason, since Brazil alone has more than twice as much forest cover as the 212 million hectares[1] in all of Southeast Asia (although it should be added that Asian logging firms are now active in the Amazon).[2] But since tropical regions contain such an immense amount of biodiversity, Southeast Asia and central Africa are probably the world's next most important area for forest conservation. The UN's Food and Agriculture Organization (FAO) reports that Southeast Asia has the highest rate of deforestation in the world, making it in all likelihood the most threatened region.[3]

Since 1960, Thailand's forest cover has roughly halved; it is now below 30 percent of the country's land area, or 14.8 million hectares.[4] But those are the official figures. Considering the extent to which conservation zones have been illegally degraded, or even cleared, unofficial estimates suggest that less than 20 percent of the country is still forested.[5] Meanwhile, some 40 percent of the kingdom's indigenous wildlife species are considered threatened, and at least half a dozen major animals have become extinct. At the same time, Thailand has gone from being a major exporter of timber to a major importer: Since 1999, it has had to annually import roughly timber products worth 50 billion baht.[6] Unsustainable management, in other words, has economic as well as environmental costs.

Deforestation in Thailand has basically followed the same pattern seen elsewhere around the world: After loggers cut down the valuable trees in an area, settlers—many of them poor and landless—move in to farm the available arable land. And the twin pressures of population growth and development have often led farmers to encroach upon standing forests, particularly where roads have been built.

Whatever the cause, deforestation is now widely recognized as a serious problem in Thailand, because forests provide a wide array of valuable environmental services that are usually taken for granted until they've gone missing. Much of the global concern over tropical deforestation is due to their role as carbon sinks—they help to sop up carbon dioxide, the greenhouse gas considered to be the biggest cause of global warming—and as vast, living warehouses of biodiversity. There are more species found on one mountain in northern Thailand than in some entire countries.

It is the immense local importance of forests, however, that has aroused the greatest concern within Thailand over their decline. In temperate countries, rain tends to fall year-round and snowmelt helps provide water during the hot summer months. But in tropical countries such as Thailand, the monsoon climate means that rain falls for only about six months of the year; people therefore depend on mountainous watershed areas, and in more recent times on dam reservoirs, to keep streams and rivers flowing year round. It is believed that healthy, living forests are able to store more water in their soil and biomass, thus helping to prevent floods during the wet season by sopping up excess rainfall, and to release more water during the dry season.[7]

Forests are crucial to the local economy in other ways. It's not only that the trees can be sold as timber or used as fuel wood; tropical forests also have a huge variety of plants and animals—everything from mushrooms to game to insects—that serve as food and medicine. And their role as climate modulators cannot be discounted. If you've ever been to the tropics, especially during the hot season when sweat seems to drip out of every pore on your body, you'll know that forests remain remarkably cool.

Thailand is thus now torn between its conflicting aims of exploiting the forest and conserving it. So is the official steward of the forests, Thailand's Royal Forest Department, which is supposed to try to achieve both goals. Ever since the RFD was first established in 1896, however, the agency's primary goal has been to profit from nature—either through rent seeking (that is, collecting money in exchange for doling out the right to exploit the resources it oversees) or through its own commercial activities. The agency is predominantly staffed by foresters trained to view forests principally as a resource that is worth more dead than alive.[8] The first attempts at conservation did not begin until the 1950s, when Boonsong Lekagul, a doctor whose passion for hunting gradually turned into an appreciation for (living) wildlife, convinced Thailand's then-ruling dictator, Field Marshal Sarit Thanarat, to set aside some land for conservation. That eventually led to a system of conservation areas based on the U.S. model in which there are national parks (set up for nature education and recreation), wildlife sanctuaries (for biodiversity conservation and research), and nonhunting zones (where some extraction is allowed). Conservation officials were generally trained in American methods, but these officials and the protected areas remain within the fiefdom of the RFD, which still retains its exploitative institutional mentality. The result is an agency that has contradictory aims—it is supposed to exploit and conserve the forests at the same time—and that is rife with corruption.

This system nevertheless went largely unquestioned until the 1980s, when satellite images showed that Thailand's vast tracts of forest had dwindled to a fragmented patchwork in the mountains and near the borders. Then, in November 1988, disaster struck. Torrential rains and floods caused a massive mudslide on a mountainside that had been denuded by illegal logging in the southern province of Nakhorn Sri Thammarat. Two villages were buried under the mud, and logs cascading down the mountain crushed many of their inhabitants. At least fifty-nine people were killed in the landslide, and the death toll from the floods rose to more than four hundred, making it the worst natural disaster of its kind in Thailand. In the ensuing uproar, Prime Minister Chatichai Choonhavan imposed a nationwide logging ban in state-owned forests that took effect the following year.

But illegal logging has continued, sometimes on a large scale, a major reason why Thailand continues to lose forests at an alarming rate, if at a slower pace than before the ban. The RFD, meanwhile, has been urged to manage its lands more sustainably, and to pay more attention to conservation. It has established a goal of 40 percent total forest cover for Thailand, 25 percent of that for "economic forests"—plantations, essentially—and 15 percent for conservation forests. The problem with this vision is that much of the land in question has already been logged and settled; and even those areas that still contain large swathes of primary forests, particularly in the mountainous north, are pockmarked with human settlements.

TRUCKS THAT PASS IN THE DUST

My wound stung, of course, but it was not serious, so after retreating back across the paddy fields, I resolved to see whether I could spot the trucks bringing the logs to the sawmill. I had found the road entering the mill, but it was not a good vantage point for photos, and it was also too risky to linger there. So I began to work my way backwards, eventually turning off onto the dirt road that led to the border crossing at Ban Nai Soi. The guard at the Army Ranger checkpoint ignored me as I zipped under the wooden barrier, just as he must have ignored the many logging trucks that had been heading in the opposite direction.

The timber trade here was really an open secret. It had already been more than five years since Thailand had imposed a domestic logging ban. To make up for the lack of wood, Thai businessmen had gone into neighboring countries—Burma, Cambodia, and, to a lesser degree, Laos—to source their timber. Burma had proved especially lucrative, for a while. It had a relatively large amount of remaining forest, and back in 1988, Burma's newly installed military junta, the State Law and Order Restoration Council (or SLORC, as the junta was deliciously known), desperately needed cash to stay afloat. Thailand's then–army chief, General Chavalit Yongchaiyudh, had flown into Rangoon and made a deal in exchange for logging concessions.

But after several years, the Thai loggers proved to be too rapacious even for Burmese dictators. The SLORC abruptly cancelled the Thai concessions and gave its state-owned Myanmar Timber Enterprise an even more lucrative monopoly over the teak trade. The SLORC, however, didn't control most of the border areas; these lay in the hands of a dozen or so ethnic minority groups who had been fighting for autonomy or independence from Burma for decades.

Across the border from this part of Mae Hong Son was the Burmese state of Kayah, where an active insurgency was led by the Karenni National Progressive Party (KNPP). They probably didn't have more than a thousand "troops" (actually just well-armed peasants), but they had been toughened by decades of war and knew the terrain well. They could attack the *tatmadaw,* the Burmese army, with quick strikes and then fade into the jungle.

The KNPP controlled the timber trade here and it had struck a deal with the local Thai logging mafia. In went the Thai road graders and chain saws, out came the Burmese teak. It was a dangerous game the KNPP were playing. By cutting down the forest, they were reducing their own cover for future battles, and they were angering the SLORC by undercutting the teak monopoly. Ostensibly, the money would help supply the KNPP with guns and food, but some of the party's "ministers" had been building suspiciously nice houses in the suburbs of Mae Hong Son.

The Thai authorities were only too happy to ignore the smuggling—for the right price. Most of the Burmese pro-democracy groups that had set up shop in Mae Hong Son also kept quiet about it. The SLORC was their enemy, and the KNPP was therefore their ally. The handful of foreigners in town working for the relief agencies were also silent. They didn't want anything to jeopardize the help they provided the thousands of refugees fleeing Burma. I had found out about it only through a tip from a Burma activist based in Chiang Mai who had simply become fed up with the KNPP's greedy ways. "There are so many logging trucks that they're blocking the road for the refugees," she complained.

That must have been an exaggeration, because I'd been down to the border checkpoint for three or four days now and, although I'd found some scattered logs and trucks, I hadn't seen any real hauling. The only things being smuggled in, curiously enough, were cows. Word had it that the livestock originated in India and were herded all the way across Burma before entering Thailand. And it must have been true because some of the cows still had little red bindi on their foreheads.

Driving on toward the border, I tried to figure out the loggers' typical schedule. Perhaps they cut the trees in the morning and brought them over by truck in the afternoon. Once the rains come in earnest, I realized, they'd have to halt everything because the road would become too messy. So they must be quite busy right now—

I nearly fell off my bike rounding the next curve. There, coming straight at me, was a ten-wheel truck with large steel prongs on the side and a huge metal claw on top. It was loaded with logs.

Fortunately, it was coming uphill and thus moving slowly. I pulled over to the side of the road and, as the truck passed, fumbled through my backpack for my camera, finally managing to snap off a shot of the truck's backside. I'd need to get a picture from the front, however. I spun my moped around and hurtled back down in the direction I'd come from, trying to catch up. But the truck was now picking up speed and kicking up a cloud of dust. I trailed along while waiting for another uphill section of the road where I could make a move past the truck. Then I heard a sound behind me.

I looked back, and was startled to discover another truck, just as big and just as loaded with logs, right behind me. Two men were in the cab, each sporting wide grins at the sight of this clueless *farang* tourist churning down the road between two diesel-belching behemoths.

Finally, we came to another ascent in the road, and the truck in front went into low gear. I edged past it, zipped to the top of the hill, and then careened down the other side, the bike slipping and sliding on the loose gravel. When I'd finally put some distance between myself and the trucks, I stopped by the side of the road, adjusted the camera settings, and then waited.

The trucks came roaring by, and I snapped away. The drivers actually mugged for the camera, and blew their horns cheerily. My pictures safely taken, I put down the camera, gave them a big grin, and waved back.

WHAT'S GOOD FOR PEOPLE MAY NOT BE GOOD FOR FORESTS

Covering environmental issues in a country as oppressive as Burma raises an intriguing issue: Is liberal democracy good for the environment? It's a question that springs up often in the developing world, especially in Southeast Asia with its wide variety of political systems.

The answer isn't obvious. We tend to assume that democracy is good for the environment because it allows environmentalism to flourish. People in Thailand, for instance, are far freer than the Burmese to express their green concerns, and to defend common resources that would otherwise suffer from exploitation or neglect by centralized governments. On the other hand, authoritarian countries such as Singapore are better able to order citizens to obey (sometimes onerous) environmental regulations. And the poorest countries, which generally have weaker democratic institutions, tend to have the most resources left pristine and intact. Half of Burma is still covered in forest, after all, while forest coverage in Thailand has been decimated. This is the result of many factors besides the two countries' differing political situations. But democratization does seem to bring deforestation in its wake; after all, logging picked up dramatically in Indonesia and Cambodia following their transitions to democracy in the 1990s.[9]

Neal Englehart, a political scientist at Lafayette College, has carried out a fascinating study of this issue by comparing the pace of deforestation in eight countries in Southeast Asia.[10] He was surprised to find that the freer a country is, the more rapidly it loses its forests. Between 1990 and 1995, the Philippines and Thailand, Southeast Asia's two most democratic countries, suffered the highest annual loss of forest coverage, and Malaysia and Cambodia were next. Meanwhile, the four most authoritarian countries in the study—Burma, Vietnam, Laos, and Indonesia—had the four lowest rates of deforestation.

This result is unexpected, but it doesn't mean the dictators and single-party systems ruling these countries have nobly set out to protect their forests. Englehart concludes that the relatively rapid demise of Thailand's forests stems from greater *economic* freedom, particularly within its timber market. In Indonesia, by comparison, the timber sector was largely a closed market; the concessions were handed out to a few well-connected businessmen, but they have not been efficient in overexploiting the country's undervalued forests. Englehart concludes that although political freedom does allow green groups and communities to mobilize in defense of common resources, that effect is "swamped" when it comes to maintaining forests because economic freedom leads to greater environmental degradation. In Thailand, as opposed to Indonesia, he argues, "privately mobilized capital distributed among a large number of competitors has created highly efficient firms which very efficiently destroy forests."

It's an intriguing argument, but it needs a caveat: The rate of deforestation also depends on the property rights regime governing the concession. In theory, it's possible for Thailand or Indonesia to design concessions so that loggers have incentives to carry out reforestation or use techniques such as reduced-impact logging, which is more sustainable than conventional clear-cutting and, according to some analysts, cheaper to operate. In practice, however, such incentives either aren't included—because concessions are often given out as rewards by politicians to their supporters—or don't work. Even loggers given long-term leases for forestland may hurriedly clear-cut it for fear they will be taken away by future administrators, or simply because the loggers are too shortsighted to adopt strategies that yield more long-term benefits.

The conclusion that the environment benefits from the presence of authoritarian governments also has to be tempered by looking at different kinds of environmental issues. By limiting economic freedom and access to valuable resource concessions, autocrats may indirectly help conserve those resources; but, because they are not held accountable, authoritarian or highly centralized governments are also notoriously bad at controlling pollution and responding to citizens' concerns about quality of life. Consider, for instance, the abysmal environmental record of Communist governments in Eastern Europe: Their pell-mell rush for industrial development, combined with their lack of democratic accountability, turned large swathes of the region into polluted wastelands. The toxic effect of these policies is now frighteningly evident in the sorry state of public health in the region, particularly in the mortality rates of people living in the former Soviet Union.

It is sometimes convenient to split environmental issues into two kinds of problems. On the one hand there is the consumption and conservation of valuable resources such as forests and fisheries (sometimes called "green" issues); and on the other, there are issues of pollution and environmental quality ("brown" issues). The latter are generally the side effects of industrial or other activities that harm people in a way not mediated by the market. In the long run, democracies can generally take steps to combat these "negative externalities," so long as aggrieved victims have recourse to sit down and negotiate with the people or companies doing them harm. But liberalism also tends to unleash the buzz-saw-like power of the market on resources that were previously either untapped or being used sustainably. Markets have the potential to help promote conservation—through mechanisms such as ecotourism, bioprospecting, and carbon sequestration—when it proves to be more profitable than logging.[11] But with a few exceptions, this potential has yet to be realized, and so it seems that emerging democracies have a tough time grappling with conservation, which usually requires governments (or some type of public sector organization) to step in and say: "The market does not rule here."

MAJORING IN JUNGLE WARFARE

The All-Burma Students Democratic Front (ABSDF) had a home in Mae Hong Son, specifically, a modest three-bedroom hill-top dwelling with a long driveway and a garage. The fighters who helped make up this militia seemed equally disarming. Maung Thwin (a pseudonym) had straggly hair and was attempting to grow a moustache, but he could easily have passed for a freshman at Chiang Mai University.

Actually, he'd been a student at Rangoon University in 1988, when a democracy uprising in Burma was brutally crushed. Maung Thwin and thousands of other students like him fled to the jungles, eventually forming a military union that often fought alongside the ethnic insurgents against the ruling military junta. Maung Thwin had seen the butchery of the *tatmadaw* up close. By comparison, life in Mae Hong Son must have been relatively sedate. When I'd first arrived in town, Maung Thwin had explained the log smuggling situation to me, and had even drawn up a map of the area—freehand, but quite detailed—pointing out where the logs were coming in, where the refugee camps were situated, even where the known SLORC bases were on the other side of the border.

The students were concerned about the potential impact of the logging in Karenni territory, fearing it would reduce cover and cause a worsening drought. But, like other groups in the area, they put ecological issues way down on their list of priorities. In fact, the ABSDF had planned to run a sawmill with the KNPP at Mae Yu, across the border in Burma, and they had even invested 50,000 baht in the scheme. But the Karenni had taken over the project, hired its own workers and basically told the students their participation was no longer needed. As it happened, the sawmill had to be moved into Thai territory anyway because the SLORC had grown angry that all this money was changing hands behind their backs.

Angry enough, apparently, to threaten war.

After I told Maung Thwin what I'd come across, he had some news for me. "SLORC troops have moved into some new positions in Kayah State," Maung Thwin said carefully. We spoke in English; many Burmese, especially the well-educated ones, were better at it than Thai. "We hear they have deployed two new battalions there, and the KNPP claims the action violates their cease-fire agreement. I'm afraid it does not look good. There could be fighting soon. . . . Our people on the border have already gone on alert."

"Is the SLORC giving any explanation?"

"They say they want to make sure there is security during the Thai election." Thais were going to the polls in a few days to elect a new national government. The campaigning was coming to such a crescendo that it was impossible to ignore it, especially if you were a journalist. Maung Thwin added, "The SLORC also says they want to stop Thais from stealing timber." Wow, this was getting more serious than I'd ever imagined. What a time to come looking for logs. A smuggling story was about to turn into a border war. I didn't know whether to be excited or scared.

Maung Thwin was still speaking—". . . so you understand that I may have to leave soon?"

The young revolutionary was calm about the prospect of going to war. But there was tension in his serious young face. He had drawn such a nice map that I wondered whether, in some other life, Maung Thwin would have been an artist, or a graphic designer. He might even have been graduating at about the same time I met him. I tried to picture him wearing a cap and gown, posing for pictures with his sister, basking in his parents' glow.

"Maung Thwin, what did you study?"

"Excuse me?"

"When you were at university, what subject did you study?"

"Political science." He grinned, apparently not at all surprised by the question.

Getting back to the issue at hand, I wondered what a military conflict would mean for the reports I would be filing with my newspaper. Not only would I have to fight for space amidst all the election news, but now, if a war really did break out, it would completely overshadow concerns about log smuggling.

"You are thinking of something," he said.

"I was wondering if all the work I've done over the last week has been a waste of time."

"No, it is good that people know what is going on. We need more environmental awareness. It is a part of democracy. This logging business has hurt many people. But under this situation . . ." He shrugged his shoulders.

I finished his sentence for him: "Under this situation, you've got a lot of other things to worry about." He just smiled.

"I know, everyone does," I grabbed his hand. "Look, Maung Thwin. Thanks a lot for your help. Take care of yourself, huh? Don't get yourself shot, okay?"

Maung Thwin laughed. "Of course not. Believe me, I will be careful. But I have to help my friends. I have to help my country." It didn't sound corny. He still had a smile on his face, but looked utterly serious.

The atmosphere was considerably more tense at the KNPP office in Mae Hong Son. This was another nondescript building, a shop house on a back street downtown. I walked in and found Jeremy reading the newspaper, intently. The KNPP tried to pretend that it was an independent government, and Jeremy's official job description was spokesman for the party's ministerial secretariat—whatever that meant. Unofficially, Jeremy (another pseudonym) seemed to be the most well-educated and polished speaker the KNPP had. He must have been in his late forties, and he carried himself with dignity. He was a native Karenni, but in a different time and place he would have made an excellent corporate public relations officer. When I asked about the report of SLORC troop movements, he slipped on his reading glasses and looked through his notes. With his lightly graying hair, the glasses made him look even more distinguished, almost professorial. But this was no academic discussion.

"Okay, here's the situation. SLORC has moved into areas they're not supposed to be in under the cease-fire agreement. Altogether, they may have four understrength battalions in Karenni territory—roughly 2,000 troops—out of the 15 total they have in Kayah State," he explained. "We've also received word they've begun forced portering." This is a particularly nasty business in which the *tatmadaw* grabs villagers, especially from minority groups, and forces them to carry materiel. The porters typically have to walk for days on end under hazardous and grueling conditions, often with little or no food to eat. If they're particularly unlucky, they are used as human minesweepers. "It's specifically mentioned in the cease-fire agreement that there will be no forced portering, and that SLORC will not expand troops in Karenni areas."

I asked what it all meant.

"It means that the cease-fire may not hold."

"Any idea why they're doing this?"

Jeremy laughed derisively. "Well, in their radio reports, they're claiming that the border may become unstable during the Thai election and . . . "

"Yes?"

"And they say that the Thais are stealing Burmese timber."

"So it's the logging that has them upset."

"Under the cease-fire agreement, we have the right to do business freely."

"Does it say you have the right to sell logs directly to the Thais?" The cease-fire agreements SLORC had signed with the minority groups were not exactly public documents. And with the KNPP, word had it the agreement was only made verbally.

"During negotiations, we were allowed to do the logging. But after the agreement was reached, SLORC said all the hardwood had to go out through Rangoon. But they only offered us a very low price. They wanted to pay us in kyat [the highly overvalued Burmese currency], at the official rate . . . but you already know all this. We've been over it."

"Yes, but now it looks like there's going to be a war over these damn logs!"

"Look," Jeremy was having a tough time concealing his own exasperation. "Logging has been going on for a long time. It picked up a bit after the cease-fire, but it's still not as heavy as before. The forests are nearing depletion." He became more passionate. "It's not just our timber they want. They want our territory. They want our people. There are lots of minerals in Karenni areas. We must have the right to sell our resources freely. They want to control the Salween River, so they can build a big dam on it. You know they do."

"Yes . . ." I intended to write something about the proposed Salween Dam, but right now I wanted to stick to the subject of logging. I put down my pen and paper and asked about a name I'd heard while investigating the log smuggling.

"Jeremy, have you ever heard of a guy named *Por Liang* Som?"

"No," he said, after some thought, "Who is he?"

"Apparently, he's the Thai businessman who's bringing the logs over. At least that's what the locals say. I came across his trucks bringing some logs out yesterday." *Por Liang* is a Thai expression that literally means "father (who) takes care." It's the common northern Thai expression for "godfather."

"I've never heard of him."

I believed him. Jeremy wouldn't have been involved in setting up the logging deals. Those were left for the ministers. And Jeremy knew little about Thai politics. He barely spoke the language.

"Look, Jeremy, I'm going to write about the troop movements of the SLORC. Of course, I have to check with some other people first. But if I'm going to do this story right, then I have to meet with the KNPP chief."

"You mean the prime minister?"

"Yes. Aung Than Lay. Is it possible?"

"To talk about what? About logging?"

"Well, yeah, and about the troop movements, and the state of the cease-fire. But about logging, too."

Jeremy thought it over for a while. The KNPP certainly wanted publicity about the border military movements because they needed to raise the alarm. But the whole logging situation could be embarrassing to them.

"I'll try to get you an appointment with him. But I can't promise. I'll do my best."

COLLABORATION IN BURMA, CONFRONTATION IN CAMBODIA

Because the preservation of tropical forests is such a big issue in the West, it's not surprising to find so many foreign environmental groups active in Southeast Asia. Most of these groups go about their work quietly and professionally, gathering research or funding projects to promote sustainable development and environmental education. But they often face a conundrum in deciding how much to collaborate with oppressive governments and corrupt forestry agencies. In democracies such as Thailand, the Philippines, and even Cambodia, the groups can combine collaboration with criticism, and work with local NGOs that dare to speak out about political issues at the risk of violent retribution. But what strategy should environmental groups take with dictatorships such as Laos and Burma, where neither NGOs nor criticism are tolerated?

Alan Rabinowitz has chosen the collaborative route, but only after stepping on a few Thai toes. An expert on mammals at the New York-based Wildlife Conservation Society (WCS), he has worked in Asia and Latin America. In the late 1980s, when he was involved in a research project on tigers in Thailand, his outspokenness regarding the corrupt mismanagement of wildlife reportedly made him *persona non grata* with some powerful Thai officials. A few years later, however, he seemed to get along much better with the government in Laos, where he came out in support of the controversial Nam Theun 2 Dam. Rabinowitz argued that although the proposed reservoir would flood a major part of the biodiversity-rich Nakai-Nam Theun plateau, much of the reservoir site had already been logged in preparation for the dam anyway. The project would help protect wildlife, he further maintained, since it entailed the establishment of a conservation area around the future reservoir and the relocation of villagers who hunted in the area.[12] His support for Nam Theun 2 enraged Thai anti-dam activists; they contended that a logged forest could at least grow back from the seeds left in the ground, but a forest flooded by a reservoir would be gone forever. They accused Rabinowitz of trying to ingratiate himself with the Lao government to further the WCS research opportunities there.

But Rabinowitz did not stop with Laos. In 1993, he became active in Burma, a country that most Western NGOs, including the World Resources Institute,[13] have shied away from because of the military junta's corrupt ways and sorry human rights record. To gain the government's trust, Rabinowitz worked hard at carrying out tiger surveys and instituting a management plan for a proposed conservation area around Lampyi Island in the south of the country. He suggested that some Burmese administrators had gotten a bad rap and noted that they had specifically asked him to make sure the park's establishment did not negatively affect the sea gypsies, sea-borne nomads who lived in the area. And Rabinowitz's hard work paid off when it gained him unprecedented access to long-isolated regions such as Mount Hkabo-Razi in the far north of Burma, where he discovered a new mammal species, the leaf deer.[14]

Rabinowitz concluded that, given the continuing decimation of forests and wildlife the world over, conservation must take priority over political niceties. And he is not alone in holding such views. Some natural scientists, particularly in the West, are quietly willing to help preserve rare and endangered species even if it means collaborating with oppressive governments. Most Asian (and many Western) activists, on the other hand, hold the equally principled view that democracy and human rights should be the top priority; among them, Rabinowitz is considered highly controversial. But in the United States, he is sometimes celebrated as a swashbuckling wildlife biologist, a real-life analogue to Indiana Jones whom the *New York Times* once lauded as "Burma Rabinowitz."[15]

Patrick Alley and Simon Taylor, meanwhile, have taken a different approach to fighting for conservation in Cambodia. Founders of the U.K.-based NGO Global Witness, their desire to fight for the environment *and* for human rights brought them to Cambodia in 1995; there, they sought to expose the involvement of the genocidal Khmer Rouge—by then a dwindling band of guerillas holed up in western and northern Cambodia—in illicit logging operations. Having found that the Cambodian government was also complicit in the rapid decimation of the country's forests, they decided on a two-pronged strategy: lobbying the Consultative Group of foreign aid donors, whose funding was vital to keeping the country financially afloat, and exposing the rampant corruption behind all the logging deals.[16]

Time and again, Global Witness managed to gather incriminating documents and expose them in the Cambodian and foreign press. Alley claims their biggest coup came in 1996 when they uncovered the infamous "million-meter deal," a secret accord signed by Cambodia's co-premiers, Hun Sen and Norodom Ranariddh, and a group of eleven Thai logging firms to extract a huge amount of timber out of the country. The resultant outcry forced Cambodian authorities to halt the export of logs, at least temporarily, and pass stricter regulations on the timber trade.

"The secret of what we've done is to provide good information to the media and lobby at the pressure points, such as donor groups," says Alley. "We've waved the stick because we don't have a carrot." At first, the donors argued that they couldn't push Cambodia to log more responsibly because it was a political issue, but by 1997 they finally accepted that forest management was also critical to development. After two more years of muckraking and lobbying by Global Witness, aid was finally made conditional on forestry reforms. Soon afterwards, Cambodia's prime minister, Hun Sen, announced a crackdown on logging. "The large concessions are still going on, but the small- and medium-scale logging have been brought under control," Alley said during an interview in 2000. "The result has been a dramatic decline in logging." Even more astonishing, Global Witness was invited by Cambodia's government to serve as an independent forestry monitor.

It has not been easy, however. A year after that interview took place, following a report by Global Witness that blamed official corruption for a dramatic rise in illegal logging, Hun Sen apparently had seen enough of independent monitoring. He accused the NGO of waging a smear campaign to depose him, and threatened to expel its monitors,[17] although he relented under donor pressure. Alley acknowledges that the tactics used in Cambodia would not work everywhere; they would not be tolerated in a closed society such as Burma, for instance, and in Thailand he felt it was better for local NGOs to speak out. But a country such as Laos, which still relies heavily on foreign aid, could well be susceptible to external environmental pressure.

THE GODFATHER OF LOGGING

My sources in the *sala klang jangwat,* Mae Hong Son's city hall, confirmed that SLORC troops had moved into threatening positions closer to the border. KNPP troops were mobilizing, but no one seemed confident that they could hold out for long against the *tatmadaw.* They were hardy, but they lacked training. The question was, did they also lack weapons? That depended on how the proceeds from KNPP logging were being spent.

I went to the nearby media center, where local journalists filed their stories, and wrote my own article about the buildup of tensions and the reports of troop movements; I also added a few paragraphs hinting that logging could be at the root of the conflict, but I didn't want to elaborate much on that point. I had to collect more information first, and then I'd write a big "scoop"—which in Thai meant a long, feature-y, front-page report. It would be a scoop in the English sense of the word, too. A potential double-scoop, I thought, my mouth practically watering.

I began chatting with a local reporter I'll call Nen. An energetic, wiry guy, with big glasses, he was hungry for the latest dirt on what was going on. He rocked back and forth in his swivel chair as he listened, but then suddenly sat upright when he heard the name of the Thai logger who was supposedly involved.

"*Por Liang* Som? Is that who's bringing out the wood? *Taung rawang, na,*" he warned. "You better be careful."

"You know him?" I stopped writing. Of course Nen would know him.

"Sure, everyone does," he said, getting up quickly and walking to the window. He pointed across the street, to the parking lot of an upscale hotel. "Look, you see over there. That's his car right there! That Mitsubishi Pajero. He always stays in that hotel when he comes into town." I went to the window to join Nen, who started speaking rapidly.

"James, *yaa yoong gahp khao, na.* Don't mess with that guy. He's a real gangster. He's involved in a lot of logging schemes along the border, and you don't get to be the *Por Liang* unless you're ready to use violence. This guy is real nasty. Listen, the story has it that he used to be in the logging business with his son-in-law. Only the son-in-law started cheating on him and the *Por Liang* found about it. So, apparently, one day he invited the guy and his daughter out to lunch. And then he just shot the guy, with his daughter sitting right there looking on!"

I must have looked dubious. "Come on. Is that really true?"

But Nen just nodded back even more emphatically.

"*Jing-jing!* Really, it's true! James, *nak khao* to *nak khao,* as reporter to reporter, I'm telling you, this guy is bad news."

Of course, the story I wrote that night didn't mention the godfather's name. The rule in Thai journalism basically goes like this: If you accuse someone of something illegal in print, not only do you need a reliable source but you have to print a response from the accused, or at the very least try to contact the accused and ask for some kind of comment. Over the next few days, I pondered how to go about asking a mafia chieftain whether he was involved in illegal log smuggling. But there didn't really seem to be a healthy way to go about it.

Meanwhile, my news story came out. It was buried on page A6, the local and regional news page.[18] Given the virtually continual strife in Burma, the story didn't exactly jump out at you, especially since it had been reduced to only nine inches of text. All the references to logging, which I'd put at the bottom of the article, had simply been cut due to lack of space. And set as it was against the building crescendo of news surrounding the imminent Thai election, it probably made little noise with most readers.

But Burma watchers would sit up and take notice. SLORC was busily signing cease-fire agreements with numerous insurgent groups, so the breaking of one such agreement would raise some eyebrows and confirm more than a few suspicions. The publishing of the story may also have convinced the KNPP that I was a legitimate journalist who could get their comments and concerns into a newspaper of record. Although I sometimes took for granted how important a journalist's role was, in the eyes of a long-neglected insurgent group located in a distant corner of Thailand, it represented useful clout.

Whatever the reason, within a couple of days, Jeremy called to say that if I wanted an interview with the head honcho of the Karennis, now was my best chance: "I still can't promise you it will happen, but he will be in Camp 5 briefly, so we may be able to meet him if we go out there."

A PRIME MINISTER WITHOUT A STATE

Camp 5 was one of the Karenni refugee camps located on the border. Getting there required a long early morning drive in the back of a pickup truck. As we headed toward the border, we raced past rolling farmland, through stately forests, and around hills still shrouded in morning mist. In some areas, the forest had been severely degraded—first logged, then settled. Tree stumps still sprouted out from fields too newly tilled to be considered proper farms. The mornings were chilly here in the highlands, and when a slight drizzle started to fall, making the dirt road even more treacherous, Jeremy and I huddled under a tarp along with the other passengers, most of whom were ethnic Karennis visiting relatives in the camp.

At a Border Patrol Police (BPP) checkpoint, the last outpost of official Thailand, I pulled the hood of my plastic rain poncho low down over my face. The BPP might let a *farang* through to the refugee camp; then again, they might not. It was better not to ask permission. But with the rain coming down, the officers decided against even leaving their guard hut and simply waved the pick-up through.

The road steadily worsened, and several times we were forced to ford some streams, all the passengers disembarking as the driver gunned the truck across the rapids. Finally we made it to Camp 5. Unlike many refugee settlements that are in disputed areas along the border, this one was definitely within Thai territory, so the Karenni inhabitants were strictly forbidden by the Thais to store weapons there or to use it as a kind of rear base. But apparently the people here had strong affiliations with the KNPP, and there was nothing to stop insurgent leaders from visiting the place to coordinate with supporters on the Thai side.

On the other hand, there was also nothing to stop SLORC troops from entering the camp and ransacking it. Although Camp 5 was in Thailand, there was virtually no official Thai presence in the camp, apart from a couple of teachers who probably doubled as spies. A Thai military response to a *tatmadaw* incursion would not come for hours, if at all. So yes, this was Thailand, but it was also kind of a no-man's land.

It was also a surprisingly pleasant place, surrounded as it was by lush jungle with soaring hardwood trees instead of the usual barbed wire or chain-link fence. The houses were sturdy affairs made of bamboo and thatch, and some had wooden pillars. Many contained small gardens, chicken coops, even flower patches winding among bamboo fencing. The place looked more like a proper village than a refugee camp. And the Karenni certainly weren't logging here.

We were led to a modest hut where we sat down to a lunch of rice, cassava curry, and some mystery vegetables. It was only after we had eaten that Jeremy again raised the delicate issue of what I wanted to ask the KNPP leader. "I assume you're going to ask about the military situation . . ." I nodded. "And do you still intend to ask about logging?"

"Yes, I have to."

"I see." Jeremy thought for a while. "Should I say you're doing a story on the general situation regarding logging in Burma?"

I wasn't sure where Jeremy was going with this. "What do you mean?"

"I mean at least you should compare what's going on in Karenni territory with logging in SLORC-controlled areas. You know they're handing out concessions to Singaporean and Chinese firms all over. There are whole forests being cut down and trucked over the border into China."

Jeremy had a point. To be fair, I would have to write about the SLORC's logging activities. But information on the situation within Burma was hard to come by when one was stranded here on the border. And comparing the amount of destruction incurred by the different groups would be impossible without seeing the damage for myself. I'd involved myself in this story because it was about smuggling logs into Thailand. Nevertheless, I assured Jeremy that I would try to be even-handed.

Jeremy sat back. He still seemed skeptical. "Well, I will try. Let's rest a bit and then I will go look for Aung Than Lay and see what can be arranged."

That sounded fine to me. I've always been of the opinion that civilization had taken a turn for the worse when it decided to junk the tradition of a post-lunch siesta. So I took a snooze, and then—waking up alone in the hut—whiled away the afternoon by exploring the camp, taking pictures, trying to chat with the few people around, even conducting a few interviews. But as the hours passed, the afternoon turned to twilight, and still I remained alone.

Finally, there was a noise outside and I went to investigate. It was one of the men I'd had lunch with, but I didn't know whether he spoke English. "Jeremy?" I asked, pointing.

"No," the man waved his hand.

"Where's Jeremy?"

"Jeremy, he say he have to go," and the man pointed off into the distance.

"Huh?!"

"Jeremy, he go down river. Be back tomorrow. Maybe." And the man turned away.

I was stunned. Then I was angry. Jeremy had gone. He must have been unable to arrange an interview. But he could have at least told me about it. Now what was I supposed to do? Stuck alone in a refugee camp where I didn't know anybody or anything.

Without quite knowing what I was going to do, I picked up my bag and made my way out into the camp, heading toward the area where I'd been told the leaders stay. It was dark now, the jungle seeming to expand and crowd in on the comparatively tiny houses. The only light came from lanterns in some of the huts, and from the stars that shone brilliantly through the opening in the forest canopy above.

I crossed a wooden bridge over the river that ran through the camp, then stopped a passerby to ask directions to the house of Aung Than Lay. I was pointed toward a slightly larger than average building, made like the others of wood, bamboo, and thatch. A couple of men stood on the steps outside, but the many pairs of sandals scattered around indicated there were more people inside.

"Aung Than Lay?" I asked, pointing inside.

"Yes?" replied the younger man.

"I'm here to see Aung Than Lay."

"You have a meeting with him?"

"Yes."

"Just a minute, then. Your name?"

I told him. "I'm a journalist," I said hopefully.

The man stepped inside, motioning for me to stay where I was. He came back after a minute or two and tried to make apologies. "Aung Than Lay can't see you right now. Important meeting."

I thought for a long moment. "Look," I finally said to the fellow, who seemed to be in charge. He appeared uncomfortable now, and was obviously desperate to get rid of me. "I realize Aung Than Lay is very busy right now. But please, can you just go in one more time and ask him to see me? Tell him I've come a very long way to meet him. Tell him . . . tell him I want to help the Karenni people," I finally blurted out.

The beefy man looked at me closely for a second, then stepped back inside. I began to wonder whether I'd just breached one of the cardinal rules of journalism. Hey, all I'd said was that I wanted to help the Karenni, a downtrodden people who certainly deserved assistance, and I reasoned that I could do so simply by telling the truth about what was going on here. To me, there was quite a difference between saying I wanted to help the Karenni people and saying I would help the KNPP. But perhaps they would interpret it as meaning that my reports would favor them in their struggle against the SLORC, and that would go against all my instincts as a journalist to be scrupulously fair. On the other hand, another rule of journalism is this: Get the story. And I couldn't do that without talking to Aung Than Lay.

I would have to fret about it all later, however, because the fellow returned and this time he waved his arm to welcome me inside. A little surprised, but mostly relieved, I slipped off my sandals and climbed the steps. Inside, I was ushered into an empty room with a large table and told to wait.

Almost silently, and a bit anti-climactically, the prime minister glided into the room. We shook hands, touching our right elbows with our left hands, a Burmese gesture of politeness. He motioned for me to sit, still not having said a word. Aung Than Lay was so gaunt that he looked almost skeletal, skin stretched tight across jutting cheekbones, veins etched clearly underneath. He had gray hair, but seemed ageless in an ascetic kind of way. Because his back was to the light, I never saw his eyes, which were hidden behind tinted glasses. But I imagined them boring into me, trying to uncover my motives. I began by explaining who I was, and what I was doing there. Then I asked about the increasingly tense military confrontation.

Aung Than Lay responded in a slow, measured voice. "The SLORC has moved into our area. It is the same as if they attack us."

"Why do you think they have moved in?"

"They say the Thais come into our area and steal teak wood. They say they have to find and arrest these Thais. But the Thais did not come and steal any wood. They come and buy it officially. The SLORC also says in the coming days there will be an election in Thailand, so they have to come and guard the border."

"What do you think they intend to do?"

"They will go to border posts 9, 10, and 11 to try and control the crossings. But we cannot let them. They say they will never allow teak wood to be taken out over the Thai border."

"Where should it go?"

"They want it to go through Rangoon. But they don't say how much they will pay for it."

This was a chance to talk about the Thai operations. "How much do the Thais pay?"

Aung Than Lay thought about it for a moment. "The Thais pay 7,000 baht per ton."

Finally, after all my digging, I had the raw, hard data I was looking for. I did a few quick calculations. At 7,000 baht per ton, the Karenni were selling teak to the Thais at a tenth of the price it was going for in Rangoon. The SLORC had obviously decided that was no way to treat a monopoly, especially one that was officially providing Burma with $150 million every year.[19]

I struggled to come up with a question while processing all this through my brain. "So, uh, you decided to deal with the Thais. That is, you . . . "

"It's okay to say we sell teak to the Thais. We have sold less than before the cease-fire, when the Thais got a concession from SLORC to log teak in Karenni territory. We want to sell our teak to anybody. We have the right to sell our property."

"Would you sell it to SLORC if they offered you the highest price?"

Aung Than Lay shrugged. For the first time, a thin smile crossed his lips. It was difficult for him to conceive of doing business with the enemy. I edged the conversation toward the consequences of the KNPP's activities. "Do you realize what kind of impact your logging has?"

Aung Than Lay responded smoothly. "I know about the damage to the environment. I had a farm, and when we cut all the trees, the water supply dried out." He hesitated briefly. "But still, it is necessary."

"There are some who say the money doesn't all go where it should . . . "

"We don't keep the money for ourselves. Our people know where the money goes. We help the local people with medicine and roads, but it is not enough. We buy rice and food, and if possible weapons to defend ourselves."

Unfortunately, I had no way to verify this statement, but I found it hard to believe that at least some of the profits weren't being siphoned off. In camp 5, I had interviewed some of the refugees; they said the proceeds from logging activities helped pay teachers' salaries, so they seemed to believe their leaders. But the people living in the areas where trees were being cut down might see things differently. They were facing the impacts directly, and if they didn't support the KNPP, they wouldn't see many of the benefits.

"Okay," I said. "But if I may speak frankly, the military situation here does not look good for you. Perhaps if you just halted logging, at least for a while, the confrontation would die down." It was a feeble attempt at peace making, but I felt a responsibility to try.

"No, we should have the right to sell, not just teak, but also minerals. We hope there will be no fighting. But we have been fighting for nearly fifty years now. We are used to making sacrifices. If we have to, then we will keep on fighting until we get our freedom. This is not a revolutionary war, it is a resistance war. They are invaders."

I could see this was getting nowhere. "Okay, but I will have to write about both sides . . . "

Aung Than Lay stiffened. "You mean about the SLORC and us?"

"No, I mean about logging and the problems it causes." There was no way I could get comments from the SLORC here. Maybe the Burmese embassy in Bangkok would say something, but it would be up to the wire services based in Rangoon to report on SLORC's response. That was just a fact of life.

The prime minister said nothing, but relaxed. He didn't seem worried about environmental criticism. Here on the border, that was well down on the list of concerns. It was ironic: Preserving trees and ecosystems meant little when war was in the balance, and yet it was the right to log those trees that had instigated the fighting. At any rate, the interview was over.

Jeremy returned the next day and apologized for having been called away. I told him what had transpired with Aung Than Lay, he filled me in on the latest troop movements, and we headed back to town. I wondered whether this camp and its people would survive the coming conflict.

MR. TIMBER

If there is one man who has allegedly been more involved than any other in the logging of mainland Southeast Asia, it is General Chavalit Yongchaiyudh. A soldier-turned-politician who seems to have had more political lives than a Siamese cat, Chavalit's precise links with the timber industry remain murky. But his name (and that of his wife, Khunying Phankrua Yongchaiyudh) has come up so often during various logging scandals, and he has been accused so frequently of benefiting from the timber trade, that he has been given the sobriquet "Mr. Timber."[20] For, as the *Nation* columnist Chang Noi so aptly noted, the ebb and flow of the logging trade in and around Thailand seems to move in tandem with Chavalit's political fortunes.[21]

It is no surprise that a military man might be involved in logging. Throughout the region, indeed throughout the developing world, guns and chainsaws seem to go hand in hand. In every country in Southeast Asia that has a significant logging industry (with the possible exception of Malaysia), there is close military involvement in the timber trade. In Burma, of course, the ruling military junta are involved in just about everything profitable. In Cambodia, numerous generals have reaped handsome rewards from handing out logging concessions in their fiefdoms. In Laos, it was the army-backed Bolisat Pattana Khet Pudoi (BPKP) that logged out parts of the biodiversity-rich Nakai–Nam Theun Plateau. And so on.

In Thailand, as well, army chiefs—although they have traditionally made most of their money from the arms trade and construction projects—have managed to parlay their formidable political power into logging profits.[22] As the country has grown wealthier and its political system has developed, Thailand's politicians and its central government have gained more administrative authority for day-to-day affairs, at the expense of the military. Even today, army commanders retain authority in the border regions, which just happen to be where most of Thailand's remaining forests still stand. It would be wrong to blame the military for all, or even most, of the illegal logging that goes on within the country—sometimes, they have actually gone in to protect forests when civilian authorities such as the RFD and the police seem unable (or unwilling) to do so. But certain generals in the border regions have been accused of using their positions to help broker logging concessions in neighboring countries.

One of the first transboundary incidents involving logging and the Thai military started in November 1987 at an isolated spot along the northern Thai-Lao border called Ban Romklao. A minor war actually broke out when Laotian troops entered a contested strip of territory to stop Thai logging operations in the area. The three-month skirmish that ensued cost the lives of more than seven hundred soldiers, and the bloody nose that the Thai military received at the hands of its tiny neighbor was considered a major embarrassment. Thailand's army commander at the time? General Chavalit Yongchaiyudh.[23]

Chavalit followed that incident in 1988 with a bold diplomatic foray into Burma. Earlier that year, a mass uprising of students and democratic activists in Rangoon had sought to topple General Ne Win, a dictator who had been in power for several decades. But a brutal military crackdown on the movement in September led to the deaths of an estimated 1,000 demonstrators. In December, Chavalit flouted international efforts to ostracize the Burmese junta—which, according to some reports, was near bankruptcy at the time—and became the first high-ranking foreign dignitary to visit the regime since its violent suppression of the pro-democracy movement.[24] During his trip, Chavalit promised to repatriate Burmese student activists who had fled to Burma and to crack down on ethnic insurgents along the border, and in his wake came a group of Thai businessmen and legislators to discuss investment deals and logging, fishing, and gem mining concessions.

"From these negotiations, 20 concessions were given to Thai companies for logging inside Burma," writes Chang Noi. "Virtually all the companies were owned by military officers and associates. Some of the executives figured among the supporters of the New Aspiration Party which General Chavalit founded a year later." The opening of Burma's forests could not have come at a more opportune time for Thai loggers because Thailand imposed its logging ban in 1989. For several years, the Thais seemed to enjoy virtual free reign in Burma, clear-cutting with abandon; but by 1993, even the SLORC became disgusted with their scorched-earth tactics and revoked most of their concessions.

By that time, Chavalit had successfully transformed himself into a politician, and as interior minister in 1993 he managed to keep the Thai-Burmese border posts open. Meanwhile, even as the flow of logs from Burma was drying up, the signing of peace accords aimed at ending its long-standing civil war led to a flood of timber entering Thailand from Cambodia. In fact, logs were coming out of Cambodia so fast and furious that the UN Transitional Authority in Cambodia, which was overseeing elections there, slapped a ban on exports effective at the beginning of 1993. Chavalit, ever sympathetic to the plight of Thai loggers (who claimed that their timber and equipment were stranded in Cambodia) sought to reopen the Thai-Cambodia border to the timber trade. But his coalition partners, alarmed by the bad press that Thailand was receiving over the issue, would not allow it.[25]

Chavalit was undaunted. After a brief hiatus out of government, he returned as defense minister in 1995, and then at last achieved his long-time goal of becoming prime minister. Once again, the logs flowed into Thailand from Cambodia, many of them allegedly harvested illegally,[26] or garnered from deals with the still-active Khmer Rouge and without the approval of the government in Phnom Penh. But Thai loggers were not picky in choosing their logging partners; after all, they had signed the infamous "million-meter deal" with Cambodia's co-premiers, Hun Sen and Norodom Ranariddh.[27] In 1996, Global Witness directly accused Chavalit and his wife of having a long history of involvement in the cross-border timber trade and helping to negotiate the million-meter deal.[28] They denied the charges.

It is not only the forests of neighboring countries that suffer when Chavalit is in power. As we shall see, during his time as prime minister, a new logging scandal emerged within Thailand, in the province of Mae Hong Son.

BORDER WAR

The big hole remaining in my logging story was proof that *Por Liang* Som was involved. Confronting him directly was likely only to get me shot. Even more difficult was trying to find out which politicians or generals were backing him. Once, I went to his hotel and attempted to ascertain his room number and to chat up the operator to see whether I could find out the names of people the godfather was calling. But the staff would have none of it. The next day, officials in Mae Hong Son received reports of shooting between the KNPP and the *tatmadaw.* Within the two weeks I had been investigating the log smuggling, a full-blown, if small-scale, war had started up. For a journalist, it seemed to be incredibly good timing.

My early stories about the fighting for the *Nation* scooped the world.[29] And my follow-up pieces contained far more detail than anybody else's. Nen wrote a smaller story for the *Bangkok Post*—the other main English-language daily in Thailand—but he had been too busy covering the election to gather much background about why the battle had broken out.

Once the bullets started flying, the wire services also began picking up the story. But they were based in Bangkok and Rangoon, so their reports inevitably contained a different slant. I was on the scene, and that was exciting. For a couple of days I felt like a bona fide war reporter—chasing down battle reports and struggling to confirm the numbers of wounded and dead—even if I did remain far from the action itself.

Jeremy, too, was surprisingly upbeat by news of the battle. I had expected him to be more nervous than ever because of the fighting, considering how outnumbered the Karenni were. But when I entered the KNPP office after the first day of shooting, he was positively gung-ho. "It looks like the SLORC is trying to take over all of Kayah State! Our survival is at stake, and we believe it is a just cause. Freedom is worth fighting for," he pronounced boldly. "I don't think there's anything the Karenni could have done to prevent this. It was bound to come sooner or later." He promised to help take me close to the fighting once the situation allowed it.

But I never got the chance. A few days after the battle had started, my newspaper called me back to Bangkok. The election was over and they needed me to cover environment stories there. I tried to explain that I was in a perfect position, with the background and contacts, to cover the unfolding situation on the border, but the *Nation* had other reporters to cover such issues. I was to fly back to Bangkok the next day. The Karenni war hadn't been such good timing after all. Now, not only would the logging story die out, but my chance to be a war correspondent was over, too.

Filled with frustration, I strode outside, jumped onto my moped, and roared off in the direction of the border checkpoint at Ban Nai Soi. I was going to see what I could see. Within fifteen minutes, however, the rain started. The road would soon turn into a mud bath, not a good place for a would-be war correspondent on a little 100-cc Honda Dream. Spent, I turned around and headed back to town.

I had one last stop to make before packing up. I headed down to the media center to file my last report, although a detailed story would have to wait until I returned to Bangkok, and even then I wasn't sure what I would write. But as it happened, I didn't even get a chance to send a news brief. After parking my motorbike, I spotted Nen standing by a white pickup and waving me over. "Jam, come on! Over here," Nen shouted. "Get in the car."

"Why? Where are we going?" I looked into the car and saw a bunch of the other local correspondents already inside.

"Nai Soi!"

"What? You're headed to the border?!"

"Yeah, we're going *over* the border to check on the refugee camps. Let's go."

I didn't need to be asked twice. I whipped off my cheap plastic poncho and threw it into the back of the pickup, which had a covered cabin that made it more like a station wagon. We both hopped in, and off we went. I greeted the other reporters excitedly. "Wow. Thanks a lot, Nen. I really wanted to get out there."

"*Mai pen rai.* No problem. We've heard that a lot of refugees are coming over to escape the fighting, and a lot of the timber is being evacuated, too."

"I wouldn't be surprised."

The weather had turned miserable. Instead of the usual monsoon storm followed by clearing skies, it continued to rain intermittently. As we turned off the main road and onto the dirt track to Ban Nai Soi, I gave silent thanks that I wasn't sliding around on my crappy Honda. Even the Toyota pickup was having a tough time negotiating the cavernous ruts and deep puddles.

It wasn't long before we found what had helped to cause all those ruts. Parked on the side of the road was a large ten-wheel truck, metal struts going up its side, piled high with a dozen or so large, lengthy pieces of timber—teak, no doubt, freshly cut in Burma. Then, beyond the truck, was another, piled equally high. And another. And another. I counted fifteen in all until, at the head of the parked cavalcade, we spotted a group of men huddled around a sport utility vehicle. Nen turned to me and flipped his thumb towards the SUV. "*Por Liang* Som," he grinned.

I reached into my daypack and scrabbled around looking for my camera. "Can you slow down just a little?" I asked the driver. But Nen immediately clamped his hand on my arm.

"No," he said emphatically. "No pictures."

I looked at Nen, stunned. "But this is our only chance . . . "

"No," Nen turned to the driver and told him, "Keep going."

"But surely if we all get out and talk to him, he couldn't do anything against us?" I looked at the other reporters, who had all turned to eye me nervously.

"Yes, he could. Believe me. You don't want to meet this guy, or write about him. You'll never feel secure again. Every time you hear a motorbike coming up from behind you, you'll wonder if there's a hit man on it aiming for your head. It's no way to live a life."

I capitulated, and sat back in frustration. Nen let go of my arm and the other reporters looked relieved. "Discretion beats valor again," I mumbled, in English this time. No one asked for a translation.

"Look," explained Nen, more calmly. "If you're worried that *Por Liang* Som is getting away without being punished, don't be. He's lost a lot of money from this whole thing."

He was right. The loggers were obviously trying to extract everything they could out of Burma, and as quickly as possible. But there must be piles of timber still left behind, not to mention trucks and road graders and other equipment that the SLORC would confiscate once they captured the territory. *"Khow khat toon neh nawn,"* said Nen. "He's definitely lost out on his investment."

We all stared at the SUV as we drove past, but couldn't see much because the godfather's bodyguards and underlings had gathered around and had their backs turned to the road. "Mitsubishi Pajero," said Oh, a stringer for one of the main Thai tabloids, obviously impressed. We drove on.

After passing through the village of Nai Soi, we reached the border checkpoint. When I had gone there on my own, the Thai guards had stopped me from going through; but it obviously helped to be in a vehicle plastered with the word *Khow* (News) all over it, along with the logo for television's Channel 7. After chatting for a few minutes, the border patrolmen waved us through. Now we were in a kind of no-man's land again. There was no Burmese checkpoint on the other side because the *tatmadaw* didn't control the area; not yet, anyway. But that was probably one of their goals for this offensive. No one really knew where the border was here. The checkpoint merely signified that the Thais were not going to take responsibility for anything beyond.

So that was where the refugees were allowed to settle, about four miles down the road. But this was no tidy village like Camp 5. It was filled with the tiny ramshackle huts covered by plastic lining you'd expect to see in a war zone. The mud added an extra touch of misery. A few people tromped around in knee-high Wellies. Others were loading up their possessions into a shiny white pickup, probably operated by some enterprising Thai or Karenni who would take the refugees into Thailand proper. But most simply huddled in their shacks, waiting for the rain, and the war, to end.

We parked our pickup and went to do a few interviews. The atmosphere was surprisingly calm. There was no sense of fear, or even urgency, among the people. Just resignation. We all tensed up a little when we heard the sound of a large truck approaching. But it was just another logging truck piled high with timber and heading for Thailand. The battle was not far away, according to the refugees, just over the hills. But there was no sound of fighting to be heard, at least not now.

"What will you do if the fighting gets close?" asked one of the reporters.

The refugee, a middle-aged man squatting next to his wife, simply shrugged. "We will go to Thailand."

"And if Thailand doesn't let you in?"

"We will just go into the forest until it is safe to come back. . . . What can we do?" He had a point. The younger people could presumably work illegally in Thailand for ridiculously low wages. The elderly would simply bide their time. I jotted down a few names and notes, but the interviews soon grew less informative. The whole trip seemed anti-climactic. I got up and wandered around.

Another pickup had arrived to take those eager to leave into Thailand with what few possessions they could carry. I asked how they'd get through the checkpoint. The driver just laughed and pointed to a wad of money. Corruption is usually viewed as an evil, but here it would help these refugees run away from the SLORC and maybe live a semi-decent life as illegal immigrants in Thailand.

Another logging truck roared down the road and through the camp. I snapped a few more pictures half-heartedly. Truth be told, photos of a logging truck do not thrill newspaper editors. I started thinking about how I was going to write this whole thing up once I got back to Bangkok. Two weeks of investigation had revealed how the log smuggling operation worked, which was a good story. But I had no proof of corruption on the part of the Karenni or Thai officials, and nothing to prove the complicity of *Por Liang* Som, much less his political backers. Nor did I have any comments from the Burmese government, apart from their bellicose official statements.

And to get my interview with Aung Than Lay, I had promised the Karenni prime minister that I wanted to help the Karenni people. How was

I going to live up to that promise and still fulfill my obligations as an unbiased reporter? Frankly, there was no way to do it, and maybe that's the solution, I decided. Forget about writing this as an objective report. Write it as an opinion piece. Don't pretend to be fair when you can't be.

I tossed that idea around in my mind as we piled back into the pickup and headed back to Thailand. But we hadn't gone far before we had to stop. Ahead, the road climbed a steep hillside before twisting out of sight through the thick forest. But the road wasn't empty.

One of the logging trucks, the first one we had seen driven through the camp, was stuck on the steep gradient and couldn't move forward. Behind it sat the white pickup, its load of refugees huddling miserably in the drizzle as they clutched their belongings. And further behind was the second logging truck, honking furiously. Looking at the scene from behind the rain-speckled windscreen of the news truck, I was struck by how perfectly it symbolized this ridiculous situation. Refugees stuck behind logging trucks. Live people having to make way for dead trees.

Before anyone could stop me, I grabbed my camera, leaped out of the truck, and snapped away furiously. I strode down the middle of the road until I had the ideal view, twisted the lens to full zoom, and kept shooting.

I half expected the other reporters to come out and take my camera. And they did come out, but this time they had their cameras, too. And they, too, snapped and snapped. And the Channel 7 cameraman hefted out his camera and zoomed in close to show the misery on the refugees' faces, the logs slightly out of focus in the background.

I turned to look at Nen, who grinned back. We didn't say anything, just nodded to each other, and kept snapping. These pics would be worth thousands of words. The following week, they were published in the *Nation,* along with a story I wrote for the opinion page that exposed the whole sordid episode.[30]

EPILOGUE: THE SALWEEN SCANDAL

The KNPP forces were soon defeated in their battle with the SLORC, whose troops took over the all-important border crossings and halted the Karennis' timber exports. But a couple of years later, during a period when Chavalit served as prime minister, a new logging scandal emerged along the border in Mae Hong Son, this time in the southern part of the province.

The Salween logging scandal, as it came to be known, had a different twist. This time, it was Thai trees that were being logged, more than 10,000 of them in the Salween conservation area alone, but they were being laundered mostly through Burma. Here's how the scheme worked: After the trees were cut down in Thailand, they were sent to the border; there, Thai firms that had been granted logging concessions in Burma stamped them as being of Burmese origin. The logs would then be floated down the Salween River and back into Thailand through official checkpoints.[31] Ironically, even the logs confiscated by Thai authorities as being of dubious origin were then sold by the Forest Industry Organization, a state enterprise, thus giving the government continued incentive to see the forests cut down.[32]

The scandal was uncovered by numerous members of the Thai press—a couple of my colleagues at the *Nation,* Kamol Sukin and Chang Noi, wrote several articles about it—who also managed to reveal details about the *Por Liang* Som, the godfather allegedly involved in the affair.[33] Poor, and poorly educated, but determined to be rich, *Por Liang* Som started off by working in a gambling den, where he learned that "the only way to win is to cheat." After a stint in the trucking business, he made his fortune by acting as an agent for logging concessionaires in Burma, where he got to know many of the powerful ethnic resistance leaders. He then went into the logging business for himself, easing his way by distributing bribes right and left. A list seized from one of his relatives contained the names of officials from the police, army, local administration, customs, immigration, forestry, and national parks departments.

Although the logging scandal was first publicized in March 1997, the logs would continue to flow so long as Chavalit remained prime minister because his administration still handed out permits allowing the concessionaires to import logs from Burma. It was only after the onset of the Asian financial crisis—triggered by the devaluation of Thailand's currency in July 1997—that Chavalit's government fell. Many of the details of the Salween operations were then exposed, including the stockpiling of 13,000 teak logs at the sawmills of one of Som's business partners. A senior member of the RFD, Prawat Thanadkha, was caught attempting to hand the subsequent prime minister, Chuan Leekpai, a bribe of 5 million baht in a bid to have the logs released from custody.[34] Prawat was subsequently punished, but Som was never arrested because of a lack of evidence. The Democrats sought to document Chavalit's links to the scandal,[35] but he denied the accusations[36] and was never charged with malfeasance. Following an election in January 2001, Chavalit has once again been named defense minister. One of his first official acts upon taking office was to make a trip to Burma.[37]

5

FORESTS AND FARMERS

ONLY THE HAUNTED JUNGLES SURVIVE

THE END OF INNOCENCE

Ajaan Pongsak Techadhammo was a conservationist monk for many years, but that doesn't make him special. Green Buddhism, a growing phenomenon in Thailand for decades, has now become a national movement, its temples scattered in forests all over the country.

Nor is Pongsak unique for having been defrocked. Several green monks have been forced to leave the clergy after taking on vested interests to protect Thai forests. Pongsak lost the right to wear the saffron robe in 1992, after a bizarre incident in which a photograph showing the monk consorting with a woman was sent anonymously to various media outlets. Pongsak is a frail and meditative man, with a visible tremor, and it is laughable to think of him in a sordid love affair. The people who distributed the picture never came forward. The woman in the photo (whose face is not clearly visible) was never identified. No evidence was offered against Pongsak, and no witnesses testified against him. The photo, in other words, was almost certainly a fake; but a day after it was "leaked," Pongsak agreed to leave the monkhood and step down as abbot of Wat Palad, an idyllic mountainside monastery on Doi Suthep overlooking the northern city of Chiang Mai. He was replaced by a habitual drunk who boasts of being a former hit man for the mob. Pongsak dismisses the incident with a shrug. "My robe is off, but in my heart nothing is different," he says. "Some vested interests wanted to build a resort and golf course at Doi Suthep. This is the reason why the picture was released."[1]

When one considers the violent fate that often befalls community activists and low-level officials who fight to protect the environment, Pongsak was probably fortunate. He has changed the color of his robes to white, moved to a remote camp on a mountain in Chom Thong district, and now goes by the honorific of *Ajaan* (Teacher) rather than *Phra* (Monk) or *Phra Ajaan* (Abbot). But people treat him with the same respect, still kneel before him, their hands folded in the traditional *wai.* And he still makes the same arguments. People should not live in headwater forests, he says, and those who do live there, mostly ethnic hill tribes, should be resettled.

Those beliefs make Pongsak stand out from the crowd, for they run counter to the convictions held so strongly by the mainstream of Thailand's grassroots environmental movement. There are anywhere from 600,000 to 10 million people living in Thai forests, depending on who's counting and the classification of forests you're talking about. Most Thai forestry activists, noting that many forest dwellers have been living in the forests since before they were earmarked as reserves, argue that they should be allowed to stay. Besides, the activists claim there isn't available land to which the forest dwellers can be relocated, or at least none of the same quality. Moving them would irreversibly destroy their communities for the benefit of city folk who want pristine forests to visit on the weekends, say the activists. They add that rural villagers traditionally lived in harmony with the forest until the cash economy came along, and can do so again if they use sustainable farming techniques and adopt more self-sufficient strategies.

The Thai government sees things differently. In the eyes of the authorities and some conservation groups, the forest dwellers are a major nuisance who destroy the integrity of the country's conservation zones and compromise the ability of forests to serve as watersheds and habitats for biodiversity and endangered species. The water issue is especially sensitive since the annual cycle of drought followed by flood seems to have worsened as the country's forests have been cut down. The Royal Forest Department (RFD), which also has vast economic interests at stake in this debate, maintains that encroachment by villagers is the biggest threat to the country's few remaining forests, and it continually pushes to have forest dwellers resettled.

Pongsak is probably the most eloquent spokesman for this conservationist point of view. "Man coexisting with the forest: That's a romantic idea, little more than wishful thinking. People still talk about it because that's the way they'd like things to be," he explains. "The hill tribe population is growing rapidly. They don't just farm to live; they farm to sell and with the support of vested interest groups. They have TVs, motorcycles, and cars." Pongsak agrees with the activists on one point: If forest dwellers returned to subsistence farming (and managed to keep population growth in check), they could live sustainably in the watershed forests. But after decades of living in the mountains, the former monk no longer believes that is a realistic possibility. And he questions whether it is fair to ask the hill tribes to forsake a comfortable lifestyle when the rest of the population is busy pursuing material comforts.

At the other pole of the moral compass on this issue is N'der; he is an elder statesman of a Karen community located in a secluded section of the Mae Khlong valley, several hundred kilometers west of Bangkok, near the border with Burma. Millions of ethnic Karen and other hill tribes live all over Thailand. But the fate of this one small group—known as the Karen of Mae Chanta—is hugely symbolic, because of all the forest dwellers living in the country's conservation areas, their situation is the most difficult to resolve. They live in an ecologically vital area that is smack in the heart of Thung Yai Naresuan, a 2-million-rai tract of land established in the early 1960s as one of Thailand's most important wildlife sanctuaries. The Karen of Mae Chanta probably number around 2,000 people (other Karen villages with a roughly equal population are located elsewhere in the sanctuary) and their presence has undoubtedly disrupted the area's flora and fauna.

On the other hand, such disruption has long been kept to a minimum because the Karen of Mae Chanta are probably the last surviving example of an ecologically sustainable indigenous culture in Thailand.[2] Long before Thung Yai was established as a wildlife sanctuary, the Karen set aside certain tracts of forest—essentially creating their own form of wilderness—out of the belief that they were inhabited by spirits who should not be disturbed. Also, like many hill tribes, the Karen practice swidden agriculture, rotating the forest plots on which they grow food and allowing each plot to regenerate before using it again. They apply these "forest farming" practices as part of their traditional spiritual beliefs, but the techniques match those recommended as environmentally sustainable by modern agricultural experts. They plant several different crops on one plot, for instance, and use various plant cuttings as "natural pesticides." After studying the Karen's beliefs and practices, Veerawat Teeraprasat, an official with the RFD's Wildlife Conservation Division who served as chief of Thung Yai for sixteen years, became an ardent defender of the tribe's right to stay in its ancestral home.

N'der and his fellow Karen have been able to maintain these traditions because their numbers have remained small in relation to the area of forest around them, and because of the isolation their remote location has afforded them. Traveling to the nearest town from Mae Chanta requires a long day's walk followed by a truck ride of several hours along a bumpy dirt road. Money is still barely used within the villages. But their isolation is gradually eroding, and the more they come into contact with the outside world, the more damage seems to be done to the surrounding environment. For instance, although the women of Mae Chanta make their own clothes, the men now trek to the town of Umphang to purchase theirs, and to pay for them they plant a few cash crops. These practices aren't as damaging as intensive agriculture, but they do have an impact on the forest. And while the men are in town, they are exposed to other conveniences of modern life, some of which they naturally want to acquire. That means finding more ways to make money out of their forest surroundings, increasing the impact. And so on.

It's hard to begrudge the Karen their interaction with the modern world because it provides them access to basic necessities such as modern health care. Around the time I visited Mae Chanta—trekking in with a group of other journalists, lawyers, and human rights workers in early 1994—a young villager died of an illness because it took so long to get him to a hospital. But improved access to health care is also causing the Karen population to grow, leading in turn to yet more impact on the forest. N'der says the Karen place no value on having large families, but children—cute, boisterous, and playful—are everywhere in Mae Chanta.

Even if you were somehow able to force the people of Mae Chanta to remain pent up in their proverbial Garden of Eden, there would still be the Border Police Patrol (BPP) unit, stationed nearby and serviced by a helicopter, to offer them the apple of modern practices. The patrolmen eat meat and drink whiskey, both of which are considered taboo in traditional Karen society, although more and more villagers are reportedly following suit. The BPP also runs the local school, which teaches only in Thai and not in Karen. More seriously, they are accused of harassing the local women, and encouraging the poaching of wildlife in exchange for the salt and medicine they are supposed to provide free of charge. "The police don't kill animals themselves because that would conflict with the Wildlife Conservation Division [the forestry unit that looks after the sanctuary]. They would rather have the villagers do it," said one young villager. "Gaur and wild cats are shot, and sometimes taken to be sold." These are endangered species and the carcasses are reportedly flown out to a market in the BPP's helicopter.

The school is another point of contention. "We like to learn Thai. But as kids we don't know ourselves yet," the Karen youth explains. "It would be good if it were taught in Karen, but we can't find a teacher." He claims that the police refuse to give medicine to parents who don't send their children to school. Paisal Surawasri, the head of the BPP unit, says the police are there to help the villagers. "They should learn to speak Thai. It's good for their future progress."

In essence, the Karen of Mae Chanta are caught between the past and the future; sticking to their traditions may mean more conflict with the Border Police,[3] which, they fear, will lead to expulsion from the wildlife sanctuary; but if they take up Thai customs, they will find it much harder to live according to their own traditions, and so they may be forced out anyway for damaging the surrounding forest.

For pressure is also being applied on the Karen by conservationists. Concerned that the villagers are destroying the forest, many Thai officials want to see the Karen resettled outside the area. Four Karen villages were removed from the neighboring Huay Kha Khaeng Wildlife Sanctuary in 1976. Veerawat says the influence exerted by outside authorities is the biggest problem facing the Karen.[4] "Normally, in Karen society, there is resistance to outside forces. They generally don't mix with outsiders. Change depends on two factors: state intervention and the strength of local community leaders. If the latter are weak, as is the case in some villages, then change comes quickly. If we want to slow down change, then traditional beliefs should be restored and community leaders like N'der strengthened." Nobody knows better than N'der that this is easier said than done. If current trends continue, he worries, "then there won't be any Karen anymore."

THE MAN-AND-FOREST DEBATE

As the experience of Ajaan Pongsak and the Karen of Mae Chanta illustrates, the debate about the role and fate of forest dwellers—known in Thai as the *khon kap paa* (literally "People with Forest," but often translated as "Man and Forest") issue—is no simple matter. It is a mighty battle over the control of valuable resources, a contest laced with ethnic tensions, hidden financial motives, and differing spiritual beliefs. And it raises some broader philosophical points: Where do people see themselves in relation to nature? Does man belong in the forest, or should he be kept separate? To what extent is development a necessity and at what point does further growth become a luxury? Does everyone have the right to the same level of development, or does it depend on where you live? And finally, who decides the answer to these questions? The answers may in the end depend on one's cultural, religious, and historic background.

Fascinating stuff, and also vital, because conserving forests is probably the biggest environmental issue in Southeast Asia and the world today—at least as far as a cause's popularity is concerned. Certainly in Thailand there are more people working on it, and more people concerned about it, than any other environmental problem. And there is good reason for the concern: Forests play a hugely important role at the global level as well as the local level, serving as they do as a home for endangered species, a storage site for carbon that might otherwise contribute to climate change, a watershed for rivers and streams, a source of valuable food and medicinal products, and, yes, a cool and pleasant place to visit.

Forests are also a key element of Asia's cultural and religious traditions. Buddhism, after all, grew out of the forests of northern India, and that natural heritage is evident in its reverence for life. The Buddha was born in a horticultural garden spot called Lumbini (after which Bangkok named its first and biggest city park), in what is today Nepal. Buddhists have traditionally viewed the natural environment as an ideal place for cultivating spiritual insights.[5] Historically, Buddhist states in mainland Southeast Asia set aside "royal parks and extensive tracts of woodland and parkland reserved for meditation and retreat," according to Jeya Kathirithamby-Wells, the Malaysian academic.[6] Monasteries are still often placed in the forests today as part of the Buddhist conservation movement. The Buddha's teachings—the dharma, or *tham* in Thai—actually resulted from his observations of nature, *thammachat.* Buddhism, in other words, is the understanding of how nature works—not just ecology but also human nature.

This signifies another crucial point. In Thailand, nature is thought to be virtually synonymous with forests, and forests are not traditionally considered to be separate from people. The distinction between nature and culture "may be less categorical in the East than the West," explain Ole Bruun and Arne Kalland, the editors of *Asian Perceptions of Nature.* "[Asian] people's approach to nature tends to be particularistic, or pragmatic, rather than governed by absolute principles."[7] Thais do not have a word for the concept of "wilderness" (to indicate that an animal is wild, Thais merely add on the word for "forest"—*paa*—as a modifier). The Buddha is said to have claimed there is no land on which man has not set foot or died, and many Thai myths involve people who live or hunt in forests.[8] The historical link between abundant resources and the well-being of the Thai people goes back at least as far as ancient Sukhothai, a thirteenth-century flourishing of Thai culture, where it was famously written that "in the water there are fish and in the fields there is rice." And indeed, such links are evident throughout the history of Thai civilizations, which have generally been based on farming, fishing, and trading, supplemented by hunting and gathering in the forests. So forests have not traditionally been thought of as untouched wilderness.

But that doesn't mean wilderness does not exist. Many areas in Southeast Asia have seen minimal human impact and can be considered de facto wilderness. And government conservation strictures are not entirely responsible. The Javanese had a similar practice to that of the Karen of Mae Chanta, and declared certain forests as *angker.* Humans were not supposed to inhabit or even approach these haunted areas, else they would go crazy and perhaps die. Such animist traditions, found among indigenous cultures around the world, remain strong throughout Southeast Asia, where they predate and have informally merged with imported religions such as Buddhism, Islam, and Christianity. Every part of nature was believed to be endowed with these spirits, says Montri Umavijani of Kasetsart University, not just the forest but also the towns and fields. And by haunting a place, it is the spirits who make it wild, not the lack of human presence.[9]

Much like the nymphs and dryads of ancient Greece, these spirits (*phii* in Thai) could be benevolent or malevolent, but either way they are worthy of respect. And that may be the best way to understand the traditional relationship Thais had with nature. It almost certainly wasn't as harmonious as many modern Thai environmentalists like to believe, but it was a wariness that served to limit encroachment on the forest and thus avoid challenging the spirits.

Consider a small event that took place when construction workers were clearing a site in Ubon Ratchathani province to make way for the building of the Pak Moon Dam. The last impediment to be cleared before the building could begin was a small shrine to a local spirit. But at that point, all the

Thai workers balked. The spirit was believed to be a powerful one; knocking down the shrine would bring very bad karma indeed. Finally, one of the *farang* supervisors of the project became fed up, climbed into the bulldozer, and proceeded to plow the shrine over, while the Thai workers looked on nervously.[10] The foreign manager presumably didn't care about the local spiritual beliefs and did not want to understand their meaning. To him, it was merely local superstition. But the Thais' respect for the *phii* can be viewed as more than just a simple fear of ghosts. It was their way of revering the sanctity of a place.

Animist spirits, in other words, serve as a kind of noncorporeal embodiment of the environmental ethic. The respect they engender for certain places and resources provided valuable benefits. Haunted mountaintops, for instance, remained unsullied and thus served as excellent watersheds, or warehouses and breeding grounds for biodiversity. Arguably, it is precisely those traditions that have beneficial side effects—whether they be the farming practices of the Karen or the traditional dietary laws among Jews and Muslims, many of which had the ancillary benefit of aiding public health—that stand the greatest chance of lasting through the ages.

Of course, spiritual and cultural beliefs go only so far in explaining the relationship Asians have with nature. Forests, after all, are home not just to scary spirits but also to more earthly creatures that have a habit of lunching on farmers' crops and livestock, or on the farmers themselves. More to the point, spiritual beliefs have proved no match for economic factors that spur exploitation. Even in ancient times, the Buddhist reverence for life was pragmatically set aside in favor of material pursuits, resulting in significant degradation, which was why many kingdoms found it necessary to establish forest reserves as spiritual retreats. In modern times, that degradation has been magnified enormously by exploding populations and access to Western technology and capital.

Some would add that with all this money and machinery came a new set of values. Columnist Mont Redmond writes

> When the Western concept of a unified, law-abiding nature came in and wiped out the particularized, local relationships Thais had had with their environment, it also removed the fears, contentment and inclination to sit still and do nothing that had been the foundations of the previous Thai 'balance.' Westerners showed Thais how to wind up the machinery of exploitation, development and economies of scale, but neglected to explain

> how and when the equipment should be turned off or allowed to run idle. Thais could go to work with a vengeance, cutting down forests and polluting rivers in accordance with the best capitalist growth models of the time for two reasons: first, they were operating with the frenzy that always accompanies a shedding of old values, and second, they had never experienced the measuring of strength that comes with the taste of victory and defeat, the taste Western man has had again and again in his struggles with nature.[11]

There is little doubt that Asian perceptions of nature—viewed by Thais as a "flow of faceless, renewable forces" (to use Redmond's phrasing)—are different from those in the West. In contrast to forest-born Buddhism, it is often pointed out, the Judeo-Christian tradition grew out of the desert, a harsh and desolate place. That doesn't mean Buddhism is a "greener" religion, or that Buddhists are more sensitive to the environment than those raised in the Judeo-Christian tradition. There are plenty of green messages in Western and Judeo-Christian folklore, the creation story in Genesis being a prominent example, and no shortage of Jewish or Christian environmentalists. "There is hardly any evidence that one religious creed protects nature better than others," write Bruun and Kalland. In fact, the evidence suggests that neither tradition is in practice a match for the forces of global capitalism.

Rather, it is in the attempt to create conservation areas where these differences in the way people view nature seem to matter most. Europeans tend to equate nature with the countryside, which helps explain their fierce efforts to protect their agricultural heritage. In the United States, nature is considered separate from man, and a vast system of parks, monuments, and sanctuaries have been established to protect wilderness. In *Wilderness and the American Mind,* author Roderick Nash explains how this came about. Europeans traditionally viewed the primeval forest as dark and terrifying, a place to be conquered and "civilized."[12] And so it was, for there is virtually no primary forest left in Europe. But in North America, although the first European settlers looked at the forests in an equally forbidding way, the view of nature among their descendants evolved differently. As they spread westward, a new ethic gradually emerged. The New World's abundance came to be seen as a sign of distinction, a source of pride, even of inspiration. The land was not only rich but also empty, or at least untouched by civilization because Native Americans, when they were considered at all, were generally looked upon as savage, just another feature of the "majestic wilderness." For a young country seeking to distinguish itself from the Old World, wilderness was one of the few things that the United States possessed, and venerable Europe could never possess. And so it came to be a vital part of American mythology, enshrined in a new, American invention: national parks.

Environmental movements tend to be shaped by the way people view nature, and although they struggle to influence economic policies, they typically have a lot of say over conservation policies. Even within these movements, there are fierce differences about how nature is viewed, and arguments about how it should be conserved. Groups in the United States have long fought over the ultimate purpose of its national parks, as exemplified by the battle between "conservationists" and "preservationists" in the early part of the twentieth-century. Conservationists, led by the forester Gifford Pinchot, argued that resources needed to be protected so they could be used by people; thus their support for the construction of the controversial Hetch Hetchy Dam in Yosemite National Park. Preservationists, led by John Muir, wanted wilderness protected for spiritual and inspirational reasons as a kind of "temple" to nature, and they fought fiercely against projects such as the Hetch Hetchy. Even today, battles rage over how to use the parks, for instance, whether to allow snowmobiles and all-terrain vehicles inside.

In Asia, another struggle informed by different cultures and traditions is now going on concerning how and why to carry out conservation. Understanding the debate, and what lies behind it, is crucial to forecasting the future of the continent's forests and coastlines.

DANCING AROUND THE DINOSAURS

The competing views between American conservationists and preservationists are mirrored not just within Thailand but also within its Royal Forest Department. Many RFD officials, particularly those who make it to the highest positions, seek to profit from their positions. They may be considered "conservationists" in the sense that Gifford Pinchot was, but they must also be considered corrupt. The way the system works is that any RFD official striving to attain a post above a certain level must pay off his superiors, the amount involved depending on the status and moneymaking potential of the position. Plodprasop Suraswadi, who was appointed head of the RFD in 1998, recalls: "When I first became director-general, I found there was a safe in my office. Every month, many chiefs of important divisions had to visit and contribute something. At every promotion or annual reshuffle, officials had to pay a minimum of Bt50,000, up to a maximum of Bt5 million."[13] They recoup this money by charging their own underlings, and ultimately this vast and corrupt pyramid scheme is supported by profiting from the land. "We have good intentions and good people, but we don't have a good system [of governance]," explains Pisit na Patalung, head of Wildlife Fund Thailand (WFT). "We have perfected only two systems, a system of corruption, and a system to deny responsibility for solving problems."

That makes genuine conservation difficult, if not impossible. Forestry officials tend to view the millions of farmers who live in reserves, many of whom hold only tenuous title to their land, not just as a threat to the forests but also a threat to their livelihoods. Some hill tribe villagers, including some of the residents of Mae Chanta, live in constant fear of expulsion because they lack Thai citizenship or even identity papers. But they do have supporters in the form of nongovernmental activists, and occasionally in the politicians who seek their votes. Most of all, they have their desperation: If they are kicked off their land, they have nowhere to go.

The result is that forestry issues in Thailand have become riddled with scandals and conflicts. Beginning in the late 1980s, the Suan Paa Kitti case involved a forestry firm, headed by an economic adviser to the prime minister and a Democrat Party financier, that was given a concession to grow eucalyptus plantations on denuded land; the firm was eventually caught cutting down 30,000 rai of pristine forest to plant its fast-growing trees. In the early 1990s, a group of forestry and military officials launched the *khor jor kor* scheme, a heavy-handed attempt to relocate as many as 6 million villagers living in degraded forest reserves to make way for privately run plantations; the scheme was scrapped following major protests. The Ta Chang case involved a couple of well-connected firms in southern Thailand that were caught logging when they were supposed to be growing trees. *Sor por kor* was a large land reform project designed to parcel out millions of acres of denuded forestland to people who had settled there, but it was eventually revealed that some of the recipients were rich and well-connected, and some of the land was potentially extremely valuable. The Salween scandal (recounted in the previous chapter) was a massive and devious illegal logging scheme that worked by laundering timber through Burma. The Dong Larn affair involved a group of farmers who settled in a protected forest and claimed they had been promised land by previous governments.

All these scandals arose from conflicts about how to use the land set aside for forests. You would think that after one or two such incidents, a serious effort would be made to resolve the issues. But that hasn't happened because the scandals themselves are typically manipulated for political reasons, much as the forests are exploited for economic benefits; a party or some other vested interest exposes the wrongdoing, and the official who's caught holding the bag is then shunted off to an "inactive post." In 1995 and 1996, three RFD chiefs were sacked in quick succession, each rocked by accusations or dismissed for political reasons. But the powerful figures behind all the shady deals are rarely exposed, much less punished. The media then soon forgets about the episode and moves on to the next scandal. Thais have a saying: "When the elephants fight, it's the grass that gets trampled." In these scandals, while the political heavyweights battle for the spoils, it's the forest that gets cut down.

Eucalyptus camuldulensis, a fast-growing species native to Australia favored by the pulp-and-paper industry but reviled by environmentalists, is the subject of many such scandals. A kind of industrial-strength tree, it is highly efficient at obtaining the water and nutrients it needs for survival, often at the expense of surrounding plants and animals. Even more cunningly, its leaf and bark litter contain chemicals such as terpenes that leach into the soil and inhibit the growth of other plants. As a result, eucalyptus grows extremely fast, even under harsh conditions. This means it can be used to begin the reforestation of degraded land; but if that's the goal, the trees must be planted in a manner that allows other species of trees to grow alongside it. In Thailand and other parts of the tropics, eucalyptus is more often planted in tightly packed monoculture plantations that are cut down after five to seven years. Such an arrangement has often contributed to—rather than cured—environmental degradation, and has so enraged surrounding farmers they've been known to attack and burn down plantations.[14]

Arguably, some progress has been made. The RFD has largely halted its attempts to force settlers off the land in favor of fast-growing tree plantations. Instead, it now provides grants and loans for individual farmers to grow certain types of trees (assuming the farmers have proper title to their land). These plantations are needed if Thailand is to reduce its dependence on foreign timber. But it's a misnomer to call them "forests." They are tree farms, and like most monoculture crops, they provide fewer environmental services than natural forests. They can't, for instance, serve as habitat to as wide a range of plants and animals, and they often make poor watersheds.

The RFD still receives so much criticism and seems so rotten to the core that it's easy to forget there are some good and dedicated people working for it who must try to "dance around the dinosaurs"; that is, work around their superiors, including many park superintendents who pay for their posts and are thus part of the web of corruption.[15] Most notable is the humble park ranger, who regularly risks life and limb battling poachers, and if he is killed, his family may not even receive welfare benefits.[16]

Then there are the more famous crusaders. Seub Nakhasathien, who first became well known for rescuing wildlife trapped by rising floodwaters in national parks where dams had been built, eventually was given the prestigious and demanding job of looking after Huay Kha Khaeng, Thailand's top nature reserve. Not only does it contain the most pristine lowland forest left in the country but it also straddles a region where fauna and flora from two different ecosystems merge.[17] His work set a new standard for conservation management in Thailand. Along with his partner, English naturalist Belinda Stewart-Cox, he helped convince UNESCO to name Huay Kha Khaeng and its neighboring wildlife sanctuary, Thung Yai Naresuan, a World Heritage Site because of their high level of biodiversity. But park and sanctuary chiefs have to be diplomats as well as managers to deal with the demands of governors and business mafias that covet the resources in protected areas, and they often receive woefully little support from higher-ups in the RFD. The budget Seub was given for running the sanctuary, for instance, amounted to about 1 baht (four U.S. cents) for each 50 square kilometers. The strain of his position gradually wore on Seub; he grew more and more depressed until the day in September 1990 when, in the wee hours of the morning, sitting alone in his cabin, he put a gun to his head and killed himself.[18]

Seub has remained a most important figure in Thailand—a martyr to the cause of conservation—but depressingly little official progress has been made towards improving the management of Thailand's conservation areas since his death. Although there are now more than a hundred such areas scattered around the kingdom, most are poorly maintained and monitored. The officers assigned the task of looking after these huge areas are given little training and suffer from low morale. Their budgets are tiny and their equipment scanty. And yet the RFD continues to convert forest reserve land into conservation areas so that it can meet its goal of turning 15 percent of the country into conservation zones. "The forestry department tries to grab more areas to be set up as parks first, and worries about management later," says Kasem Snidvongs, a well-respected economist and the former top-ranking civil servant in the Ministry of Science, Technology, and Environment. He argues that the agency is so difficult to reform that it would be better simply to move the conservation agencies—the National Parks Division (NPD), the Wildlife Conservation Division (WCD), the Wildlife Research Division, and so on—out of the RFD. The opposition to such institutional reform, however, is strong.

WATERMELONS VS. BANANAS

Even if conservation were managed with the utmost professionalism in Thailand, there would still be a conflict between its ecological goals and the economic aspirations of the people living in and around protected areas. Although this issue shows some similarities to the conservationist vs. preservationist debate in the United States, there is an added class element that does not seem present in the American discourse. The presence of natives or settlers in parks and sanctuaries has seemingly never been a major issue (at least not on the national level) in the United States, where it is generally assumed that such areas need to be free of people if they are to be considered "pure wilderness." That may now be changing. Mark David Spence recently came out with an illuminating book, *Dispossessing the Wilderness,* that describes how Native Americans were relocated from such fabled U.S. parks as Yellowstone, Yosemite, and Glacier.[19] The Americans had "made little islands for us and other little islands for the four-leggeds," observed Black Elk, the fabled spiritual leader of the Oglala Sioux, whereas his people "were happy in [their] own country, and were seldom hungry, for then the two-leggeds and the four-leggeds lived together like relatives, and there was plenty for them and for us."

In Asia, the traditional view of man's role in nature is probably more similar to that of the Sioux; but the modern view, particularly among officials in the RFD, is closer to that of the Americans who displaced them. The issue of whether people should be allowed to live in conservation forests—and if so, under what rules—has therefore created deep divisions among Asian environmentalists. There seems to be a similar split throughout the developing world, at least in democratizing countries. Even in Brazil, where more than 80 percent of the populace live in cities, it is viewed as quite natural for people to live in the forests, according to Samyra Crespo, a program coordinator at the Instituto de Estudos da Religiao and an expert on the country's environmental movement. "Environmental NGOs in Brazil are split, but the mainstream thinks it's possible to combine people with the forest," she explains.[20] A *pavos de foreste* (forest people) movement has been created to support forest dwellers and help them live sustainably, spearheaded in the Amazonian region by the Grupo de Traballo Amazonico, a coalition of organizations representing farmers, indigenous peoples, and environmental groups.

In Thailand, as well, the environmental democracy movement defends people living on forestland. But a vocal minority of green groups claims that the growing population and development of forest dwelling communities has turned them into a major source of environmental degradation. They argue, in other words, that Black Elk's vision of indigenous people living fraternally with nature has broken down. As a result, the Man-and-Forest debate has become so bitter that it has taken on ideological tones, and some of the farmers' defenders criticize the concept of wilderness as foreign and imperialist.

A broad spectrum of views surround the Man-and-Forest debate, but it's simplest to split them into two camps. On one side are the grass roots activists and rural development groups who form the mainstream of the environmental movement and who generally argue that it would be wrong to relocate farmers away from conservation forests. These groups maintain strong links to anthropologists and social scientists in academia, and they promote participatory democracy. Because so many of them belong to nongovernmental organizations, they are often labeled "the NGOs," but I prefer to think of them as the environmental democracy movement. Although they are concerned with protecting the forest, they focus on community development, and their first priority is to defend the rights of villagers.

On the other side are a group of conservationists, ecologists, and wildlife experts who are probably smaller in number but whose influence is growing with the middle class. They are oriented more towards the natural sciences and are primarily concerned with the welfare of the watershed forests and their nonhuman inhabitants. Although sympathetic to the plight of the farmers, they argue that it is dangerous to allow people to stay in the conservation areas because, now that so little pristine forest is left, the disturbing presence of people can have an insidious impact. Many of these people also belong to NGOs and private foundations, but let's call them "preservationists" to avoid confusion.

This nomenclature is not great, but it is better than the "fruity" epithets that are sometimes hurled around behind the scenes. The environmental democracy advocates are occasionally accused of being "watermelons" for being green (i.e., environmentalists) on the outside, but red (i.e., leftists) on the inside. Their preoccupation with politics is said to have blinded them to the reality that the activities of the rural poor have a serious impact on the environment. The preservationists, meanwhile, are sometimes derided as "bananas" for being yellow (Asian) on the outside, but white (Western) on the inside. That's because the protection of wilderness is often considered a preoccupation of the *farang* (a term that not only refers to Westerners but is also the Thai word for "guava"), particularly the American variety. NGO critics also sometimes refer disparagingly to the preservationists as "deep green," but despite the often bitter terms used in this debate, the two camps hold many views in common; therefore, "green" is more suitable as a label for the entire environmental movement.

How many people live in Thai conservation areas? There is little agreement on even a rough estimate. It is typically suggested that around 1 million families, or from 5 to 10 million people, live on forestland, but that holds only if you include all the reserve land that has long since been denuded. A 1995 RFD survey counted roughly 156,000 people living in national parks and wildlife sanctuaries. Another 436,000 people resided in important watershed areas that were slated to be declared conservation zones.[21] But these numbers are almost certainly outdated. With new parks and sanctuaries continually being prepared and declared, the problem is growing ever more serious.[22]

The situation is most serious of all in the mountainous north of Thailand for a variety of social, geographical, and historical reasons. Historically, Thais are part of the Tai ethnic group that migrated from southern China more than a thousand years ago and settled in the territory that has become their heartland: the Chao Phraya River basin. Their staple crop is rice, and the fertile central plains of the lower basin form their "rice bowl," an immensely rich agricultural region. It is also home to Ayutthaya, the once-flourishing historical capital of Siam, and to Bangkok, the current economic and political locus of Thai civilization. The Chao Phraya, the River of Kings, is therefore of tremendous economic and historical importance to Thai society. But its source lies in the highlands of the north, which the Thais have long ruled but which they settled in only small numbers.

During the twentieth century, the highlands have been increasingly settled by other ethnic hill tribes, in particular the Hmong, many of whom are still migrating into Thailand from strife-torn regions in neighboring Laos and Burma. A great deal of primary forest is left in the northern mountains—mostly because of its traditional inaccessibility and the relatively low levels of development—but that is changing quickly. The activities of these hill tribes in clearing the highland forests have raised concern and resentment among lowlanders near and far.

This situation reveals how environmental conflicts can become intertwined with a whole range of other issues, making them vastly more complicated to resolve. This is not just a fight between conservation and development; the conflict involves various ethnic groups, lowlanders and highlanders, and local, national, and even global interests. Most ominously, it's taking shape as a battle between human rights and national security: In the 1970s, many of these forest areas were considered security threats because they hosted armed insurgents; today, security concerns are increasingly being invoked due to the forests' value as watersheds.

What's surprising is how little attention the Man-and-Forest issue receives, particularly in the international press and among international environmental groups. The omission may be partly due to the incremental nature of the threat as well as the difficulty of presenting the damage poor farmers do to the forests as a simple black-and-white issue. We usually think of conservationists and human rights groups as allies, and normally they are, but in this debate they are just as likely to be combatants. That makes it a sensitive issue, one that many people would rather ignore, or at least try to resolve quietly. It is far easier to make a fuss about nasty loggers and big, bad transnational corporations (which do deserve a large part of the blame for deforestation, particularly in the poorest of countries such as Laos, Burma, and Cambodia). But the impact of poor farmers encroaching on forests is a growing issue everywhere in the developing world,[23] and the danger in not raising it is that it may not receive the attention it requires.

A MARRIAGE OF CONVICTION OR CONVENIENCE?

We already know where the RFD stands on the Man-and-Forest issue. Ever since the logging ban took effect, it has argued that encroachment by farmers is the biggest threat to Thailand's remaining forests. RFD officials stress the practical reasons for keeping forests and people separate, and maintains that they are necessary to protect wildlife and watersheds. But it's a hard sell considering the agency's well-deserved reputation for corruption and mismanagement. The RFD made several attempts to relocate groups of villagers out of conservation areas in the mid-1990s, but they were predictably botched. In one relocation, there were reports that some of the resettled people were left so destitute they ended up eating tree bark; and in another, the villagers were left without clean water to drink.[24] Further efforts were stymied, thanks to denunciations by nongovernmental activists—not just pro-democracy and human rights groups, but also many environmental groups. It may seem strange for a majority of greens to oppose a project designed to protect the forests, but it is quite revealing about the nature and grassroots origins of the mainstream environmental movement, not just in Thailand but also in other young democracies such as India, the Philippines, and Brazil.

Some Westerners may not even be aware that Asia has active green groups, but in the long run they are likely to have much more of an impact on the region's policies than foreign pressure. They tend to be broad coalitions of mostly middle-class NGOs that have teamed up with groups of grassroots farmers and fishermen. They are concerned not only with how resources are used but also with *who uses them*. As suggested by the term "environmental democracy," they work to empower small-scale farmers, fishermen, and other largely disenfranchised groups. That entails fighting to gain control of the resources vital to rural livelihoods, which is why these groups make up such a large part of the green movement.

On the practical side, many NGOs work in the field to merge development with environmental concerns in their efforts to come up with sustainable alternatives to destructive practices. In the highlands of Chiang Rai, for instance, Tuenjai Deetes of the Hill Area Development Foundation has worked with the hill tribes that are attempting to switch from swidden techniques to a more settled form of agriculture. By learning how to build terraces, the villagers reduce soil erosion, use the land for an extended period, and increase their long-term economic prospects. Another good example is a group based in the southern province of Trang called *Yadfon* (Raindrop), which has worked with coastal fishing communities to preserve mangrove forests and sea grass beds, leading to a dramatic increase in fish catches and greatly improving local livelihoods (see the next chapter).

There is a link between the practical work of NGOs and their equally important political work. They fight frequent battles—mostly rhetorical ones—with the centralized state agencies that monopolize control of resources such as water, forests, and minerals. The main forum in Thailand for these battles is the court of public opinion. In the West, and in some developing countries such as India, the real courts (that is, the legal system) are arguably the most important advocacy tools for environmental groups. But in Thailand, judges are rarely kind to environmental claims. So nongovernmental groups instead lobby VIPs or seek attention through protests, perhaps closing down a major thoroughfare. "Protests are our way of holding a press conference, since we can't afford the real thing," explains the media activist Ing K.

As with the Assembly of the Poor, the guiding principle that unifies the environmental democracy movement is that the management of resources should be decentralized to give local people more of a say. Local people, according to this theory, can take care of forests, fish stocks, and water supplies, and also protect them from the predations of big capital if necessary, far better than some distant bureaucrat making decisions in Bangkok. Locals have more incentive to do so because they live with the consequences every day. You can take issue with some of these assumptions—there are important national and global interests at stake that local stakeholders may ignore—but it has a powerful logic. Similar arguments have led to battles between central governments and local interests in federated countries around the world.

In attempting to translate the Thai environmental democracy movement into an American context, it appears to be an ironic conglomeration of ideas from two completely opposite U.S. groupings: the environmental justice movement and the anti-environmental wise-use movement. The latter, also known as the property-rights movement, is largely based in the western United States, where local landowners have fought for the use of federal lands and against "takings" by government regulators. Similarly, Thai environmental democracy advocates are plainly out to protect their supporters' economic interests, but they also argue that decentralization will help the environment. And whereas the wise-use movement seeks to maintain control of resources to which American ranchers and landowners already have access, the environmental democracy movement in Thailand is still trying to gain a legal foothold—most notably in the form of a Community Forestry Act—over public lands and resources.

So there is a class element to the environmental struggle in Thailand that makes it similar to America's environmental justice movement. This is a largely urban-based phenomenon that seeks to make sure the environmental burdens of industrial society don't fall unduly onto the shoulders of ethnic minorities, the poor, or the disadvantaged. It fights against the siting of toxic waste dumps or heavily polluting factories in poor neighborhoods, much as the Thai environmental democracy movement battles against the declaration of national parks in areas inhabited by hill tribes.

Thailand's NGOs are essentially trying to wrest control of the rural economy from the many powerful business cartels in Thailand run through patron-client relationships and backed up by the threat of violence. That is a polite way of saying that many rural areas (and many urban areas, too, for that matter) in Thailand are effectively run by the local mafia, known as the *ittiphon meurt* (dark influence). The godfathers (*jao por* or *por liang*) typically operate businesses such as road contractors, whiskey distilleries, mining concerns, rock quarries, and timber concessions. They have the local politicians in their pockets, or perhaps have their own sons run for Parliament. Under such repression, environmental issues can serve as a "legitimized arena of resistance," writes Philip Hirsch, an Australian geographer and the editor of *The Politics of Environment in Southeast Asia.*[25] "The *chao baan* [villagers] are often discouraged from speaking out publicly on issues such as socioeconomic disparities, for fear of being labeled a 'Communist' and possibly suffering for it," explains Apichai Puntasen, a professor at Thammasat University. "They can get away with being more outspoken if they frame it as an environmental issue."[26] The tolerance goes only so far, however: Protests about projects such as dams, rock quarries, and industrial estates can grow so heated that Thai community activists opposing them are gunned down in cold blood.

When villagers come out against, say, a factory that is polluting the local river, they have numbers in their favor, but typically not much organization. NGOs help provide that, along with valuable advice and access to the media. Like the Assembly of the Poor, the environmental democracy movement can be seen as a kind of amorphous left-wing opposition. During the last decade, thanks to greater activism and stronger organizing efforts, its reaching out to workers as well as to farmers and fishermen has grown increasingly effective at putting its agenda on the table. But the movement still lacks cohesion and remains too distrustful of the electoral system to enter politics formally, at least for the moment. None of that should minimize the tremendously important role that NGOs serve in protecting the environment. Together with their villager allies, they form the most effective lobby against the many poorly conceived development and industrial projects that have caused so much destruction in Thailand.

NGOs and villagers also form the most effective watchdog against environmental abuses, such as the illicit dumping of pollutants. That is actually supposed to be the task of the Department of Industrial Works (DIW), but it is often slow and ineffective (and reportedly weakened by corruption). Its monitoring and enforcement duties are hamstrung by its primary role, which is to promote industrial development—once again proving the folly of asking agencies to serve both as promoters and regulators. So there is a very good reason why the environmental democracy movement has come to form the mainstream green movement: Protecting the environment and protecting the rights of villagers usually go hand in hand.

But sometimes mutual protection sows division within the movement and allows observers to see whether the marriage between conservation and rural democracy is one of conviction or convenience. The Man-and-Forest debate is an example, and most NGO activists have fought against the resettlement of groups such as the Karen of Mae Chanta, or the Hmong, who live on the same mountain as Ajaan Pongsak, out of conservation areas. Many even fight against the declaration of new parks and sanctuaries. Environmental democracy advocates see it as part of a broader struggle to protect the rights of indigenous people, particularly in the north of Thailand, where they consider the hill tribes to be an oppressed and disenfranchised minority group. They have concluded either that the damage caused by villagers to forests is exaggerated or that defending villagers is a greater priority than protecting the forest.

Many are convinced that, ultimately, defending villagers is also a good way to defend the forest. It's not a coincidence, they point out, that Karen villagers are being targeted for resettlement in precisely those areas where they have conserved the forest. And the environmental democracy movement believes it has a solution that will help both: community forestry. For more than a decade now, environmental democracy activists have been fighting to have a community forestry bill passed through Thailand's Parliament that would grant farmers some legal rights over local forests and give them more say about how they are used. It's a policy that fits in nicely with their efforts to promote community development (essentially, a rural-based civil society) and local participation. But the RFD has fought them tooth and nail.[27]

Defenders of forest dwellers argue there are major practical obstacles to resettlement. Most important, they claim there simply isn't enough fertile land available to which people can be relocated, and that in practice relocation simply does not work. "I don't know of any case where communities that have been resettled have been better off, have a better life, a better quality of life," says Witoon Permpongsacharoen, the Thai activist and antidam campaigner. "When you break apart those community links that bind people to their home, they lose something forever." They add that letting villagers stay and participate in a community forestry system would give the villagers more incentive to look after the resources for the long term. By remaining in the forests, they assert, local people can serve as watchdogs to make sure other vested interests—state agencies or local businesses—don't abuse the area's resources. At least one high-ranking official agrees. Petipong Pungbun, a well-respected technocrat who is permanent secretary at the Agriculture Ministry, which oversees the RFD, explains: "After we let people get involved, I would hope that they would represent an electoral force for conservation. That is the only force which can fight the power of money."

At times, however, the argument in support of forest dwellers can become dangerously close to ideology—an abiding faith in the ability of farmers to live self-sufficiently and sustainably, and unremitting opposition to any form of resettlement. Much as deep greens in the West base their visions of a sustainable future on the very Judeo-Christian pursuit of a "new Eden," one that is largely devoid of people, environmental democracy advocates in Asia have their own vision, based on their own spiritual beliefs, of man living in harmony with nature.

Indeed, if Huay Kha Khaeng is the Thai conservation movement's greatest success, then neighboring Thung Yai is surely the symbol of environmental democracy, thanks to its impressive political history. In May 1973, a student ecological club from Ramkamhaeng University exposed some military rulers who had gone illegally hunting in the wildlife sanctuary. The revelation not only kick-started Thailand's environmental movement but helped spark a student protest that eventually overthrew military rule and installed what is considered Thailand's first truly democratic government. Following a murderous reactionary backlash in 1976, Thung Yai served as a major guerilla base for Thai students who were forced to flee to the jungle. For six years, political expression was severely restricted until mass protests started against plans to build the Nam Choan Dam in the wildlife sanctuary. The defeat of that project still stands as the Thai environmental movement's greatest victory.

Defeating attempts at Karen resettlement would no doubt be considered another landmark achievement in Thung Yai's democratic history. But what a price to pay if it comes at the expense of the sanctuary's ecological integrity. Arguably, this is where the environmental democracy movement can make its greatest contribution. The Karen had great forest-protection lore, but were in danger of losing it. Green pressures are forcing the Karen to rediscover them, to re-invent themselves as conservationists. Some other local groups, not just hill tribes but also ethnic Thais, are doing the same through conservation Buddhism. And everywhere in Asia, advocates of environmental democracy are working to fulfill a similar dream.

THE PRESERV-ASIAN-ISTS

The question is, will that be enough? Are defenders of forest dwellers too optimistic in dreaming of people living in harmony with nature? It is not only the RFD that thinks so. An increasing number of Thai citizens and organizations, most notably the Dhammanat Foundation, are worried about the impact villagers are having on conservation areas. These ecologically minded groups caution that people should heed Ajaan Pongsak's warning about seeing only what we want to see; the defenders of forest dwellers, say the preservationists, are inclined to paint an overly romantic portrait of the humble villager and other indigenous people; in their fervent desire to see people living in harmony with their surroundings, they make it seem as if forest dwellers are already well on their way to doing so. Exaggeration about Thailand's hill tribes is rampant, say the preservationists, adding that, with the exception of the Karen and Thai Lue, even calling the hill tribes "indigenous" is a misnomer, since many of them migrated into the country in the period following World War II. Indeed, new immigrants are arriving all the time.

But when it comes to conservation, it doesn't matter whether forest dwellers are indigenous or migrants, native or nomadic. What matters is their impact on the landscape. In an essay titled "What Black Elk Left Unsaid,"[28] the anthropologist Roy Ellen notes that the impact of preindustrial societies—often idolized as "noble savages" yet treated as "primitive"—is mainly a result of their access to technology. Native societies living in harmony with nature, in other words, generally evolved in isolation from the modern world, and they would presumably have to stay that way to remain so. As we saw with the Karen of Mae Chanta, that isolation is difficult to keep in rapidly developing countries. So protecting Thailand's endangered species, argue the preservationists, will require reining in encroachment and cracking down on poaching by villagers.

Their most critical contention of all is that Thailand's water supply is now threatened by the continuing degradation by hill tribes of headwater forests in the highlands. "The migratory hill tribes that have settled in headwater areas need to be relocated. Resettlement should be done slowly and take many years—the way they have always migrated. And many sectors will need to be involved to make sure they are provided with proper facilities such as land, water and houses," says MR Smansnid Svasti, the outspoken founder of the Dhammanat Foundation, who is also convinced that land suitable for resettlement can be made available. "Yes, we can say it is a human right to be resettled properly. But to say that they should not be resettled at all because of human rights is silly. With our rivers threatened, it is human *life* itself that is at stake."

The preservationists have allies among lowland farmers, who see their water supply shrinking year by year. The factors behind increasing water stress in Thailand are varied, and their relative importance is subject to dispute. Some scientists claim there is little evidence that forest degradation has much impact on stream flow; this assertion is ironic since virtually all the stakeholders in Thailand—villagers and officials, environmental democracy groups and preservationists—are convinced from first-hand experience that it does. What is undisputed is that in the uplands and the lowlands irrigation to support agricultural activities has intensified in recent years, and this has led to tension. Increasingly restive farmers' groups in Chiang Mai province's Chom Thong district blame their dwindling water resources on the activities of the hill tribes. In recent years, they've held loud rallies where fiery speakers call on the government either to resettle the highlanders or to exert greater control over their activities. Although the demonstrations have so far remained peaceful, roads up into the mountains are typically blockaded, and there is potential for ethnic violence.

The "deep green" sensibility that Thai preservationists have adopted in their efforts to keep forests free of human disturbance is remarkably similar to the inspirational ideals of American preservationists, except that it is framed within a Buddhist rather than a Judeo-Christian tradition. It is essentially a call for wilderness: "The forest should be forest. People should live with people," intones Ajaan Pongsak. To him and his followers, the forests remain haunted, if not by spirits, then by a green Buddhist spirituality. And perhaps a forest needs to be haunted if it is to remain a genuine jungle—wild and awesome. Otherwise, it becomes merely a resource.

What about the people involved, the ones who are to be moved? Many groups claim to speak for the farmers living in protected areas, but what do the farmers say they want? Finding out was often difficult. Even those who knew what they wanted were sometimes reluctant to say it publicly, for fear of upsetting the authorities. Nevertheless, one point that was abundantly clear was the desire for more secure land tenure rights. Nobody wants to live with the constant fear of being resettled, or even expelled from the country. Beyond that, however, it is hard to say. People are naturally fond of their homes and generally reluctant to move, but many willingly do so if it means better opportunities. Some villagers living in especially remote protected areas would seemingly accept being located to a more convenient place if they could be assured of receiving good land and proper facilities—but that's a very big "if" indeed.

An even thornier question is deciding what minimum level of development is necessary for people. It's like asking someone how rich he or she wants to be; you never really know when you'll be satisfied. But the question must be answered because, assuming villagers stay in protected areas, they are going to face greater constraints on development than people outside. They are already supposed to be living under strict rules, but enforcement varies from park to park, and in some places it's virtually nonexistent. A Community Forestry Act will probably decide the matter, but several different versions of a bill have been drawn up. Predictably, the various stakeholders involved disagree on a host of issues: whether hunting can be allowed; whether trees can be cut to build houses; who will decide whether the rules have been broken and what the punishment will be. A version of the bill has passed in the lower house of Parliament, but was blocked in the Senate over the issue of whether community forests will be allowed in national parks and wildlife sanctuaries, or only in forest reserves.

Some of the issues raised are even more profound. The vision of farmers living sustainably in the forest basically assumes they will adopt (or return to) a more traditional, self-sufficient lifestyle. But even if that's achievable, some preservationists ask whether it's fair to force them to live under such strictures. Wouldn't they be better off resettled outside protected areas rather than serving as a living exhibit in an open-air museum? Environmental democracy groups respond that these are loaded questions—loaded with the assumption that everyone holds the same modern, urbanized vision of development that has produced the preservationist ethic itself. We tend to assume that other people want the same things we want, that they want the same form of development we have; but the two sides in this debate have very different notions about what form development, and conservation, should take.

The fundamental question surrounding the Karen and other inhabitants of protected areas confounds conservation everywhere: Is saving endangered species or ecosystems worth sacrificing people's homes, culture, livelihoods, and even lives? We'd like to avoid this zero-sum decision, and sometimes there are ways to do so, but implicitly it is still being made all the time. On a political level, the answer depends on the relative strength of the stakeholders; that is, who's got more clout, the Karen and their defenders, or the RFD and the preservationists? On a personal level, it depends on those making the decisions, and how much they care about the people involved. When you meet people such as N'der and his fellow Karen, you become more sympathetic. It's human nature to care more about people you know, and less about people you don't. That, after all, is the root cause of conflicts between local and national and global interests. They are conflicts of scale, and crop up all the time in environmental affairs. Your opinion reflects your reference frame, whether you're looking at the local situation, or the big picture.

A practical, sustainable solution to the forest dweller dilemma will require not only compromise between the goals of environment and development, but also between local, national, and global interests. It would be a mistake to bow to ideology, of one form or another, and make that the basis for a broad national forest policy. The Man-and-Forest debate involves so many factors—the culture of the forest dwellers, their wealth, history, traditions, legal status, degree of geographic isolation, land tenure status, the strength of their leadership, the importance of the forest at stake and its nonhuman inhabitants—that each situation needs to be resolved on its own merits.

Thailand needs a community forestry law that establishes a system of checks and balances between the competing interests. It should set clear limits on the extent to which forests can be disturbed, but be flexible enough to allow differences in the way people interact with them. Working out the details of such a compromise is devilish, but guidelines can be crafted from the sustainable development projects already underway among hill tribes and other forest dwellers. The Doi Sam Mun Highland Development Project in Chiang Mai province,[29] carried out by the RFD's own Watershed Management Division, is typically held up as a model for integrating highland agriculture with forest protection, as is Tuenjai's work in Chiang Rai; and the buffer-zone concept also holds promise. These schemes focus their attention on areas surrounding conservation zones to make sure that people have the wherewithal to avoid exploiting protected forests. A buffer-zone project carried out by the Thailand Environment Institute in Chaiyaphum province, for instance, works closely with villagers living near the Phu Khiow Wildlife Sanctuary, since their activities are crucial in determining whether it will be preserved.[30]

Each of these projects has many components, but three are common to just about all of them. First, community development is crucial to get people working together; otherwise the behavior of a few can undermine the efforts made by everyone else. Second, they all involve establishing community forests as an alternative to exploiting protected forests. Third, education is vital. Training adults in sustainable practices is obviously necessary, but they are usually too old to change the way they look at the world. Ultimately, progress is made by teaching kids the importance of the forest at an early age. One hears numerous stories, for instance, about poachers who have stopped their illegal hunting at the insistence of their own animal-loving children.

Would a series of successful, nation-wide community forestry programs do away with the need for wilderness, or at least the need to resettle people currently living in conservation forests? My colleagues and I at the *Nation* argued about these issues long and hard in trying to establish an editorial line for the newspaper, and I came to realize that the American model of national parks isn't suitable everywhere. I also understand the European critique that creating preserves devoid of people is itself an artificial act. But I still believe it's important to protect wilderness, for practical reasons—there is so little of it left—and spiritual ones: In an age where humanity's impact is everywhere, it is a sign of our humility.

Forced resettlement, on the other hand, should be avoided, used only as a last resort for flagrant law breaking. It's generally better to use carrots instead of sticks. If city-dwellers are so keen on maintaining healthy forests, perhaps they should pay forest-dwellers to protect them. New York City, for instance, is spending $1.5 billion to inhabitants of the rural Catskills to protect its watershed, which is far cheaper than the $10 billion it would cost to build a new waste water treatment plant. And now dozens of other American cities are considering following suit. When a good community forestry or watershed rehabilitation program won't work, voluntary resettlement is a much better solution: Squatters should be tempted to move from the most remote forests by offers of new homes. And it seems feasible. Forest and biodiversity conservation is now considered so important that funding should be available. Making the move voluntary would help ensure that resettled forest dwellers are offered decent land and infrastructure. But if Thailand really wants to protect its forest, its first priority should be to reform the RFD—or perhaps, more realistically, follow the suggestion of reformers such as Kasem Snidvongs and remove the conservation agencies from the forestry department altogether. In fact, the Thaksin government plans to reorganize its ministries, and perhaps put many of its resource agencies into a newly created Ministry of Natural Resources and Environment, by October, 2002. That could help make Thai policy more coherent, but much depends on the details, and how the ministry is run.

My faith in the importance of wilderness was dismissed by some Thai activists as a result of my U.S. upbringing. But the urge to keep some sanctuaries as free as possible from human interference may not solely be an American dream. Rather, it is a post-industrial dream. The author Roderick Nash speculates that people first saw themselves as separate from nature when they stopped hunting and gathering and relied instead on agriculture. This perception has surely been strengthened by the move from agricultural to industrial societies, and with it comes an ever-stronger urge to maintain wilderness as a means to escape from civilization. The dichotomy is a familiar one. People who live and work in nature tend to take its beauty and abundance for granted, but those who visit it occasionally on trips away from the city find it awesome and inspiring, at least in limited doses.

AN ASIAN VIEW OF NATURE

In the end, the people living in and around a forest have the most direct impact on it. This is why the greatest sign of hope for conservation in Thailand is the increasing pride local people feel for their parks and sanctuaries. Support groups and conservation clubs have sprung up around many of the most famous protected areas—including Huay Kha Khaeng, Khao Yai, Phu Kradung—and even some of less well-known ones such as Mae Wong National Park. This is a new phenomenon. Most such groups generally did not exist even ten years ago, and they are made up of common people, not just of activists and the wealthy.

Stewart-Cox, the English naturalist who helped to have Huay Kha Khaeng named a World Heritage Site, has also been impressed by the outpouring of support for the sanctuary. When she went for a visit on World Heritage Day one year, she was flabbergasted to find that the vendors from a Bangkok market had loaded up two huge vans with freezer boxes full of food for the trip up north, and then slaved away all day so that hundreds of people could celebrate (just outside the entrance to the sanctuary) by having fresh seafood for dinner. "Forget the World Conservation Union. Forget the World Bank. If the women who run the Huay Kwang fresh market can donate so much energy and effort to celebrate world heritage at Huay Kha Khaeng, then we're winning," exclaims Stewart-Cox. "Yes! Yes! There's hope!"[31]

Wilderness, in other words, may have universal appeal, in much the same sense that the naturalist Edmund Wilson sees a human predisposition toward "biophilia." Certainly, many Europeans recognize the value of wilderness, even though it is difficult to find on their crowded continent. After all, when European governments had a chance to preserve wilderness without affecting their own people—as during colonial rule in Africa—they went ahead and did so. And a growing number of Asians, or at least Thais, also seem to believe in its importance.

The preservationist position in Thailand is almost always backed up with practical arguments. The spiritual arguments are rarely stated publicly; they would probably come across as insensitive, not to mention weak, when placed against the fate of poor villagers. But as the musings of Ajaan Pongsak and his followers suggest, that inspirational support for wilderness is present in Asia, and it raises interesting questions: Is this a continuation of the tradition to set reserves aside for spiritual purposes, or is modern-day prosperity helping to create a new ethic that sees humans as separate from nature? And as Asian societies industrialize and urbanize, will support for wilderness evolve and grow as it has in the United States?

Considering the strength of the environmental democracy movement, and the legacy of Asian and Buddhist myths about the forest, one would suppose that a modern Asian vision of nature would entail people living in the forest and, ideally, in harmony with it. In that sense, it would be closer to the European ideal of the countryside than to the American ideal of an uninhabited wilderness. But the rise of Thailand's park support groups, combined with the activism of groups such as the Dhammanat Foundation and the Chom Thong farmers, indicates a competing view of the human relationship with nature. Thai preservationists will have a tougher time selling their vision than the American preservationists had, however. The hill tribes and other forest dwellers are semi-marginalized in Thailand, but they have much more of a voice than the native Americans did when the United States was establishing its park system, or than the Africans during European colonial rule. Even the more authoritarian governments in Asia and the developing world may find it difficult to establish unpeopled sanctuaries. Creating wilderness in a democracy is awkward.

A couple of things seem clear, though. There will probably always be several competing views of nature within Asian societies. And within mainland Southeast Asia, virtually all these views are likely to be influenced by green Buddhism, its increasing popularity being another sign of hope for conservation. In Thailand, this movement is especially promising because it resonates with all of society, including both sides of the Man-and-Forest debate. Preservationists are naturally drawn to it because the spiritual ethic is so similar to their own beliefs. And environmental democracy groups find it appealing because monks are a key component of civil society; they do important work in every type of community, urban and rural.

It is fascinating to see how old religious traditions are evolving new, green meanings. The most famous such practice is ordainment. Traditionally, this ceremony is practiced only on people who become monks. But now green activists move through a threatened forest and ordain many of the stately old trees by marking them with colorful fabrics to show they are sacred. And ordination does actually seem to prevent loggers from cutting them down. Walk through a degraded forest and you're likely to find that the only big trees left are those adorned with colorful scarf-like fabrics. Another tradition (probably a fusion of Buddhism and spiritual beliefs) that has been adapted for conservation is the *seub chatta* (prolong life) ceremony, which is typically used to help people who are gravely ill, but is now conferred on rivers threatened by dams, pollution, or too much fishing. At one such ceremony for the Nan River, local monks cordoned off a portion of the river where fish will be fed and bred, and fishing prohibited.[32] Another ancient Buddhist ritual, the *tawd phra paa* (merit making) ceremony, which was traditionally used to make offerings and to cleanse oneself, has now been adapted to help raise funds for the cause of Buddhist conservation.

Could the spiritual sentiments that Thais feel for the forest help overcome the ideological differences between the two green camps over the Man-and-Forest debate? If they are to reach a compromise, they will first have to gain respect for the other's spiritual beliefs: Preservationists cannot simply dismiss the rural activists' dream of people living in harmony with the forest as "romantic wishful thinking"; and farmers' rights advocates can't simply denounce wilderness as a kind of "deep green imperialism." They need to recognize how much they have in common; indeed, both visions seek to protect existing forests and bring degraded ones back to health. That sounds a lot better than viewing the debate on a purely practical level, where it resembles just another conflict between different interest groups and their competing designs on the forest.

If the history of other countries that underwent industrialization is any guide, spiritual appeals should be effective in inspiring the majority of people who remain outside the environmental movement to support green causes in the future. For although the practical reasons to protect forests are powerful, they rarely carry the day, which is why so many countries have cut down their forests with abandon once they have had the means to do so. Perhaps these practical arguments need to be combined with a movement based on inspiration to gain the strength necessary to resist the tide of deforestation that engulfs industrializing societies. And so it's reassuring to realize that people everywhere do feel that spiritual urge. It is expressed differently from culture to culture, and from era to era. But whether each effort is marked by the preservation of wilderness, by an attempt to create sustainable practices, by the fight to revive harmonious traditions, or by a humble shrine, it is a testament to the resilience of that human bond with nature, and thus a sign of hope.

6

THE COAST
ON THE EDGE

THE FISHERMAN'S SALON

Considering how small and ordinary it is, the fishing village of Chao Mai receives a lot of important visitors. This sandy hamlet of several hundred Muslim villagers in Trang province, tucked away quietly on the Andaman coast just next door to Haad Chaomai National Park, is home to a most engaging couple. On any given evening at the modest wooden home of Yahed and Meeya Hawa, you may run into visiting actors and artists, environmentalists and journalists, photographers and professors, officials (both the governmental and nongovernmental kind), and just plain passers-by. From all over Thailand, they drop by the Hawas' bamboo hut for a pleasant bout of eating, singing, and conversation.

Bang (Uncle) Hed, as he is known, has a thin frame and a skin beaten copper by the sun. He is a forty-six-year-old fisherman by trade and the host of this bamboo salon. He is also a formidable naturalist, schooled entirely in the field, whose advice on local species is sought by professional researchers. *Ja* (Aunt) Ya is not just a fisherman's wife. Her grass roots work on behalf of women and small-scale fishing folk has flourished. She now travels all over the country to meet with officials and high-ranking ministers. She met and lobbied Chavalit Yongchaiyudh when he was prime minister, and even made it all the way to Beijing for the landmark UN Women's Conference in 1995.

It's no wonder, because Meeya has the ideal qualities for a nongovernmental activist. She's clever, has impeccable grass roots credentials, and her buck-toothed mouth, which usually sports a grin, also hides an unusually sharp tongue. She uses it often to lash the officials whom she claims are either too corrupt or too inept to protect the villagers' interests. "The government always plans from the top," she complains. "They bring development to the village like instant noodles, 'just add hot water.'" (Asked what she thought about the reported plan by women's rights activists to carry out a topless street protest in Beijing, Ja Ya replied with a laugh, "The Chinese police are stupid. Here they had a chance to see women from 180 countries bare their breasts, and they were afraid of it!") Together, Bang Hed and Ja Ya epitomize that much-abused term "local wisdom." Yahed claims he knew the first day he saw Meeya that "she was the one." Two decades and five children later, she still blushes like a schoolgirl. But her explanation is slightly different: "No one else would take me because of my big mouth," she grins.

It was only a decade ago that the road connecting Chao Mai to the outside world was paved, and Ja Ya is not happy about the way it happened. "They should have educated the people first, prepared them, then bring in the physical development," she argues. The local youth race their motorcycles along the tarmac straight-aways, then as crash victims the boys are raced to the hospital. But she and her husband can also appreciate the advantages of modern technology. Standing in the bow of his long-tail boat with the wind whipping against his face, Bang Hed's grin spreads from ear to ear as he lets the engine rip in a race with his mates.

The road has helped provide villagers with better access to markets, health care, and other advantages offered by the outside world, but it has also brought the attention of developers who want to build resorts and shrimp farms. That in turn has led to further encroachment on local mangrove forests, a source of food for the villagers and a vital breeding ground for marine life.

Virtually all the resources that have supported Chao Mai villagers' traditional livelihoods have suffered. Fish stocks have dwindled, partly as a result of the big trawlers that make illegal sweeps close to shore under the cover of night. The trawlers also tear up the sea grass meadows, an important feeding ground, and damage the coral reefs. Meanwhile, local fishermen are under increasing pressure from conservation officials seeking to protect the nearby marine reserve. And the villagers complain that wealthy investors from town have been able to gain land holdings within the boundaries of Chaomai National Park, but local families have trouble even having street addresses assigned to their houses.

Not all the problems can be blamed on outsiders, however. Villagers themselves often sell their land to tourism and aquaculture investors, use unsustainable fishing practices, operate small pushnet trawlers that "bulldoze" the seabed, and encroach on the mangroves or cut them down for the charcoal factories. "Villagers are selling off their land, their birthright, to buy cars," Bang Hed warns in his soft nasal voice.

That said, the situation in Chao Mai today is better than it was in the 1980s, when most of the villagers were destitute. "Everyone was poor then, and we had to work in the factories," Meeya recalls. Some even resorted to eating canned fish, which was particularly demeaning. Desperation helped make people receptive to the message of a local NGO, Yadfon (Raindrop) and its director Pisit Charnsnoh, who urged the fishing communities to preserve the mangrove forests and sea grass meadows. The results were striking. By rehabilitating the coastal ecosystem, they also resuscitated the fish stocks and thus their own livelihoods. People were no longer forced to the cities and fishing ports to look for work, and the communities began thriving again. "People are about 70 to 80 percent better off than they were ten years ago," says Meeya. Even the dugongs, an endangered sea mammal similar to the manatee, began to reappear around the villages once the sea grass they feed on was restored.

There were several keys to success. With the help of researchers from the local university, Yadfon demonstrated the importance of a healthy sea habitat to supporting fish stocks by carrying out studies on the sea grass meadows. The village headman recounts, "We told [Yadfon] that we didn't want to find a way to bring back the sea grass, we wanted to bring back the fish! But then they showed the link and we slowly learned the connection between the sea grass and the fish."[1] That convinced some fishers to stop using destructive gear, and—thanks to lots of meetings, peer pressure, and raising awareness through the local network of mosques—others gradually came around. Incentives for cooperation were dreamed up: the bulk purchasing of fuel and gear, for instance, helped reduce costs. And it helped when the fishing quickly improved. "This is the beauty of a tropical ecosystem where nature can heal itself fairly rapidly due to high productivity," explains Yadfon's Jim Enright. The villagers couldn't keep the big trawlers out of the bay on their own. But the return of the dugong drew the attention of the media, which in turn pushed the Trang governor to pass a provincial ordinance making the bay a conservation zone; and so the villagers were given another legal weapon. The media presence also helped convince local people what they were doing was important.[2]

Chao Mai reveals the importance of local participation; the work carried out there is a classic example of how civil society in the rural sector can be mobilized to make livelihoods and communities more economically viable and sustainable. It could also be argued that, by learning from field tests and trying out different approaches to come up with resource management schemes that fit local circumstances, the folks at Chao Mai were practicing a rough-and-ready form of "adaptive management"—an incremental, almost experimental approach towards environmental management currently expounded by policy experts.[3] Unfortunately, corruption, politics, and bureaucratic strictures make it difficult for the Thai government to follow suit. Nevertheless, other fishing communities up and down the coast are now learning from Chao Mai's successful methods through NGO extension work.

This is encouraging news because the zone where the land meets the sea may be the most important area on Earth. Coasts are immensely abundant in all forms of life, particularly humans. Two billion people, more than a third of the world's population, lives within 100 kilometers (60 miles) of the sea.[4] Fourteen of the world's fifteen largest cities can be found along coastlines. In Asia, coastal urban areas alone are home to half a billion people. And even those who have never been near the coast end up affecting it: Most of the roughly 20 billion tons of waste a year produced by humans (and all the animal waste, eroding soil, and so on) sooner or later ends up flowing into the sea.[5]

This chapter will focus on fishery and aquaculture issues because they offer an object lesson in environmental management. Other coastal resource issues, which must be factored into any coastal zone management plan, are spread out among other chapters. Tourism development is discussed in Chapter 2; toxic pollution in the Gulf of Thailand in Chapter 8; and the international shrimp-turtle and dolphin-tuna trade disputes in Chapter 9, which also covers what may ultimately be the most serious threat to the coasts: the rise in sea levels due to climate change. But this potential danger seems impossibly distant in Chao Mai.

For Bang Hed and Ja Ya, their problems remain intensely local. They live with constant insecurity because it's so difficult to obtain title to their land without the backing of a wealthy investor. Meanwhile, the investors will help only those villagers who are willing to sell—a classic Catch-22. The couple has been tempted to give in. One of their neighbors sold out, and a shrimp hatchery run by noisy diesel engines that belch black smoke now operates next door. Elsewhere in the village, a Malaysian investor has built a small resort at the pier where fishermen used to moor their boats. Bang Hed has been warning about this development for years: "If we can't park our boats there, then we will have to park them in the national park, and we will be accused of encroaching." Meanwhile, a fence erected by a neighbor who manages a small bungalow operation (on behalf of another outside investor) has blocked the path leading from their house to the beach.

It's all especially troubling because they had hoped that tourism would let the village develop in a less disruptive way than industry. "If we can control it, tourism should be okay," says Yahed. "But if it's controlled by outsiders, then there will be problems."

FISH WARS AND THE BLUE REVOLUTION

I've described some of the battles waged over land in Southeast Asia, but the battles at sea may be even more ferocious. Governments attempt to regulate fishing in the region by imposing rules on the type of gear that can be used, awarding fishing concessions, and restricting how closely big trawlers can come to shore. In Thailand, they're not supposed to cast their nets within 3 kilometers of the coast. But law enforcement on the open sea is scarce, and small-scale fishermen all over the country complain of encroachment by big trawlers. Fed up with a lack of response from the authorities, many of these communities are organizing their own coastal defenses. Armed guards equipped with radios or mobile phones patrol the fishing grounds at night. If they spot illegal trawlers, they alert the Fisheries Police and sound the alarm at home. A fleet of long-tail boats will then sail out to confront the often heavily armed intruders, and occasionally a pitched battle will ensue.

So who owns these rapacious trawlers? It's often hard to tell, as many go unregistered and employ low-wage Burmese crewmen. Somphol Jirojmontree, chairman of the Commercial Trawler Associaton at the Trang port of Kantang, says his vessels are too big to catch fish in the 3-kilometer zone. "Trawlers and push net boats which catch fish near the coastline are the small ones operated by villagers, not by me," he contends.

The sea grass beds are not the only ecosystems that have suffered from trawlers' transgressions. Coral reefs are a vital feature of tropical coasts; they provide not just a popular tourist attraction, but they are also one of the most biodiverse habitats on Earth, home to as many as 3,000 marine species. An ecological marvel, reefs result from a symbiotic partnership that has helped this riotous variety of life to bloom in a barren environment. Coral is an animal whose body lives within a limestone exoskeleton. It gains some food by waving tentacles that reach out from this rock-like protective exterior, but most of its meals come from a kind of algae, known as zooxanthellae, that lives within its tissue. These plant-like organisms use the coral's waste products to manufacture nutrients from sunlight through photosynthesis, and they can grow to outweigh their hosts' tissue by as much as three to one. It's a wonderful example of how separate species have evolved mutual bonds of nourishment. And it's particularly remarkable because the clear tropical waters in which reefs thrive are mostly devoid of nutrients; that's why the sea is so transparent in areas such as Phi Phi.

Half the world's reefs are found in the Indian Ocean basin, including along Thailand's famed Andaman coast; but here a majority of the coral is now dead, according to Sombat Poovachiranon of the Phuket Marine Biological Center.[6] Because it requires a suite of stable conditions to thrive, coral is struggling everywhere. Pollution and sediment runoff cloud the waters and block some of the sunlight used for photosynthesis. Tour boats drop anchors on the reefs, and tourists step on them. Fierce storms, especially at low tide, can devastate a reef. Perhaps most ominously of all, considering the portents of global warming, is that a mere 1-to-2-degree increase in ocean temperatures for an extended period causes coral to "bleach"—they expel the zooxanthellae, turn white, and, if conditions persist, die from a lack of nutrients. But in Southeast Asia, the most visible damage is from fishing. There is nothing I like better on holiday than to snorkel or scuba dive, and the tragic site of reefs blasted away by dynamite fishing or strangled by nets is all too common.

Establishing marine reserves, such as the one that protects Australia's Great Barrier Reef, can help with a range of coastal issues, particularly fisheries. They are easier to monitor than complex rules on fishing gear or catch limits, and have proved surprisingly effective. A study conducted at the University of California at Santa Barbara found that "no-take" zones showed average increases of 91 percent in the number of fish, 31 percent in the size of fish, and 23 percent in the number of fish species present, all within two years.[7] Crucially, these benefits extend outside of reserves, which serve as a kind of "safe house" from which marine creatures can spread out to repopulate surrounding waters. So the fishermen of Chao Mai village, for instance, should benefit from conservation at the nearby marine park.

But law enforcement is so lax in Southeast Asia that the potential benefits of marine reserves are often squandered. It's difficult to estimate the extent of illegal fishing activities, but by most accounts they are rampant. Go to just about any Thai marine park, and you're bound to see trawlers with their nets cast, sometimes in broad daylight. As has happened so often in Thailand, the state has failed to perform its role as a natural resource manager. And out on the open sea, it is difficult for civil society to help. So the fish wars continue to rage, and the oceans become another vast and tragic commons.

Reining in the trawlers isn't easy, because fishing is a hugely important industry. Worldwide, marine captures alone generate roughly $80 billion in revenues, and provide more than a million jobs.[8] Asian fleets sail the world over and catch nearly half the world's fish.[9] Developing countries as a whole account for 65 percent of global marine landings. Much of this catch is exported, yielding $16 billion in earnings every year. But perhaps even more important is the catch that is eaten domestically, since fish is the main source of protein for most of the developing world.

But the industry faces a basic problem, not just in Southeast Asia but globally: too many boats chasing too few fish. The World Wide Fund for Nature (WWF) estimates the world now has two and one-half times as much fishing capacity as needed to achieve sustainable yields. Viewing the sea as a virtually limitless resource that is easy to exploit, governments have poured money into their countries' fishing fleets, providing an estimated $15 billion in subsidies every year.[10] Thailand, for example, sells trawlers cheap fuel.

There is also huge waste. Roughly 60 percent of the fish caught by Thai vessels is considered bycatch,[11] and goes to make fish meal used as animal feed. For shrimping boats, the bycatch level can be as high as 80 to 90 percent—a major reason shrimpers were targeted by a U.S. trade action in 1996. People may not like to eat these species, but other fish do. So removing them in large numbers from the food chain can damage marine ecosystems.

This is just one of many reasons—including most notably the overfishing of target species—why one fishery after another around the world has collapsed. Assessing the status of fish stocks and the factors leading to their decline is notoriously difficult (it is said that "counting fish is like counting trees . . . only they are invisible and move around"). But the crisis facing global fisheries is increasingly well-documented,[12] and Thailand has done its share of the damage. After largely fishing out its home waters, the Thai fleet now sails to Indonesia, Burma, and the Philippines to obtain much of its catch. They have often been granted concessions, but disputes with foreign authorities and boats are common. Rarely a week goes by without an armed clash involving Thai fishing boats, or reports that Thai fishermen have been imprisoned or released from foreign jails.

If fishermen act like ruthless predators, that is because our attitude toward fish is still primarily that of hunter-gatherers. Just as our ancestors thousands of years ago turned to agriculture, growing their own crops and herding livestock, the demands of a booming population are increasingly being met through aquaculture—farming fish and shellfish in controlled environments. Aquaculture is now the fastest growing segment of the world's food business. Boosters such as the FAO (UN Food and Agriculture Organization), have heralded the potential of a veritable "Blue Revolution" that would increase food security and provide more protein to the poor, much in the same way the Green Revolution raised crop yields in the developing world by adopting the techniques of intensive agriculture. Donor agencies have therefore provided financial support to aquaculture development, and the FAO has called for a doubling of production between 1995 and 2010.[13]

Lured by voracious demand from Japan and the West, shrimp farming has become especially popular in tropical countries. But the industry is still in its adolescence and experiencing growing pains. Like the Green Revolution before it, the rise of intensive aquaculture has created serious social and environmental problems. The farming of carnivorous species has been especially damaging and has led to disputes about the impacts of salmon farming in the West, mirrored in Asia by the controversy about shrimp farming, an industry that in some countries has collapsed in its own filth.

Asians have been farming fish for centuries. Freshwater species are still cultivated in paddy fields throughout Southeast Asia, and farmers whose land is flooded by seawater during certain parts of the year often alternate growing rice and raising seafood.[14] In Indonesia, the traditional *tambak* system entailed breaching paddy dykes once the rice harvest was collected to let in seawater at high tide so that larval fish and shrimp are captured and then tended until maturity. Similarly, the *gei wai* system used in Hong Kong entails digging channels around mangrove islands, then stocking them with fish, shrimp, and crabs at high tide. You can still see it in action at the Mai Po Nature Reserve in the New Territories. But this *gei wai* system is a museum piece, the last remaining example of the technique; it sits just across the bay from the rising skyscrapers in the Chinese boomtown of Shenzhen.

These traditional methods cultivate a variety of species, and have about as much in common with new monoculture techniques as Shenzhen does with a traditional Chinese hamlet. Developed in Taiwan in the 1970s, modern shrimp farming relies on specially built ponds into which seawater and fresh water are piped to create the necessary brackish conditions. The ponds are stocked intensively with larvae of a single species, raised from hatcheries, and are aerated continually by specialized machines. Shrimp are omnivorous scavengers in the wild, but on farms they are fed high-nutrient meal largely made of ground-up bycatch to speed their growth. Numerous biochemical inputs are added to the ponds, including an array of antibiotics to stave off viral epidemics.

These epidemics have been the bane of the industry. Since hatchery-raised shrimp are genetically similar to one another, disease spreads quickly through the farms. Along with other environmental problems, they led to the collapse of the prawn farming industry in Taiwan in the late 1980s. Taiwanese investors then began exporting the technology. Their capital and expertise has helped the industry spread rapidly to Thailand, the Philippines, Malaysia, southern China, India, Vietnam, Cambodia, Indonesia, Bangladesh, and Sri Lanka, which all have the tropical climate, extensive coastline, and water resources needed for farming shrimp. Along with government subsidies, the industry has benefited from financing by multilateral lending agencies, including the Asian Development Bank and the World Bank,[15] which for instance gave India US$425 million for aquaculture development in 1992.[16]

Although prawn farming in Asia has suffered from the same viral and environmental problems that caused the industry to collapse in Taiwan, farmed shrimp has become one of Asia's most important food crops because of the hard currency it earns. Between 1982 and 1994, cultured shrimp rose from accounting for 5 percent of the world's shrimp supply to 30 percent,[17] and Asia now supplies around three quarters of the world's farmed shrimp.[18] Most is exported to developed countries. Japan's seafood imports account for nearly a third of the international fishery trade, and of those imports, shrimp rose from 29 percent in 1986 to 46 percent in 1991. The United States has the second largest shrimp market, and consumption has been rising quickly. In 1995, the average American ate 2.5 pounds of shrimp every year.[19]

The goal of "feeding the hungry" by farming shrimp has therefore turned out to be a pipe dream. To improve food security, say the experts, countries need to focus on herbivorous aquaculture. It may be feasible to grow shrimp and salmon on vegetable-based feeds such as soy,[20] but currently shrimp feed is usually made from processed bycatch; protein is being gathered from the oceans by developing countries and then sold to well-fed consumers in Japan and the West, packaged as shrimp. And since the input of fish products is roughly from two to four times the volume of fish output for shrimp and salmon farming, the raising of such carnivorous fish leads to a net protein loss.[21] Because fishery resources are depleted rather than augmented, claims about increased food security are dubious.

Throw in the damage done to coastal ecosystems, and it's not surprising that critics compare shrimp farming to the ravages caused by cattle ranching in the tropics. Shrimp farming has become so lucrative in Asia and some Latin American countries—in 1996, the total farm-gate value of cultured shrimp was estimated at $6.2 billion, and its retail value maybe three times that figure[22]—that the FAO is eager to promote it in Africa and other tropical regions. But its social and environmental impacts have now led the World Bank to halt loans for shrimp farming unless they are part of a broader coastal zone management plan. To see why, once again we need merely examine Thailand's experience.

SHRIMP FEVER

Aisha Petphrom has seen the Blue Revolution up close, and it's not a pretty sight. Thailand's Gulf coast in particular has been ravaged by shrimp farms. Much of the shoreline, once covered by mangrove forests, has been turned into ponds, many of which now lie naked and abandoned, remnants of a dream gone bust, the soil rendered sandy, salty, and infertile. Ruined scenery is the least of the problems, however. Aisha, a forty-seven-year-old landholder in the Trang village of Thung Kilek, has watched the shrimp farming business ruin lives.

On the macro level, shrimp farming would seem to be a huge success for Thailand. Spurred on by government promotions and foreign assistance, farmers found that although operating costs were high, they could achieve spectacular returns by using intensive stocking and feeding methods. The result was a shrimp farming boom, euphemistically known as "shrimp fever," as farmers rushed to convert their land into ponds or to dig farms in publicly owned mangrove forests. By 1989, Thailand's cultured shrimp production exceeded its wild catch.[23] By 1994, Thailand became the world's largest shrimp exporter and accounted for one-third of the world's market share;[24] by the following year, it was the world's largest producer of cultured shrimp. When foreign revenues from farmed shrimp reached $1.3 billion in 1995,[25] shrimp became the country's second largest export commodity after rice, and from 80 to 95 percent of these exports came from farms.[26]

But the collapse of the Taiwanese industry was no anomaly. Similar problems became evident in China in 1993, Vietnam in 1995, and then Thailand a year later, resulting in a decline in Thai shrimp production from 259,000 metric tons in 1995 to 210,000 metric tons in 1996 and 215,000 metric tons in 1997.[27] On the micro level, the situation is even more serious. Ponds are often overstocked, poorly cleaned, and weakly managed. Only an estimated 17 percent of the feed applied to ponds is converted into shrimp biomass.[28] The remainder settles onto the bottom and mixes with chemical additives to form a toxic sludge that contaminates the ponds. Even when it's flushed into the sea, polluting coastal waters, it often returns to haunt the farms. Under the "open system" initially preferred by Thai shrimp farmers, seawater is regularly pumped back into the ponds, which in effect become fouled with their own waste. These unhygienic conditions have led to a much less metaphorical type of shrimp fever: epidemics of the notorious "white spot" and "yellow head" viruses.

Farmers often turn to antibiotics in response, but these may be counter-productive. Besides making the water toxic, the cocktail of chemicals added to the pond remain as harmful residues in the shrimp. In 1993, Thailand's Medical Sciences Department reported that 24 percent of exported shrimp were contaminated with tetracycline, a common antibiotic, although that figure had been reduced to 4.7 percent by 1997.[29] "Japan has identified over twenty antibiotics used in the farmed shrimp industry and has banned shrimp farmed with these antibiotics," writes Alfredo Quarto of the Mangrove Action Project. "Meanwhile, the U.S. Food and Drug Administration only looks for residues of between just two to six antibiotics, and as yet, bans no shrimp cultivated with their use."[30] But foreign consumers have less to worry about than local ones, because Southeast Asian governments are usually more concerned with exports than with domestic public health. Seafood sent abroad is randomly inspected, by both the exporting and the recipient countries. Seafood sold domestically in Thailand is rarely checked.

The problems with disease make the economic returns from shrimp farming extremely volatile; because of its high risk and high potential, Enright compares it to a form of gambling. One poll found that 62 percent of prawn farmers have experienced years with drastic production losses, often due to disease and poor water quality.[31] "In Asia," adds Quarto, "the average intensive farm has been found to survive only 2 to 5 years before serious pollution and disease problems cause shutdowns . . . [and] in Thailand, over half of the shrimp ponds have closed down in the first decade of Thailand's entry into the great race for world dominance in the shrimp export market."[32] Because agro-industrial conglomerates such as Charoen Pokphand have thrived by vertically integrating their industries, the bulk of their profits come from providing feed, hardware, and consultant services to the producers. In other words, small farmers absorb most of the risk associated with shrimp aquaculture.

"Since the shrimp farms came, the quality of life in our village has gone down," confirms Aisha, bedecked in a colorful headscarf. She has 500 rai (about 80 hectares) on which she grows rice, rubber, and oil palms. Four people, from Trang and Bangkok, have offered to buy her land at a price of around 25,000 baht (then about $1,000) per rai. The potential profits from shrimp farming were also alluring. A recent study found that farming tiger prawns yields on average thirty-five times more income than growing rice[33]–at least in the first year or two. So it must have been tempting for Aisha. But she is a conservative woman; her family has lived in Thung Kilek for two hundred years and she was suspicious of the get-rich-quick schemes put forth by the investors and feed merchants. So she held on to her land. Echoing Meeya, Aisha says, "I have enough in my life."

Eight of the one hundred families in her village did sell their land, much of it in the mangroves, or invested in shrimp farms. For them, says Aisha, "the money came easy, and went easy. They have to spend it because it makes them happy. They buy a new car, or motorcycles for their kids, or go gambling, or take on a mistress."[34] Then the businesses fail, but the farmers can no longer go back to their old crops because the land has become too salty. Many end up as laborers, she says, or homeless. Sometimes the farmers borrowed land title deeds from relatives to secure loans because the banks quickly realized that land converted into ponds makes for lousy collateral. So when money to pay off the debts runs out, conflicts break out among the families. "It took four years to prove it, but I made the right decision not to sell," says Aisha. That doesn't mean she entirely escaped the ills caused by "shrimp fever." The land around the ponds is often affected by salinization, she says, causing rubber trees to sicken and paddy to wither. And her village can no longer use well water even for washing. Wastewater from the ponds fouls the coast and damages the fisheries.

The most significant environmental impact has been the rapid decline of Southeast Asia's formerly abundant mangrove forests. Mangroves, the only trees able to grow in salt water, are a vital part of tropical coastlines because they protect the shore from erosion and serve as breeding grounds. "Scientists have estimated that three quarters of the tropical world's marine catch is dependent upon mangrove forests for food or habitat for some part of their life cycle," says Quarto. They are also home to numerous endangered species, including the dugong, Olive Ridley sea turtle, and the crab-eating monkey.

Most mangrove areas in Thailand are demarcated as forest reserves, but they are generally treated as an open-access resource, and the RFD has been unable to protect them. As a result, Thailand's mangrove forest coverage shrank from roughly 370,000 hectares in 1961 to 170,000 hectares in 1996.[35] Along the Gulf coast, only from 10 to 20 percent of the country's original mangroves are left, according to Suraphon Sudara, a marine scientist at Chulalongkorn University, while 60 percent remain on the Andaman coast. It's difficult to say how much of this is due to conversion by shrimp farmers, but most analysts believe it is the biggest factor, at least in recent years. In Trang, asserts Pisit, prawn farms are estimated to be responsible for 64 percent of the 15,000 hectares cut down during the last decade or so, even though mangrove soils are highly acidic and shrimp ponds built from them must be treated with lime.

These land-based issues make the controversy about shrimp farming slightly different from the one about salmon farming, since the latter are usually grown in netted pens offshore (although salmon fry are nurtured in freshwater ponds). A major criticism of farmed salmon is that they spread disease and parasites to fish in the wild, and those that escape from the pens may ruin the genetic diversity of wild salmon. It is not clear whether prawn farming shares the same problem, although the wastewater from ponds could spread viruses; but there is no doubt that the loss of genetic diversity among wild shrimp (stemming in part from the decline of mangrove forests) threatens the industry's long-term sustainability. There are more than 120 species of shrimp worldwide, but 6 species now dominate the cultured shrimp trade, and just 2 species–the black tiger prawn and western white shrimp–account for 80 percent of that trade.[36] And hatcheries are still dependent on wild shrimp as a source of genetic brood stock. Enright recounts how, when a pregnant female shrimp is caught at sea, hatchery personnel are immediately alerted by radio; they then race out to buy it, often paying hundreds of dollars for one animal.

Shrimp farming is thus a case study in how economic and environmental issues become inextricably intertwined. Veerapun Preechakul, another Trang farmer who has studied in the United States, has sought more sustainable techniques by building his ponds on former paddy fields rather than in the mangroves. Most important, he is careful not to overstock his ponds with larvae, which are tested for viruses before he buys them. He uses fewer chemicals and antibiotics than before. He has also moved toward a closed system that requires less water exchange, although he still has to release some effluent into the nearby river. He doesn't dump pond sludge into the river, but piles it on his land and treats it with lime.

"Other farmers' shrimp ponds last for only four years," says Veerapun, "but I hope to use mine for ten years or longer." That still wouldn't qualify as sustainable, but Veerapun appears to be successful; his modern, comfortable house and expensive car give him the air of a suburban professional. He admits, however, that "sometimes I lose as much as a million baht because of disease and bad prices in the market." Clearly, aquaculture is still in its experimental stage. Compared to the millennia we've been farming crops, intensive shrimp farms have been around for only a generation. "The fact is," says Enright, "we still don't know which methods are sustainable."

Thailand needs to find out in a hurry, though, because during the last few years the industry has taken a dangerous turn. Prawn farms are increasingly being built in inland areas (causing national production to grow after it leveled off in 1996–1997), including the central provinces that form the country's rice granary, where farmers can truck in salt water and benefit from well-developed freshwater irrigation systems. Barring collapse, growing shrimp is far more profitable than growing rice, but it uses more water, and salt from the brackish ponds can seep into the water table and surrounding fields. So this trend has sparked a major outcry, particularly among nearby rice farmers, who have seen their harvests reduced.

SUSTAINING LIVELIHOODS, ADAPTING MANAGEMENT

When shrimp farming was confined to the coasts, the government passed some fine-sounding regulations that banned the conversion of mangrove areas and restricted wastewater discharge. As usual, however, the laws were rarely enforced, and they made more marks on paper than they did on the landscape. The issue was taken far more seriously once the shrimp farms began cropping up in the central plains, where the economic impacts of salinization are easier to see than, say, the long-term implications of mangrove deforestation. Under pressure from powerful rice-growing interests, the Thai government in 1998 banned shrimp farming in all freshwater areas of the country. The ban was not widely enforced, however, and the inland expansion continued. An influential lobbying effort joined by Charoen Pokphand, a key supporter of the current Thaksin government, sought to have the ban revoked, but it has so far been stymied by the Chat Thai party, a coalition member that draws support from rice farmers in the central plains.

The situation is reminiscent of the logging ban imposed a decade earlier, and it reveals a clear pattern in Thailand in which lucrative resource-based exploitation—fishing, logging, aquaculture, tourism—takes off in a frenzy with little regard for environmental concerns and externalities. If rules such as prohibiting trawlers from fishing near the coast are adopted, they are weakly enforced. The damage builds until the resource is depleted. Companies and officials look for resources elsewhere, often abroad, whether by building dams or seeking timber and fishing concessions. Finally, if a crisis or public scandal emerges, the government enacts a sweeping ban on the activity in question. Thai resource governance tends to follow a pattern: It first adopts a laissez-faire approach, until the negative consequences become all too clear; it then suddenly switches to the precautionary principle in the form of a ban. What's missing is a steady, moderate approach to environmental management.

Along with Malaysia and Singapore, Thailand, Vietnam, Indonesia, and the Philippines have enough capable technocrats to devise and carry out reasoned policies. Occasionally, they are allowed to do so, as with Thailand's demand-side management energy program. But usually they are stymied by widespread corruption, a sectoral approach to resource management, and rivalry among state agencies. Their voices are drowned by the clout of vested interests. Politics supersedes science or even economics, more so than in developed countries. These weaknesses, combined with a general dearth of research capabilities, make it even more difficult for developing countries to use adaptive management, which is based on "social learning," and incorporates the results of previous policy measures into governing institutions. If planning development based on other countries' environmental mistakes and successes is too much to ask for, governments should at least learn from their own. But in Thailand, prawn farmers rushed into the mangroves, and then inland areas, before most understood what their long-term impacts would be. Loggers, fishers, tourism developers, and condominium builders exhibited the same recklessness.

Social learning comes slowly, especially when compared to the rapid rate of economic change. But pilot projects and grass roots efforts like the one at Chao Mai can help persuade stakeholders to adopt precautionary measures, ideally (but all too rarely) before frenzied development exhausts local resources. That needs to be followed by effective management plans which, especially in areas where the land meets the sea, balance the needs of multiple users and the demands on multiple resources.[37] Zoning seems especially important. It can be used to keep shrimp farms separate from freshwater agricultural activities, and behind (rather than in) mangrove forests, which can then be used to filter moderate levels of organic wastewater. Shrimp farmers have promised to reduce impacts by moving to "closed systems," but there is confusion about what this term means, and few believe the process can become completely closed at a reasonable cost.

If aquaculture can be done sustainably, without damaging surrounding areas and thus precluding other activities, it can become a valuable component of the rural economy because, as Steve Rayner, an environmental sociologist at Columbia University likes to point out, sustainable livelihoods ideally should allow villagers access to multiple survival strategies. The problem with mainstream intensive development is that it prevents such flexibility. Development experts typically visit a place and decide that the people there are all farmers, or fishers, or hunters, or laborers, when in fact they might rely on more than one—or even on all—of these occupations. By trying to make them more efficient at one thing—even if it means tempting them with subsidies, urging them into debt, and creating environmental problems—the villagers often lose the ability to do the others. They become locked into a kind of occupational monoculture. If farmers use too much pesticide on their paddy, they can no longer raise fish in their fields. Prawn farmers can't forage in the mangroves because of encroachment, nor can they turn to other crops because of salinization.

And so, by resorting to inland shrimp farming in the country's rice bowl, Thailand is playing with fire. In return for short-term gains, it risks ruining the country's historic heartland and breaking the back of its rural economy.

7

THE PIPELINE
A SNAKE IN THE WOODS

LANGUISHING IN LIMBO

On the morning of July 21, 1994, a Burmese refugee named Nai Sai Mon was going about his daily chores when a hundred troops from Burma's 62nd Light Infantry Battalion suddenly showed up. Having just been relieved of their duties guarding nearby Three Pagodas Pass—the main crossing point between Burma and Thailand in the Kanchanaburi area—they proceeded to ransack Palai Thumpai, an ethnic Mon refugee village located in Burmese territory about 10 kilometers from the Thai border. Nai Sai Mon was tied up, beaten, and burned with cigarettes while the soldiers interrogated him. He watched while several of his fellow villagers were held down as troops wrapped plastic tubes around their noses and mouths. The tubes were then filled with water and held in place until the refugees nearly drowned.

Knowing that Palai Thumpai was part of a larger Mon refugee settlement called Halockhani, the *tatmadaw* troops then began moving up the long, thin valley. They grabbed twenty-year-old Suwin and about fifty other refugees to use as human shields. But along the way, Mon rebels lurking in the hills started shooting at them. A firefight ensued and the Burmese troops were driven back. Most of the captured villagers managed to flee, but Suwin was stuck in the middle of the battalion. Returning to Palai Thumpai, the soldiers set fire to half the houses before leaving. "They kept repeating all the time, 'You must not stay here. We will come back and if you are still here we will kill you. Wait and see. Wait and see,'" recounted Nai Mit, another resident of Palai Thumpai who was also tied up, beaten, and interrogated. The soldiers then took Nai Sai Mon, Suwin, and fourteen other village leaders away with them to serve as porters. "If we walked too fast, we were beaten. If we walked too slow, we were beaten," Suwin said. "Finally, after four days, I managed to escape while the soldiers were fighting in a Mon village." Nai Sai Mon was eventually released once the troops reached their home base near the coastal town of Ye. The two men fled into Thailand and joined up with thousands of other Mon refugees who had promptly abandoned Halockhani and rushed for the border.[1]

Everyone was mystified about what lay behind the attack, especially since it came at an incredibly inappropriate time for Burma's ruling military junta. The Association of Southeast Asian Nations (ASEAN) was holding a summit in Bangkok, and the SLORC—which was eager to join the regional grouping—had been invited for the first time to send a representative. That angered human rights groups, which wanted the brutal regime isolated, but it was another indication that ASEAN was moving toward "constructive engagement"—the pursuit of closer relations with Burma despite the ruling generals' refusal to recognize the victory of Aung San Suu Kyi and her party in the democratic elections of 1990. ASEAN claimed the policy would help Burma evolve into a more open and developed society, but critics claimed it was merely a cynical attempt to grab the resources that Burma was selling. At any rate, the Halockhani incident spurred on the protests outside the summit. Burma's foreign minister, instead of arriving smiling and triumphant as he had on the first day, was met by a firestorm of reporters' questions and cameramen once news about Halockhani and the subsequent refugee exodus broke.

But there was never any explanation from Rangoon for the original attack. The battalion may have needed porters, some speculated, or perhaps it sought revenge for a skirmish with the villagers in June that had left one Burmese soldier dead (Nai Sai Mon later said this was the subject of his interrogation). Or it may have been part of SLORC's ongoing attempt to pressure leaders of the Mon insurgency—which, like other ethnic-based militias in Burma, had long been fighting for autonomy from Rangoon—into signing a cease-fire agreement.[2]

Thai authorities were also eager for the Mon to put down their arms. Whereas they had previously sought to maintain the insurgencies as a buffer between the two countries, Thailand's National Security Council, led by General Charan Kullavanijaya, had devised the new policy of constructive engagement. Thailand and its fellow ASEAN states are concerned about security and the close links Burma has formed with China. Thai businessmen, meanwhile, have long eyed Burma's resources and determined they could do business with the SLORC. The big losers have been the ethnic insurgents and refugees stuck on the border—the Mons, the Karen, the Karenni, the Shan—who have been squeezed from both sides.

The refugees who fled from Halockhani quickly became a symbol of the ethnic groups' dire straits. Allowed to move only a kilometer or two into Thailand, they became stuck on the border, in a virtual no-man's land, perched in a muddy ravine quickly dubbed New Halockhani. Under drenching monsoon downpours, and with the help of relief workers, they erected makeshift shelters. But the sanitary conditions were miserable, and the camp was threatened with outbreaks of disease. Although a few Thai Border Patrol Police (BPP) were on hand, the camp was vulnerable to new attacks from the *tatmadaw.* But the refugees kept pouring in, eventually totaling 6,000. Refusing to go back into Burma, prohibited from moving farther into Thailand, they were stuck between the proverbial rock and a hard place.

The crisis attracted the attention of journalists from foreign news agencies, but I had a head start on the story because I had already investigated reports that the Thai military were harassing refugees in camps all along the Burmese border with Kanchanaburi province and ordering them back into Burma. The repatriation campaign contrasted with Thailand's relatively hospitable record toward refugees, and my sources were convinced it all had to do with the pipeline.

For Thailand, the most crucial resource Burma has to offer—more important than the timber, the fish, the gems, and the minerals—is its natural gas. Foreign oil exploration firms had found a couple of commercially significant deposits under the Gulf of Martaban: first at a field called Yadana and then farther south at Yetagun. With its booming energy demand, Thailand was the obvious (perhaps the only) market for the gas, and notions of importing it via pipeline had been floating around since 1990,[3] the year I started working at the *Nation.* In 1993, Thailand approved an energy development plan that entailed buying Burma's gas, and then asked Thailand's state-owned oil company, the Petroleum Authority of Thailand (PTT), to begin negotiating a gas sales agreement. That set off a complex series of talks between the PTT, Burma's state-owned Myanmar Oil and Gas Enterprise

(MOGE), the French oil firm Total (also partially state-owned at the time), which had had signed a contract to produce the gas, and America's Unocal, which joined the Yadana operation the following year.[4]

The Yadana project wasn't important only for the gas it would provide. It was also a huge diplomatic symbol, since Thailand and Burma have traditionally seen each other as enemies. It would be only the second international pipeline in Southeast Asia—the first began transporting gas from Malaysia to Singapore in 1992—and would require unprecedented cooperation between the Thai and Burmese governments. So it came to represent the promise of constructive engagement. But it also symbolized the policy's potential peril. Burmese democracy activists complained that proceeds from Yadana would go directly into SLORC's coffers and help prop up the junta for decades. They also worried that building the pipeline within Burma would result in human rights abuses among the local population.

I had long been interested in the issue because oil and gas pipelines have become serious environmental issues. In the Amazon basin, pipelines are the subject of fierce protests because of all the problems they pose for people whose land they run through.[5] It's not just that forests are cut down to make way for them; pipelines also open up remote areas to all kinds of unwanted interlopers: illegal loggers, wildlife poachers, wildcat miners, and the like. In particularly isolated regions, pipeline builders may introduce diseases that can decimate indigenous peoples—that is the fear, for instance, concerning the proposed Camisea project in Peru.[6] Leaks can also lead to disasters, either by spilling oil into the environment or, if the pipeline is transporting gas, causing explosions. And many are built in areas subject to armed insurgency where they attract a heavy military presence and become obvious targets for saboteurs. That was certainly the situation with the Yadana pipeline, which would have to pass through densely forested and militarily contested regions along the border. As with dams, the pipeline project revealed close links between environmental and human rights concerns.

The first goal was to pin down exactly where the pipeline was going to be built. Because the operators would want as short a route as possible within Burma, it had to cross the border either in the provinces of Kanchanaburi or Ratchburi, where the Electricity Generating Authority of Thailand (EGAT) was building a power plant that would run on the gas. In 1993, the PTT suggested the pipeline would follow the route of the World-War-II-era "Death Railway" (made famous by the movie *Bridge Over the River Kwai*) and cross the border at Three Pagodas Pass. But in early 1994, Total wrote to a French refugee worker describing the actual route; based on that information I published an article revealing that the pipeline would run about 40 kilometers south through the border village of Ban I-Tong.[7]

It was then possible to examine what the authorities had done to clear the route. In Burma, that required cutting down pristine tropical forest. And human rights groups such as Earth Rights International (ERI) documented how—despite protestations of ignorance from Total, which was building the Burmese section of the pipeline—several Karen villages were forcibly relocated by SLORC troops before the signing of the contract.[8] Just as disturbing were reports of a new "Death Railway" being built on the Burmese coast. Tens of thousands of villagers were being conscripted to refurbish the old rail line between the towns of Ye and Tavoy, and human rights groups worried that it would be used to help build the gas project. Another EGAT press report declared that the line would "be used to transport troops to guard the pipeline." The conditions at the work camps were said to be so appalling that many of the workers had simply fled.

Because journalists were not allowed to travel to the region in Burma, it was difficult to corroborate what was going on there. But in early 1994, I teamed up with *Nation* photographer Prathai Piriyasurawong to visit refugee camps in Thailand around the border towns of Sangkhlaburi and Ban I-Tong. By tracking down the pipeline's route in the Huay Khayeng forest reserve, we confirmed that the project was going to have serious environmental impacts on the Thai side, too. And we documented that it was not only in Burma where forcible relocation was going on; refugee camps in Thailand were also being moved, and the 9th Division of the Royal Thai Army was ordering the frightened inhabitants back into Burma.[9] The push toward repatriation served two purposes: to remove potential security threats to the pipeline and to push the Mon and Karen insurgent groups into signing cease-fires. We collected and published the horrific stories of the Burmese villagers who had been conscripted to work on the Ye-Tavoy railroad,[10] forced to work not only without pay but often without food or water. Those who refused were hit or shot. We even photographed an unfortunate baby named Pu Twe whose head had been stuck in a fire by a Burmese soldier. And we learned about the systematic abuses that occur in Burma: Many of the refugees were simple farmers who had fled because the soldiers had taken half their rice crop. The "Burmese way to socialism," it seems, had led to Thailand.

THE BACK WAY TO THE BORDER

So five months later, when the Halockhani incident occurred, I went back to Sangkhlaburi, a funky little town draped around the northern edge of the

Khao Laem reservoir. All manner of buildings—from rickety huts to comfortable lodges—are perched on its steep hillsides. Half-built homes and the dead trees sticking out of the lake give the place a half-finished feel. The town's Karen and Mon populations, which probably outnumber the Thais, mostly live in crowded wooden shacks in the western part of town, where flooding is more likely to occur. A few even live on boats anchored on the reservoir. The town's impressively long and tall wooden bridge, its signature sight, separates the old part of town from the newer buildings on the Thai side. With the border and a brutal war not far away, there's a conspiratorial air to the place. Like Mae Hong Son and Mae Sot to the north, Sangkhlaburi contains a strange mix of country living and frontier intrigue.

The situation was a mess, literally. Much of the countryside, certainly the roads and trails, had turned into mud under the relentless downpours. As we traveled to the site, now called New Halockhani, even the souped-up four-wheel drive pickup that was the informal transit service from Sangkhlaburi often become stuck and we'd have to tramp into the camp by foot. But things got more serious when the Royal Thai Army's 9th Division announced that New Halockhani would be closed to all visitors on August 10.[11] The refugees would have to start returning to Burma, the military ordered, claiming there was now peace in the region, just eighteen days after the conflagration had started.[12] The Mon National Relief Committee feared it would no longer be able to send in rice, water, and other supplies. But the refugees still refused to return. They had moved to Halockhani from a safe Thai camp earlier in the year only because they'd been promised the SLORC would not harass them.[13] Now they feared repeated attacks if they went back to the Burmese side. So it was to be a contest of wills between the Thai army on the one hand, the stubborn refugees and their sympathizers on the other.

Stranded in a Sangkhlaburi guesthouse the day before the blockade was to take effect, I discussed with "George," a veteran Burmese journalist who used to work at the *Nation,* what was going to happen. Would the Thai army try to force the refugees back across the border the next day? There was only one way to find out, I suggested. We should go the back way to New Halockhani, avoid the army checkpoints by trekking to the camp through the forest, moving in the opposite direction of the Burmese refugees entering Thailand. George talked it over with his photographer, an American named Mick Elmore, and I discussed it with mine, a large and amiable young man named Kittinun Rodsupan, or Kod for short. We agreed that it seemed a better option than sitting in town and watching the rain fall and invited some of the other frustrated journalists in town along. If we ran into trouble, there might be safety in numbers. In the end, eight of us agreed to go, half Thai and half *farang;* we represented the *Nation,* the *Bangkok Post, Manager,* a German press agency, and various other publications.

I went off to find a guide, and finally made a deal with a doe-eyed young Mon named Ban. He wanted to go by night because he feared running into Thai patrols. But we vetoed that idea, and set it up for the next day. Instead of trying to obtain permission to enter the camp by wrestling our way through some tedious bureaucratic jungle, we'd try our luck with the real thing instead. So we went to bed that night, each of us praying to our own gods for good weather.

One of us must have sent the message to the right deity because the incessant rain finally stopped in the early morning hours. We were in high spirits as we hit the trailhead, surrounded as we were by lush green paddy overlaid by a heavy mist. Best of all, our guide said it would take only two hours to reach our destination if the rain held off (others had warned it would take all day). This would be a cakewalk, we hoped.[14]

We found out otherwise soon enough when Ban tried to cross a rain-swollen "stream" and got swept away by a deceptively fast current. It wasn't clear whether he could swim, but he managed to stay afloat until he caught an overhanging tree branch and dragged himself ashore on the other side. Now we were stymied. We could have swum across, but what about our camera gear? After much experimentation with fording the river and cutting down bamboo poles, we realized that we could just link hands and help each other through the hard parts. All the to-ing and fro-ing took about half an hour. From the subtle change in expression on our guide's face, he must have realized that Bangkokians may pay well, but he would have to earn every baht.

Nevertheless, we felt pretty good about our river conquest—until we found out five minutes later that the trail crossed the same river again. This time, we managed to hack our way through the jungle to a log that crossed the waterway, another time-consuming exercise. By now the difficulty of the task we had undertaken began to set in; our thoughts were punctuated by a soft drizzle that had started to fall. The trail then began to climb through spectacular scenery.

But not all of us could appreciate the stunning vistas, Kod in particular. He sat down frequently on the mud-covered trail—sometimes intentionally, more often not. After one of his initial slips, he confided (a bit late, I thought) that he had never been trekking before. I looked at him, and gulped. Photographers tend to be the toughest of hombres, but it would have been hard to come across a person more ill-suited for an arduous jungle hike. Kod was large even by *farang* standards, his considerable height complemented by an equally impressive girth. I'd played basketball with him at work, and in stature he was the virtual Shaquille O'Neal of the *Nation*'s court. He must have weighed over 100 kilograms, and his camera bag appeared almost as heavy. I looked at his shoes, and saw with relief that at least he was wearing sneakers. There was little I could do to comfort him, however, apart from trying to appease the increasingly insistent rumbles emerging from his belly with the only offerings I had: dry "Mama" noodles and dried mango. "Don't worry, trekking is just like walking. You'll be fine," I said weakly.

We pushed on. The trail headed down toward another stream-turned-torrent, which we again had to cross several times. But we had finally got the hang of it and begun working as a team to help each other across. By now, we were also grateful there were so many of us.

We would occasionally pass Burmese villagers, illegal immigrants crossing the border. Presumably, they were on their way to Sangkhla or surrounding villages to engage in a little trade. We could only stare in awe at the heavy loads the men carried, and at the sturdy women trundling along little children. They stared back in equal amazement at the sight of these multi-hued city-slickers trudging in the opposite direction. They must have wondered what on Earth we were doing. After all, who ever heard of people trying to sneak from the comfort of a town *into* a refugee camp? We just said hello, and occasionally asked how far it was to New Halockhani.

The illegal immigrants were an interesting part of the story. One of the Thai military's claims in forcing repatriation was that, with Burma allegedly more peaceful, the refugees were now really nothing more than your ordinary economic migrants.[15] When I asked Kamol Rangsiyanun, Sangkhlaburi's district chief, if it wasn't cruel to send the Mon back to Burma in the midst of a civil war, he claimed that "this is economic war" being waged against Thailand.

Most people probably saw through the claim. But I had been collecting the quotes and stories of the refugees so that I could write an article emphasizing that they were in fear, running away from political repression. If they had wanted to seek their economic fortune in Thailand, they could easily have left the camps (and many probably did), just as these migrants were doing. Of course, the distinction was often murky in real life—if a farmer leaves the country because SLORC takes half his rice crop, is he an economic migrant or a political refugee?—but it seemed an important one, at least to Thai officials.

Then came the most exhausting part of the journey: a hike up a long, winding, rock-strewn stream. We became quieter and our heads began to hang. Then, just after midday, about six hours into our trek, we heard a faint sound, familiar but out of place amid the jungle noises. A Royal Thai Army helicopter soon passed overhead and headed toward New Halockhani. I had to suppress a giggle as we ducked under some bushes to avoid being seen. It seemed unnecessarily dramatic, as if we were in some bad Vietnam War flick. But, anyway, we needed the rest.

The worst thing about the helicopter's passing was realizing that something was going on at the refugee camp and we didn't know what it was. The second worst thing was not being sure whether we *wanted* something to be happening. On the one hand, it would have made a great scoop for us; on the other, the situation may have turned violent, even deadly. The helicopter flew away a couple of hours later, piquing our imagination further.

We were getting close now and had gone completely silent at the request of our guide, who said we were close to a police checkpoint. We hoped we hadn't strayed into Burma, which would have been awkward, not that the border seemed to make much difference here. About half an hour from our destination, we ran into a Mon villager who offered us some fruit. When we sat down for a rest along a riverbank, we suddenly heard a shout from across the stream.

We looked over and our hearts sank into our mud-covered boots.

There, coming out of the jungle, were four fatigue-clad Thai soldiers carrying automatic weapons. They were obviously surprised to see us, and a couple were even smiling. But the two senior officers quickly began interrogating us one by one. I looked over at George, who stared back at me with a nod. He didn't have to say anything. As the only *farang* who spoke Thai fluently, as the one who had organized the trip, I had become the *de facto* leader of this expedition, and now I had to act like it. It was a seminal moment for me: I had to take responsibility for other people's lives.

Given the circumstances, there was only one thing I could do: pretend we were stupid tourists.

"What are you doing here?" the leader of the soldiers asked. He wasn't smiling.

"*Ma thiaw,*" I grinned. "We're on a tour."

"Where did you come from?"

"Bangkok." I was being a bit cheeky, but I wasn't sure what else to say.

"Who's your guide?" This was the truly dangerous question. If they had found the answer, they would have taken Ban away and done who knows what with him. All I could think of was to pretend not to understand him. The word for guide—*phu nam*—sounds like the word for water—*nam*—so I just pointed to the river several times and looked quizzical. He started asking the others, but everyone just pretended to be confused, or else said nothing at all. That's when I knew for sure that we had a good bunch of people.

Fortunately, one of the soldiers had seen the Mon villager we had met at the refugee camp that morning, and they didn't think about asking what Ban was doing there; they just assumed he was another refugee. I took out my wallet and showed them my name card, but was rather miffed to discover they apparently hadn't heard of the *Nation,* much less of me. "Where are you going?" asked a balaclava-clad soldier, who was turning out to be quite a hard case.

"Oh, we heard there was a village around here."

"You mean Halockhani? It's empty. The villagers have all fled to a Thai border checkpoint." And so we finally learned that the refugees had not yet been pushed back across the border.

They then started discussing whether to send us back. But at this point, they met their match in Kod. He was tired. He was hungry. He was half soaking wet and half covered in mud. There was no way he was going back the way we had come.

"Can't you carry us?" he asked plaintively, "Maybe send a helicopter in to pick us up?"

The soldiers looked at Kod's bulk in amazement and then at each other in distress. They started complaining about how little equipment they were given, pointing to holes poking through their big, black boots. We knew we had won.

"It's very dangerous around here," the hard-case soldier said in a last bid to send us away. "There are Mons, Karens, and Burmese shooting at each other, even Indians and repatriated illegal immigrants." He spoke of them as if they were criminals. If it's so dangerous, why are you demanding that the refugees return, I wanted to ask. Instead, we promised that we would look after ourselves and slowly, quietly, began walking away.

We regrouped a bit further down the trail and babbled away in excitement over our encounter. We quickly paid off Ban (adding a generous tip), who departed sporting a wicked grin. Perhaps he was just relieved about being rid of us, but I like to think that he had enjoyed the adventure, too, in hindsight at least. Finally, we reached Halockhani, a raft of blue plastic sheets covering ramshackle huts. To some, it may have looked like a cramped and muddy hellhole. But to the refugees, it was a home of sorts. And to us it looked like heaven. If the Mon were surprised to see eight mud-covered, waterlogged, camera-clicking city-slickers come traipsing into camp, they hid it behind delighted smiles.

There was no doubt about who among us was happiest to be there. Kod dashed, or rather shambled, straight to the nearest little food stall, plopped down in the mud and, to the astonishment of the nearby refugees, proceeded to gobble down a dozen packs of cookies. We all had a good laugh, but we were laughing with him, not at him. Kod had more to be proud of than any of us. And it was nice to be able to laugh again.

The BPP stationed at the camp greeted us with equanimity. On the whole, they seemed pleasantly surprised to have new company, although they warned us that next time we would have to arrive with proper permission. We wolfed down some *kwitiaow* (noodle soup) and the latest news. A fact-finding group from the Parliamentary Committee on Human Rights had been forbidden to enter the camp. The district officer and a soldier from the 9th Army Division had arrived in the helicopter and warned the refugees that they had one more week to return voluntarily or things would get "a little difficult."

Then we heard the really bad news: A *Reuters* reporter had also been on the helicopter. We cursed our luck. Even with what little news there had been, we'd been scooped. But on second thoughts, the trek to the camp had demonstrated something useful. It had shown just how porous the border was, how easy it would have been for the refugees to become genuine illegal immigrants, if that's what they had wanted to do.

Besides, we journalists could now walk with a bit of swagger. We had reached the camp the *hard* way. And so we had our own story to tell, which no one else could scoop.

"THIS INTERVIEW IS DEFINITELY OVER!"

The situation in the camp was serious. From ten to fifteen people were crowded into each hut. Malaria, diarrhea, and respiratory infections were widespread. The jury-rigged clinic staffed by Mon volunteer medics was thick with patients, most of them young children, some beginning to suffer from malnutrition. The Thai authorities had also blocked off the flow of supplies and apparently planned to keep squeezing the refugees until they left. Before leaving Sangkhlaburi, I'd asked Kamol, the district officer, how the Mon would get enough drinking water, which had previously been trucked in. He replied, à la Marie Antoinette, "They can drink rain water."[16]

The refugees had indeed set up tanks and rigged hoses so they'd have enough to drink—assuming the rains didn't stop. "We don't have enough food and water to stay for a long time if they block supplies," said Nai Mon Chusa, the camp secretary. "We have enough rice to last one or two months, and we can collect roots and bamboo shoots from the forest, but we have nearly run out of beans and sardines."

To make matters worse, Thai authorities had begun deporting Burmese illegal immigrants caught elsewhere in Thailand up to New Halockhani.[17] That not only confused the refugees' status but further depleted the supplies of Mon relief officials. I had made friends with the BPP here, and they seemed sympathetic to the refugees' plight. But the army was calling the shots, as they do in border regions all over Southeast Asia. It's a geographical certainty: The farther out on the periphery you are, the less control the central government has. The Thai military's ability to blockade the refugee camp—from journalists, from doctors, even from a parliamentary fact-finding commission—was an indication of its clout. Not a word was to be heard from Prime Minister Chuan Leekpai, just two years after the restoration of democratic rule.

In Burma, too, local military units far away from the capital have plenty of autonomy. It wouldn't be surprising if Rangoon had had no idea that its 62nd Light Infantry Battalion planned to attack a tiny village of Mon refugees. The extremely poor timing of the Halockhani incident, which well could have been just to settle an old score, may have been an indicator of just how out of touch the border generals were. But the lack of uncertainty annoyed me. A thought began to take root: Why not ask the local SLORC commander?

A couple of weeks passed, and it was clear the refugees were losing the battle to stay in Thailand. They had been kept largely incommunicado, and the media soon lost interest. No domestic pressure group rallied to their cause. Thailand's Mon community proved to be either weak or apathetic. I had gone with a *Nation* news team to the 9th Army Division's headquarters outside Kanchanaburi for a press briefing, but my questions seemed only to anger the commanding officer, Major General Chalong Chotigakarm. So I suggested to Nong, the *Nation* photographer who had accompanied me on this trip, that we head up to Three Pagodas Pass and see whether we could talk to the *tatmadaw* chief. I'd done something similar earlier in the year at Ban I-Tong when I interviewed the Burmese commander at the checkpoint across the border, and it had gone reasonably well. Nong looked skeptical, but on we went.

It was mid-afternoon by the time we arrived at Three Pagodas Pass. We crossed to the Burmese side and walked around the decrepit market with its cheap teak furniture for sale. I spotted a Thai officer who seemed to know his way about, and, after screwing up my courage, asked him where I could find the SLORC commander. He stared at me, and then asked around for a *khon suay* willing to take me there. I couldn't figure out why I'd need a "beautiful person" as a guide, but Nong explained that *suay* with a neutral tone didn't mean "beautiful," but rather "unlucky."

Not a good sign, but we piled into a car and went to the edge of town, up a steep driveway, and parked in front of a modest wooden house. There was a dark middle-aged man in a torn white T-shirt standing outside with a hose. We got out of the car, and I was introduced to him. His English was poor, but he wore military pants. This apparently was the SLORC commander.

He seemed surprised by our visit, but he smiled and invited us in. Our guide left, quickly. Inside, we were met by another soldier, this one in full uniform and much more imposing, tall and handsome but very stern looking. I never did get his name, but he was the subcommander. We sat down and tea and cookies were brought in. We introduced ourselves. I told him who I was, where I was from, and inquired whether we could ask some questions.

The subcommander conversed briefly with his superior in Burmese, and then picked up a large, green phone that looked strangely bare. It had no dial, or touch pad, just one button. But he started talking into the handle. Meanwhile, the commander introduced himself. "I am Tin Kyaw," he said. "I am a colonel."

"Are you the head of the military force here?" I asked.

"Yes, I am head."

"The 62nd battalion?"

"No, the 61st." These two battalions often seemed interchangeable, and I wondered whether they were part of one big unit.

"I guess you know about the Mon refugee situation at Halockhani?" I tried to start off diplomatically. "The Thais say it's okay for the refugees to go back, that you will leave them alone. Is that true?"

"Yes, if they are peaceful and don't make war, there will be no problem."

"So you will not interfere in their camp?"

"We may have to inspect them sometimes. That's our duty."

"See, I guess they're afraid of that."

"We get our orders from Rangoon," he said with some difficulty. "If the Mons have guns, we will have to arrest them. That's war."

That sounded a bit ominous. And the commander seemed disturbed, too. He started talking in Burmese to his fellow officer, who again picked up the phone to have a hushed conversation. I sipped my tea while trying to figure out who he was calling. Nong, who was sitting next to me—he wasn't allowed to take pictures—leaned over and whispered, "Jam, I think the interview is over."

I looked over at the colonel. He was fidgeting, trying to smile. He had thrown on a shirt but hadn't buttoned it. His hair was uncombed. I couldn't believe how unkempt he was. But I had come in unannounced, probably disturbed the guy after his nap or something. What had he been doing with the hose?

The subcommander just stared at me, malevolently. But I had come so far. I had to find out just one more thing.

"So why did the Burmese soldiers attack Halockhani? Was that on orders from Rangoon?"

"Yes, yes. I get orders from Rangoon."

"So it wasn't just because of an incident before? It wasn't because of an old . . . conflict?"

"Yes, we have problems before."

Maybe he didn't understand my questions. "I'm sorry, I'm a little confused," I said slowly. "The Thais say it was because you had a fight before with the Mons there . . . "

He cut in forcefully. "Whatever the Thai soldiers say, that's the truth." Now he was moving his chair about, speaking more emphatically with his subordinate.

Nong leaned over to me and hissed, "Jam, the interview is definitely over!"

He was definitely right. But all I could think of doing was to drink my tea again. The aide picked up the phone once more. But before he could talk, a dark, plump officer in a Thai uniform walked in and began speaking rapidly in Burmese to the soldiers, who got up to greet him. It was a member of the Thai border unit; *that's* who they'd been calling. He'd come to take us back to Thailand.

I tried to say goodbye and thank you, but we were quickly ushered out and driven back to the border. I attempted to make conversation with the Thai border guard. "You speak Burmese very well," I said. "Did you grow up around here?"

"Yes." But he was sporting one of those scary Thai smiles, the kind that says, "I'd actually like to kill you." He was livid. Then he burst out, "Do you realize what you've done?! You're not allowed to go and disturb them!"

"Oh, I didn't realize. I didn't think there would be any problem."

"There's a problem! We never would have let you cross if we knew you were going to do something crazy like that."

"I'm sorry. I was just doing my job. You know, trying to be a fair reporter and talk to both sides."

He glared at me. "You're lucky you didn't get thrown in jail."

"Come on! They wouldn't have done that."

"Yes, they would. They have."

With that, I shut up. We arrived at the border and thanked him again. I'd clearly ruined his afternoon.

Leaving Three Pagodas Pass, I felt a bit naïve for having gone in so casually, but also giddy that we had got out unscathed. And hey, in the process, we'd given Colonel Tin Kyaw a little taste of a free press.

A GLOBAL CAMPAIGN FOR THE '90S

The Mon refugees held on for a couple more weeks. Phra Wangsa Pala, the monk who had the difficult job of heading up relief operations for the Mon, came under increasing strain as he was forced into becoming the refugees' chief interlocutor with the Thai authorities. In a last desperate bid to stay, the refugees called on Thailand's Mon community to speak up on their behalf, and make appeals to the king. But there was no response, and by September 9, 1994, the last of the refugees had returned to the old Halockhani, on the Burmese side of the border.

In December 1994, SLORC launched Operation Natmin (Spirit King) with two objectives: destroying rebel forces in the pipeline area, and securing the pipeline against attack. Then, on February 2, 1995, Thai and Burmese authorities signed the gas sales agreement for Yadana, which would eventually provide SLORC with $400 million per year. Meanwhile, the PTT began negotiations to buy additional gas from the Yetagun field, operated by Texaco. And in June 1995, the Mon signed a cease-fire agreement (I was in Mae Hong Son looking into the teak smuggling story at the time).

Activist groups such as ERI and SAIN were gathering as much information as they could on the activity surrounding the pipeline. They had linked up with an impressive worldwide network of civic and political groups pushing for democracy in Burma, and together managed to put increasing pressure on corporations and institutions that had links to the ruling junta. It was a spirited campaign, reminiscent of the anti-apartheid movement during the 1980s, and it eventually forced many institutions to divest from firms that did business in Burma. Through shareholder activism, it also urged foreign companies to close their operations in Burma, with some success.[18] And ultimately, it pressured the Clinton administration into imposing a ban on all new U.S. investment in the pariah state.[19] All in all, it was an impressive example of what can happen when civil society is globalized.

Aung San Suu Kyi supported the investment ban. It gave her a bargaining chip with the SLORC. But unlike the fight against apartheid, where the frontline states had encouraged keeping South Africa economically isolated, Burma's neighbors opposed the ban. They maintained that political openness would come only through economic development and greater interaction with the outside world—the constructive engagement argument.

Burma pro-democracy campaigners in the United States took a further step in October 1996 when they filed a class action lawsuit in the Los Angeles federal district court accusing Total, Unocal, and MOGE of "egregious human rights violations."[20] The plaintiffs included a dozen anonymous Burmese victims who claimed they suffered direct harm—from forced labor, portering, assault, rape, and the death of family members—as a result of the project, explained Katherine Redford, ERI's then-director. Total and Unocal both denied using forced labor on the pipeline project, said they paid workers above average wages, and claimed the pipeline will be a boon to the local economy.

Indeed, the charges are not centered on the construction of the pipeline but on the measures used by the Burmese military to protect it. Ka Hsaw Wa, ERI's cofounder and current director, has compiled extensive documentation of the alleged abuses from interviews with the victims. Plaintiffs and witnesses reportedly had to be smuggled out of Burma. "Nobody is arguing that company officers went out and [committed these offences] themselves, but they can be held responsible if they were done in furtherance of the joint venture," said Redford. "We think the companies know what has been going on, but we just have to show they should have known what's going on because of SLORC's history of human rights abuses."

Piecing together what did actually happen was difficult, particularly for me, since I wasn't able to travel into Burma and investigate. But other correspondents, such as Ted Bardacke of the *Financial Times,* were invited to visit the project site by the companies and able to make unofficial forays into the area. He concluded, along with most observers, that forced labor was not used to build the pipeline. But the Burmese troops guarding the pipeline have been accused of using it. Distinguishing between the project and the support operations was a murky issue. Does the railroad support the pipeline or not, and was it intended to?

Everyone knows you can't do business in Burma without the military's cooperation. According to Bardacke, wages for the pipeline workers were at first being handed out, and allegedly siphoned off, by the local development committee, headed by a general. Total then sought to pay its workers directly. But the soldiers had apparently screened which laborers would be allowed to work on the pipeline so that they could still take kickbacks.

As for the environmental impacts in Burma, the biggest fear of Nai Shwe Kyin, leader of the New Mon State Party, was that an earthquake might cause the pipeline to explode. "Southern Burma is an earthquake zone," he warned. "When I was a child, there was a big earthquake, 8.5 on the Richter scale." The local community would then bear the brunt of the impacts but receive few benefits from the project, he added.[21] Total claimed that the pipeline, which was placed underground, is built to withstand earthquakes, and its route chosen to minimize forest destruction. Nevertheless, a swathe of jungle was cut down. For maintenance and security purposes, gravel roads were also built alongside the pipeline, but they reportedly aren't open to the public. That means they won't help the local economy, but it also means that villagers can't use them to log, hunt, and settle in the forest, although the soldiers may be able to.

At first, environmentalists in Thailand showed surprisingly little reaction to the pipeline. An exception came in early 1995, when the environmental group Terra expressed concern that the project would have a major impact on Thailand's western forest complex—a series of parks, sanctuaries, and reserves that combined to form the largest conservation area in mainland Southeast Asia.[22] The NGO called on the builders to reveal its social and environmental impacts.[23] But for the most part, Thai groups were quiet.

In early 1996, I obtained a copy of the project's environmental impact assessment (EIA), which had been prepared on the PTT's behalf by a Thai firm called Team Consulting Engineers. The details of the pipeline's 260-kilometer-long route in Thailand were finally revealed: from Ban I-Tong it would follow the old mining road to Jet Mit, then plunge into a pristine section of the Huay Khayeng forest reserve, slated to become Thong Pha Phum National Park. The area is classified as a 1A watershed forest—the highest rating given.[24] Somphongse Tantisuvanichkul, the PTT's project director, said the 6-kilometer stretch was the shortest route possible through the forest. The pipeline would be built underground, he added, and without roads to service it. But the Office of Environmental Policy and Planning (OEPP), which vets large development projects, rejected the EIA, saying it did not sufficiently address the issues of landslides, erosion, and the possible effects on Sai Yok waterfall, a popular tourist attraction.

Despite these revelations, it wasn't until the end of 1996 that the issue began to flare up in Thailand. That's when local groups in Kanchanaburi—the same ones who had defeated the Nam Choan Dam project in the 1980s—began speaking out against the pipeline because they had not been properly informed about the project. A new set of environmental problems then emerged. As with the Mon, concern was expressed over what might happen should there be an earthquake. And besides passing through the 1A watershed forest, the pipeline would cut across a 5-kilometer stretch of Sai Yok National Park. "There is another direct route for the pipeline to reach Ratchaburi without passing through the forest," said Boonsong Jansongrassamee of the Kanchanaburi Conservation Group. "The PTT must make clear why this route does not appear on the project's plan."[25] The alternative route would have followed the border from Ban I-Tong until it reached Suan Phung district in Ratchaburi, then headed east towards the power plant. But that route had reportedly been rejected for security purposes. So in a roundabout way, the conflict in Burma led to greater damage of Thai forests.

More dramatic was the discovery that the pipeline could threaten two rare and endemic species that live near the chosen route. One is Kitti's hog-nosed bat, whose only known habitat is in Sai Yok National Park. The size of a bumblebee, the bat is listed in the *Guinness Book of World Records* as the world's smallest mammal, according to Suraphol Duangkhae of Wildlife Fund Thailand (WFT), and the World Conservation Union has declared it one of the world's twelve most endangered species. Concern was also raised over the regal crab, a forest creature found only in Thong Pha Phum district. Its vivid markings—a blue back fringed with white strips and striking scarlet legs (the colors of the Thai flag)—have made the crabs a popular tourist attraction every rainy season when they come out of their holes and turn the forest floor a brilliant red.

Most of all, these discoveries highlighted just how weak the project's EIA was. It listed hundreds of species that lived in the area, but generally failed to explain what the project's impacts on them would be. "[The] environmental impact assessment carried out by Team Consultants does not even mention Kitti's hog-nosed bat," said Sanguan Thanachaisaitrikul, a forestry official. The RFD was subsequently asked to survey the area and check how the project would affect the two species. It found that although the pipeline does not pass directly through the bat's habitat, some of the bats were present in the general vicinity. As for the crab, in December 1996, the PTT agreed to alter the pipeline's route slightly to protect its habitat.[26]

That was not enough, however, to prevent the OEPP from citing inadequate information about key wildlife species when it rejected the shoddy EIA for a second time.[27] The decision evidently surprised the PTT, which began throwing its weight around. Songkiat Tansamrit, the oil company's public relations director, said construction of the pipeline had to begin in March 1997 if it was to be completed on schedule. Otherwise, he said, the PTT would have to pay hefty financial penalties. So it was not surprising that when the National Environment Board chaired by Prime Minister Chavalit Yongchaiyudh met on March 24, it approved the project despite the EIA's obvious deficiencies. Deputy Prime Minister Samak Sundaravej than accused environmental groups and local activists in Kanchanaburi of colluding with Burmese dissidents to oppose the project.[28]

Construction of the Thai pipeline went smoothly until January 1998, when protestors began a last-ditch bid to halt the project by staging a sit-in along the pipeline route in Kanchanaburi. The activists had received a copy of the gas sales agreement, which had never been made public, and discovered that the penalties for delays would be waived if caused by "acts of government," or *force majeure.*[29] A three-day work suspension to try to iron out differences did little to calm things down. Tension rose as the demonstrators moved to block workers from cutting down trees in the Huay Khayeng forest. Groups of villagers supporting the project began mobilizing—the PTT was accused of instigating them, but denied it—and it looked as if there might be a repeat of the violence that engulfed the Pak Moon Dam conflict.[30]

THE GREENS ARE SNAKE-BIT

In February, Ted Bardacke and I went to Kanchanaburi to visit the construction site. We talked a bit with the protestors at their camp along a newly expanded road, and then went to meet Phinan Chotirosseranee, the Kanchanaburi housewife who was leading the rag-tag group of protestors. She was returning from a two-day foray deep into the jungle, having successfully defended a stand of thick forest from the workers' chainsaws, at least temporarily.

It should have been a moment of triumph. Her exploits had made Sunday's newspapers, and her colleagues at the protesters' camp were abuzz about the photos splashed across front pages showing students wrapping their bodies around stately old trees. They eagerly waited to greet Phinan as the fifty-one-year-old environmentalist struggled up the final hill. But upon reaching the top, Phinan immediately collapsed into the arms of her friend and burst into tears. Perhaps the arduous trek had been too much for her, or perhaps she had simply become overwhelmed at the enormity of the task she had set for herself, but over the next few moments all the despair borne of a battle fought against tremendous odds poured out in the sobs that wracked her body.

"May mii kray ma chuay raw jring," she cried. "Nobody has come to help us!" Finally, her emotions spent and her energy so drained, she had to be helped into a car and taken back to camp. There, she silently brooded over events. Her colleagues left her alone as they quietly went about their affairs. Once Phinan had revived a bit from her journey, she attempted to explain why her spirit had seemed so broken earlier in the day. "I came up that hill, and I saw my friend and just broke down. We have fought this project for over a year, and we're always retreating, retreating. Now we only have this small stretch of forest left . . . ," she said, and her voice trailed off.

If the protesters' cause looked doomed, so did the forest, despite the claims of the PTT that it would grow back as good as new. A trip to the construction site revealed just how much devastation building a pipeline can cause. The pipe was only a few feet in diameter, but each segment weighs over 5,000 kilos, and the equipment needed to put it in place is massive. As a result, the firm installing the pipeline, Tasco Mannesman cut a swathe from 20- to 25 meters wide through dense forest—a bright orange strip of earth running through brilliant green flora. "You can see it's good forest here. It's like this all along the pipeline route [for 13 kilometers]," said Neung, a Ramkhamhaeng student who had surveyed the area extensively and, despite being a protester, guided us around the construction site as if he worked there. He pointed to a stream: "This is a 1A watershed, but it will soon be gone. It's a real shame."

We watched as a bulldozer dragged off the heavier logs. A huge mechanical claw gripped a pile of foliage and dumped it onto the back of a truck. Dust was everywhere. It lay centimeters thick underfoot, and huge clouds billowed up with each passing vehicle. "Soon this place will be a desert," Neung sighed. The biggest threat to the forest was not the route being cleared for the pipeline but the poachers and encroachers who would inevitably use it once the contractors were gone. "It has already started," said Phipob Thongchai, another leader of the protesters. "The 9-kilometer road leading to this camp used to be a footpath. Now that it has been widened, one of the encroachers with land along it has brought in some earth-moving equipment. He's a police general, so who's going to stop him? It's really disgusting."

Phinan testified how quickly the encroachers are to spot an opening. "We first came here on December 7. By December 14, the road had been built. Then we set up camp. By January 6, people had already come here to mark out their land. They were just standing around, picking out parcels. That's how bad it is."

The protesters then urged that the pipeline be constructed along the paved road that winds from Thong Pha Phum up to Ban I-Tong, but the PTT said the road was too curvy and that the pipeline could suffer from landslides. Phipob and Phinan suggested that it would be better to expand the road than cut a new route through the watershed forest; but the government and the PTT thought it was too late for any route changes, and the protesters' cause was hampered by entering the debate so late in the game. Although the pipeline route had been well known and documented in newspapers for years, Phinan said she and her colleagues had learnt it was passing through Kanchanaburi only at the beginning of 1997. "We were a bit selfish. We had thought it was going to pass Ratchaburi—our MPs said it wasn't coming through Kanchanaburi—and so we didn't pay much attention to the issue," explained Phinan.

This highlights a deficiency in many local environmental groups. They can be parochial in their views, to their own detriment. Although the pipeline's impact on Burma has made it a source of controversy the world over, the Thai protesters barely mentioned the subject. They apparently believed that raising the human rights issue would be counterproductive. "Most Thais don't see the importance of the Burma issue," explained Phinan. "Korn [Dabaransi, the former industry minister] once said to us, 'It's a foreign issue, so let's not talk about it.' In fact, a lot of people seem to think it's a good thing to take resources from our neighboring country, but we know what is happening to the minority groups there."

Maybe so, but opposition to the pipeline within Thailand still seemed fractured and disjointed—a very different situation from the battle over Nam Choan Dam, where the environmentalists won their struggle thanks to their ability to forge a broad alliance among civil society groups. Diversity is useful not just for nature but also for governmental institutions and environmental movements, too. The broad coalition opposing the dam project applied more effective pressure than the narrow group of greens that fought the pipeline. Added Phinan, "We had a lot more time with [Nam Choan]; we fought for nine years before we won."

Phinan's analysis points to another problem with many local groups: They are often reflexive, reacting to events instead of being proactive. That's why it's so important for the state to ensure that proper EIAs are carried out. Although some Bangkok-based NGOs and think tanks—such as Terra, Wildlife Fund Thailand, the Seub Nakhasathien Foundation, and the Thailand Environment Institute—are more farsighted and try to influence policy, there are fewer of them than in most developed countries, and they have less clout. Had the Thai pipeline opponents taken up the fight earlier and linked themselves more closely to the international Burma democracy groups, you can't help but wonder whether that would have helped their cause. But green politics is often seen through a nationalist prism; both sides in a development dispute feel the need to drape themselves in the flag.

The pipeline protest was finally brought to a peaceful end by a government promise to set up a national review committee,[31] this one to be chaired by Anand Panyarachun. As prime minister, Anand had approved steps that moved the gas deal along in its early stages. During the public hearing, which had some hostile give-and-take between opposite sides sitting in the same room, he defended the decision to purchase gas from Burma as a sound policy decision. But a lot of dirty laundry about the project came out. It was announced that EGAT's Ratchburi power plant would not be completed on time,[32] meaning there would be delays in receiving the gas and Thailand would be forced to pay penalties. But this time the PTT did not seem to mind and said the "fines" would go towards future gas purchases. "The only money we will lose will be the interest," declared the PTT's Jira Chomhimvet. The implication was clear: A delay caused by construction woes was considered more acceptable than delays aimed at preserving forests.

This was just one of the PTT's deceptions that were unveiled.[33] The oil company had claimed the pipeline's route was the best for both countries, but it turned out that Burma had insisted on selling the gas at Ban I-Tong, forcing Thailand to build the project through a 1A watershed. The PTT also said it would limit the width of the pipeline track to 12 meters in forested areas, but video taken by the activists showed trees being knocked down even beyond the normal 20-meter limit. The PTT maintained that once the pipeline was laid underground, the forest would be able to grow over it, but it was revealed that trees with deep roots could not be allowed above the pipeline. So farmers will have to take care and forests will have to be pruned. The PTT, in other words, said whatever was necessary to see the project completed.

Given the power plant delays and the reduced energy demand following the onset of the economic crisis, the Yadana committee could conceivably have asked the PTT to shift the route of the pipeline; but the committee declined to recommend the project be delayed or the route altered.[34] Anand did criticize the project for a lack of transparency, and the committee's report chided the EIA process for not including public participation, adding that "there was only public relations." It called for a much-needed review of the process by which EIAs are carried out and approved. The limitations of the system became obvious during the hearing. Because consultants who perform the assessments are usually paid by the company eager to build the project, they have an incentive to underestimate impacts in their reports.

After misleading the public so frequently over the Yadana project, the PTT could find it a struggle to build other major pipelines, such as one planned between Thailand and Malaysia. In June 2002, the $500 million Trans Thai Malaysia project, which would send gas from a Joint Development Area offshore through southern Thailand into Malaysia, had been stymied by protests from villagers in a Muslim majority region who fear they will suffer negative impacts from the project.[35] In a way that's a pity, because gas does have certain advantages—particularly environmental ones—over other fuels, and it will be an important energy source for Thailand and the region in the coming decades. Some of the impacts and controversy over Yadana in Thailand could have been avoided with a different route and a more forthright PTT. But like its cousin EGAT, the PTT has no one but itself to blame.

In a regional sense, the project's greatest impact could be to tie together Burma and Thailand, two ancient antagonists, for decades to come. In the United States, meanwhile, the lawsuit against Unocal could end up setting an even more important precedent. The case against MOGE and Total was thrown out because they were considered foreign entities, but the courts did declare jurisdiction in the case against California-based Unocal and its top executives. In August 2000, although a federal judge found that Unocal clearly knew of and benefited from atrocities committed by the Burmese military in connection with the pipeline's construction, it dismissed the case, arguing that the oil company couldn't be held liable for those acts.[36] A California state court, however, accepted the case, which is now moving ahead.[37]

Another intriguing question is how the Yadana project will affect Burma now that it is up and running (gas from the Yetagun field is exported along the same overland route). The fears of democracy advocates that it will single-handedly support the treasury of the State Peace and Development Council (SLORC's new name) have yet to materialize. Under the gas sales contract, Burma receives relatively few profits in the project's early years because it has to pay off MOGE's 15 percent stake in the production consortium.[38] There were also reports that the military borrowed a good deal of cash to pay for imports when the country ran out of diesel fuel in recent years.

Nor has the project exactly jump-started development in Burma. Unocal was due to build a so-called "3-in-1" project, in which another pipeline would be built to send gas from Yadana to Rangoon. There it would be processed into cooking gas, fertilizer, and electricity. But the project has never taken off, in part because the returns would come in the highly overvalued local currency. Unocal claims the pipeline to Thailand "has brought significant benefits in health care, education, and economic opportunity to more than 40,000 people living in the pipeline area." With access to the region so limited, it's hard to confirm or refute that claim, but it certainly must be balanced by the effect of an increased military presence in the area. Whereas the economic benefits of building a pipeline seem mostly short-term, the soldiers are now there for the long haul.

That partly explains why there are reasons to question the argument for constructive engagement. The issue of whether trade can be used to help liberalize oppressive regimes has cropped up in regard to many countries—including China, Cuba, and Iraq—besides Burma. Never mind that the policy's main advocates tend to have vested interests in such dealings, it's a legitimate debate. Along with the other "Asian tigers," Thailand is a good example of how trade and economic development can lead to democratization. But political progress also seems dependent on the creation and growth of a middle class, which means that the benefits of development must be reasonably dispersed. "Constructive engagement won't help create a Burmese middle class unless the economy is opened all the way," says Nai Shwe Kyin, the Mon insurgent leader. "But the SLORC does not dare make the kyat convertible because it has no legitimacy. The ensuing turmoil would cause the people to rise up."

The weakness in the constructive engagement argument is that when foreign investment is aimed at exploiting resources, the profits are often highly concentrated. True, some jobs are created for miners, loggers, oil-platform workers, and the like. But if the resource isn't used sustainably, and value isn't added on domestically, those jobs can be short-lived. Meanwhile, the officials, generals, and dictators who control how resources are sold off receive the windfall benefits in the form of "rentier revenues." If you believe in trickle-down economics, perhaps that can still help a country to grow; but all too often, the money is spent on buying foreign chalets and imported weapons. And you end up with basket cases like Nigeria and Burma.

Investment in manufacturing and services may stand a better chance of helping a country to develop, both economically and politically. Despite the potential for abuse with these industries, they seem to provide more dispersed benefits. Ironically, these are precisely the types of investments that have been stunted in Burma by foreign activism and U.S. sanctions, which have indeed hurt the country. In practice, it may be impossible to make a clear distinction between these forms of investment since the producers of goods and services in developing countries often rely on cheap resources to give them a competitive advantage. But if a poor country is going to exploit its resources, it should at least try to add some value to them rather than just export them wholesale.

But neither sanctions nor engagement seem to have pushed Burma towards democracy, and the debate between the varying approaches may be academic. Unlike China, Cuba, and Iraq, Burma held democratic elections in 1990 that were won by the opposition, led by Aung San Suu Kyi. She has called on foreigners to refrain from investing in Burma until the ruling junta agrees to political pluralism. So the question for outsiders becomes quite simple: Should we do as she asks, or should we cooperate with the generals?

8

TOXICS
MERCURY RISING

THE RIDDLE OF THE SEDIMENTS

My biggest scoop ever started with an anonymous fax. Pages and pages long and addressed to me, it was lying on my desk one afternoon in late 1995. It included excerpts of an environmental study being carried out in preparation for the expansion of Mab Ta Phud, the center for Thailand's petrochemical industry and arguably the country's most important industrial estate, located in Rayong province on the booming Eastern Seaboard.

The researchers had made some disquieting findings. In testing the sediment on the sea floor around the estate's industrial port, they had come across extremely high levels of cyanide in the outer approach channel for ships. Alarmed, they carried out a new round of tests. This time, the quantity of cyanide measured was negligible, but a significant amount of mercury was found in the sediment off the port's western wall. The rest of the fax listed the dozens of firms that operate in Mab Ta Phud. It was signed simply, "friends of the environment." There was no indication at the tops of the pages where the fax had originated, and to this day I still don't know who sent it. Its mysterious nature was intriguing, as was the gravity of its contents.

The resulting investigation into an arcane world of petrochemicals and toxic compounds led me on several wild goose chases and into many dead

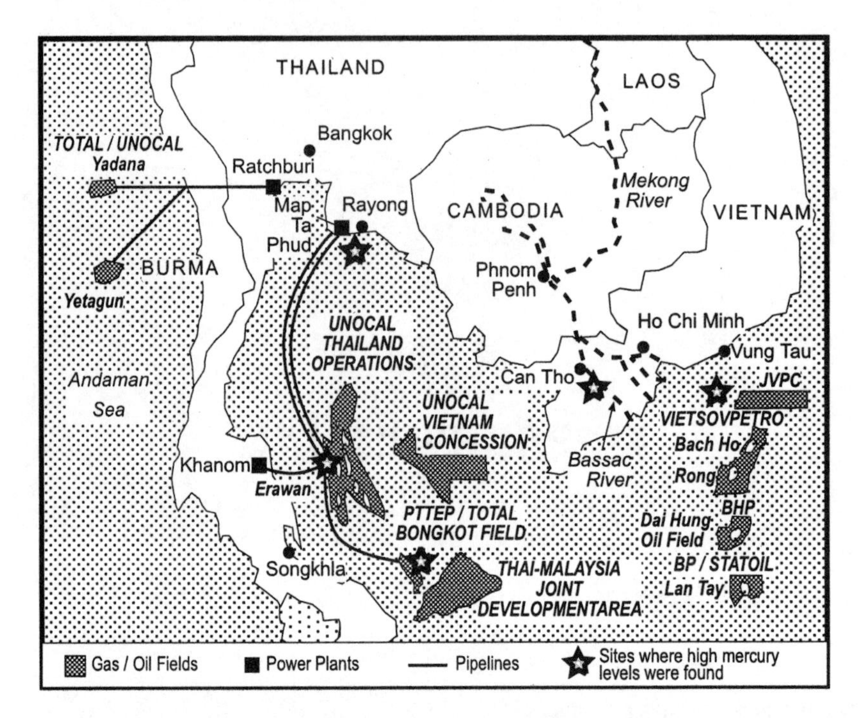

Mercury Contamination Sites

ends; I also made an embarrassing blunder or two. But my investigation also exposed a potentially serious problem that, if not handled carefully, could pose a threat throughout Southeast Asia. Mercury is a neurotoxin that builds up in the food chain. When ingested, it can lead to reproductive problems, including birth defects, and chronic damage to the central nervous system. High concentrations can cause a loss of hearing and vision, and ultimately death. And mercury was involved in one of Asia's most infamous industrial disasters at Minamata in Japan, in which the Chisso Corporation—a manufacturer of acetaldehyde, used to make plastics—illegally dumped mercury into the bay for more than a decade. Chisso concealed its actions and continued its dumping even after 1956, when people in the area started coming down with "Minamata disease" from eating contaminated fish. The Japanese government failed to acknowledge the source of the problem until 1968. Ultimately, more than 2,000 patients (less than half of whom remain alive today) exhibited symptoms severe enough to be considered official victims, but another 10,000 sufferers have mild symptoms.[1] The disease has no cure.

The situation at Mab Ta Phud was nowhere near as serious as the Minamata disease tragedy, but it was still worth checking out. First, I had to verify the results with the researchers. Three consulting firms were performing the environmental impact assessment: Siamtec International, Asian Environmental Protection, and L. G. Mouchel. So I went to one of their offices and talked to a consultant who confirmed that his team had found mercury and cyanide in the sediments. Researchers had taken samples from the seabed and sent them to various labs. Each lab produced a different result. But beyond that, he wouldn't say much, especially when I declined to reveal how I'd found out about it. Although clearly agitated about the results of the environmental impact assessment (EIA), he refused to hand over the hard data. "We need to do further testing," he explained. I needed to find exact results because they were crucial backup for my story, but in the meantime, I had to look into what might have caused the contamination.

That proved a frustrating exercise. Thailand is a relatively open society, but every country has limits to the public disclosure of information. And industries everywhere are loath to reveal what types of toxins they use and emit. In the West, there are laws that support the public's right to access information—such as the Toxic Release Inventory mandates in the United States—that require companies to disclose what chemicals they are releasing into the environment. But nothing like that existed in Thailand, at least not at the time my investigation began.[2] So during the next couple of months, I contacted anyone who might be able to shed light on the situation: customs officials, industrial estate authorities, environmental officials, and public health experts. I learned that mercury and cyanide compounds are used in a variety of industrial processes, including those used at downstream petrochemical plants at Mab Ta Phud.

In theory, that narrowed down the suspects, but in practice it was of little help since I had no way of checking the factories' wastewater emissions. This type of investigation requires technical capabilities that, at least in the developing world, are not widespread—one of the more frustrating aspects of covering industrial waste issues. Perhaps that explains why so few environmental groups in Southeast Asia focus specifically on toxic waste. Information regarding toxic substances is concentrated in the hands of governments and corporations, and they tend to be secretive. The lack of assistance from civic groups made investigation that much more difficult.

There was one promising line of inquiry, however, that involved following up on a public dispute that had broken out in 1989. That was the year the Petroleum Authority of Thailand (PTT) opened its gas separation plant at Mab Ta Phud to process gas piped in from Unocal's Erawan field out in the Gulf of Thailand. Unbeknownst to the PTT, the gas was contaminated by mercury, which quickly damaged the equipment. The new facility was shut down for more than a month while repairs were made and mercury removal units were installed. The state-owned oil company threatened to sue Unocal over the incident. But Unocal claimed it had fulfilled its responsibility by informing its overseer, Thailand's Department of Mineral Resources (DMR), about the presence of mercury in the gas. Apparently, the DMR had neglected to pass on the information. Legal action was never taken. Unocal later expanded operations to several nearby offshore fields, and Total began producing gas from the Bongkot field further south in the Gulf. That led to some interesting questions: Could mercury be leaking from the PTT and Unocal facilities? And what do they do with their contaminated wastes?

Unfortunately, it was impossible for me to "stake out" the gas separation plant. And going through the "proper channels" would simply have left me with a lot of public relations spin. Instead, I sneaked into PTT headquarters on Phaholyothin Road and talked directly to the people in the health department. It turned out that numerous tests had been carried out at the plant following the 1989 incident to determine the level of contamination in the soil, the plants, and the fish. The unit of measurement for these tests is usually parts per million (ppm). So if you want to see how contaminated a fish is, take a gram of tissue and find out how many micrograms of mercury it contains. Different test subjects, however, have different standards for what is considered "safe." Fish, for instance, are considered safe to eat by the World Health Organization (WHO) if they contain less than 0.5 ppm of mercury. According to the PTT's health department, fish in the plant's compound were not found to have mercury levels higher than the standard, although some had levels as high as 0.37 ppm. Some of the watercress found on the grounds had mercury levels as high as 6.17 ppm, and soil samples revealed levels that reached 29.45 ppm, almost thirty times higher than the recommended standard.

Those tests had taken place in 1990. So the question was, what had they done with the mercury-contaminated soil? Buried it, said the PTT health staffers. They'd dug up the mercury-contaminated topsoil and covered it with supposedly cleaner soil from underneath. That didn't sound like a very practical solution. Maybe some of that mercury had leached into the ground water and run into the sea? But they assured me that all the water from the factory is collected in equilibrium ponds and does not escape the compound untreated. A molecular sieve was placed 5 meters underground to cleanse ground water runoff. As for Unocal, they said they kept their contaminated sludge in Songkhla, the southern province, which is a staging area for Thailand's offshore gas producers.

To be fair, the options for disposing toxic waste in Thailand were severely limited at that time. Incredibly, despite having embarked on a massive industrialization program, the authorities had never[3] managed to build a full-scale industrial waste treatment plant—an oversight that typified the unplanned nature of Thailand's development. Many developed countries, including the United States and Britain, also failed to build adequate treatment plants as they industrialized. But Thailand had their examples to learn from, and failed to do so, illustrating once again why so many countries follow the environmental Kuznets curve.

Finally, in 1994, a joint-venture hazardous waste treatment company was established, the Industry Ministry owning a 25 percent stake. Genco, as the firm came to be known, was supposed to process 500 tonnes of toxic waste per day by 1996, and then twice as much in a few more years. That would cover just a fraction of the 1.6 million tonnes of hazardous waste the Pollution Control Department estimated Thailand would produce by 1996, but it was a start. Genco, however, immediately ran into the same problems that had scuttled previous industrial waste treatment projects. To keep costs low, it wanted to avoid building its project in an industrial estate, but building it elsewhere inevitably involved running into "Not in My Backyard" opposition. Such "Nimby" problems confound proposals for waste treatment facilities around the country—and, indeed, around the world.

Genco decided to build its plant and landfill in a rural section of Rayong province called Pluak Daeng district. Villagers protested vociferously and argued that the plant would contaminate the local watershed, the source of fresh water for one of the Eastern Seaboard's most important reservoirs. Logistically and environmentally, they argued, it made much more sense to locate the toxic waste treatment facility in Mab Ta Phud, where much of the waste originated. Genco hated that idea because the industrial estate's success in attracting firms had made land there scarce and expensive. The dispute became so heated that one of the protest leaders, Thong-in Kaewwattha, was gunned down mafia-style on his own doorstep. Ultimately, the industrial authorities gave in and agreed to build the waste treatment plant at Mab Ta Phud, although another landfill site will sooner or later have to be found.

In a sense, the effort to build toxic waste treatment plants has suffered from the same dilemma that hampered the construction of mass transit rail projects in Bangkok. Because these infrastructure projects are privatized, the priority is to make a profit. But the effort to keep costs down can lead to greater environmental impacts and controversy—whether they be over-crowded streets in Bangkok or contaminated watersheds in Rayong. On the other hand, keeping the costs down helps make sure the projects are used. If it leads to lower fares on the Skytrain, more people will ride it. And if it yields lower costs for waste treatment, more companies will agree to dispose of their toxic refuse properly.

That is an important point, because apart from simply storing toxic waste, the main means of disposal in Thailand is apparently "midnight dumping." Nobody knows how much hazardous material has been illicitly dumped. But considering how much is produced and the lack of options for proper disposal, the implications are scary.

FIRST THE SCIENTISTS, THEN THE SOOTHSAYER

Dumping may have been behind the contaminated sediments at Mab Ta Phud. That would explain why high concentrations of certain compounds were found during one set of tests but not during others. But the cyanide and mercury could also have originated from wastewater running out of the industrial estate. They could even be a natural phenomenon.

Once the EIA was completed and sent off to the evaluation committee in early January, I called Kasemsri Homchean, the environmental chief for the Industrial Estate Authority of Thailand (IEAT) and a member of the EIA evaluation committee. She was skeptical of the findings. "Why did the data change so much? Something must be wrong," she said. "They need to clarify why the results are so different [from one set of tests to another]."[4] But the consultant who carried out the tests stood by their validity. "The testing methods were checked. There was no error," he said. "The sediment samples were taken at different times of the year, under different tidal conditions. Both sets of results could be right." I asked him again for a copy of the hard data, but he hemmed and hawed and still refused to hand them over.

Kasemsri noted that the mercury levels found outside Mab Ta Phud's drainage canal were not as high as those found near the port's western wall, suggesting that the waste water from the factories in the estate was not the main source of the mercury. "We are worried about the cyanide findings, but not the mercury results. They are nothing new," she added. "Researchers at Chulalongkorn University carry out a continuous water quality monitoring program around Mab Ta Phud and they don't find mercury levels over the standard. The levels are the same as those found naturally."

But I soon came across other evidence that mercury contamination was indeed becoming a growing problem at the industrial port. And contrary to Kasemsri's statement, the proof was in the water. What's more, it had been gathered by another government agency, the Pollution Control Department (PCD). Unfortunately, the PCD official I talked to said he couldn't release the information publicly until I had met with his boss, the deputy director general, Sirithan Pairojboriboon, who was out of the country for a few weeks. Still, the presence of mercury in the sediment and increasingly in the water around the port was highly suggestive. I had yet another vector to check, however: the fish. This was a crucial issue because the biggest threat of illness from mercury poisoning would come through eating an excessive amount of contaminated seafood.

The difficulty with mercury and other persistent organic pollutants (POPs) is that they build up in the food chain. Once they enter the aquatic environment, they are digested by tiny organisms or siphoned out of the water by filter feeders such as shellfish. Many of these organisms also convert the various mercury compounds that are discharged into methyl mercury, a form dangerous to humans. (Mercury comes in many forms—organic, inorganic, elemental, and complexed—some more dangerous than others, but for simplicity's sake I have decided not to distinguish between them in this book.) These creatures are eaten by larger, carnivorous species, which then accumulate ever larger concentrations of the heavy metal. That means the largest concentrations are often found in fish at the top of the food chain—sharks, swordfish, tuna, and barracuda in marine environments; pike and bass in fresh water ecosystems—and if we are not careful, humans.[5] This bio-accumulative nature of persistent organic pollutants has spurred the UN to help negotiate an international POPs treaty that limits their use in pesticides and other manufactured chemicals. And not before time: the U.S. Food and Drug Administration now warns pregnant women not to eat shark, swordfish, king mackerel, and tilefish because of the potential mercury threat to the unborn fetus.[6]

Asians may be especially susceptible to contamination because of the relatively high proportion of fish in their normal diet. According to a WHO study, people who eat fish from two to four times per week have on average four times the concentration of mercury in their bloodstream compared to people who don't eat fish at all. Researchers led by Mike Dickman from the

University of Hong Kong, who carried out a study suggesting there is a link between the consumption of mercury-tainted seafood and male fertility problems, argued that the WHO health standards should be re-examined in Asian countries. "In our opinion, the WHO guideline for maximum mercury concentrations allowed in foodstuffs [0.5 ppm] is too high for parts of the world like Hong Kong where seafood is a dietary source of protein."[7]

I needed to find out who was checking the fish for mercury. The Fisheries Department seemed the most likely option since they inspect seafood exports and conduct testing at their Eastern Fishery Development Center in Rayong. They refused to divulge any information, however, for fear it would hurt business. Next, I talked to the Food Analysis Division at the Ministry of Public Health, which is supposed to inspect food for domestic consumers. Prakai, an official at the agency, said that some fish they had tested had mercury levels "almost" as high as the 0.5 ppm standard. But the last such report had been filed five years earlier, after which the project was no longer funded—further evidence that the Thai government is more concerned about the safety of its exports than the health of its own people.

Finally, I came across someone who had information and was willing to speak on the record. Piamsak Menasveta, a respected marine scientist at Chulalongkorn University, had carried out mercury studies on fish in the Gulf of Thailand, and he expressed concern about his results. Fish caught around Unocal's gas platforms were showing elevated mercury levels. The normal level for fish in the Gulf of Thailand, he said, is about 0.1 ppm. Studies conducted near Unocal's Erawan field in 1990 found fish with levels of 0.3 ppm. And two studies of twenty-five species carried out since then revealed that some had sustained mercury levels higher than the standard of 0.5 ppm. "A study in 1994 found greater mercury levels than previously," he said. "The problem exists and is increasing."[8] The gas platforms are hundreds of kilometers south of Mab Ta Phud, so it was not clear whether the two situations were related. But Piamsak had also conducted studies on the Eastern Seaboard at Bang Sarae, about 40 kms west of Mab Ta Phud, and found that some fish there also had mercury levels that were higher than the standard.

Unfortunately, Piamsak was legally barred from revealing the latest data because he had been hired by Unocal to carry out the studies. So I called Unocal and spoke to Konthi Kulachol, the company's public relations manager. The company was aware of the problem, he said, and had been monitoring it since 1990. It had filed a comprehensive report to the DMR. "Consumption of fish from the Gulf, as represented by the fish in our study, is safe for human health," he said. But the company would not release the data to the public.

Piamsak said the industry minister had the right to reveal the information to the public under the Petroleum Act. "I would like the Industry Ministry to acknowledge the data and set up some countermeasures before the situation becomes truly dangerous," he said. But ministry officials refused to comment. After many phone calls, I finally managed to speak to the minister himself, Chaiwat Sinsuwong, who had a reputation as a reformer. But apparently the issue was too hot for him. He said he had already stated his views on the matter and would make no further comments—his way of saying no.

After months of digging and sleuthing, I had some tantalizing testimony that a toxic waste problem was brewing in the Gulf of Thailand, but still no hard data connecting it conclusively to Map Ta Phud. Before going public, I needed to get my hands on that EIA. So I headed over to the office of a fellow I knew had a copy. My source was out, his secretary informed me, but I could wait in his office until he came back. Wait I did, for ten minutes, then twenty. There was no sign of the guy. I looked around his office. Papers were scattered across a huge table. Perhaps the report I wanted was right there. I took a closer look. Why yes, there it was, sitting right on top of a stack of papers: *Final Report: Environmental Impact Assessment: Second Phase Expansion, Map Ta Phut Industrial Port.*

No one was watching me. Feeling distinctly furtive, I opened it and scanned the pages until—Aha! It was a "sediment analysis report" from a company called United Analyst and Engineering Consultants. They'd done the lab work. Here were tables of sampling results for various pollutants. My adrenaline was really flowing now.

I scanned down the list until I hit cyanide. Wow, the numbers were high: 4.46 ppm in one place—that was from the port's outer approach channel; 3.45 ppm in the port basin; and 3.89 ppm in the estate's main drainage canal. That could mean it was coming from the wastewater. I quickly started scribbling the numbers down. Still nobody was in sight.

What about mercury? There it was, but the numbers were low. I flipped through a few more pages until I came across a table showing data from all the sampling points. And there I found a measurement of 0.28 ppm at the outer west revetment. There are no standards for heavy metals in marine sediment, but that figure was at least four times higher than all the others, which meant something was going on. More scribbling. By now my notes were a mess.

I decided to be bold. There was a photocopy machine out in the hall. Maybe the office workers wouldn't mind if I used it. I grabbed a page from the report and went to see what would happen. The copier wasn't being used, and nobody seemed to notice as I sidled up to it. But I couldn't get the darn thing to work! A light was blinking. What did it mean? I opened a few panels and pressed a few buttons, but nothing happened. I became more frantic, and cursed my lack of expertise in office equipment. Here I'd been a technology editor and I couldn't get a stupid photocopier to work when I needed it most—

"Can I help you?"

I nearly jumped out of my skin. Someone was right behind me. It was the secretary. She was smiling. I returned the smile, but it couldn't have been very convincing. "Uh, yes," I said, "it seems like he's not coming back today, so I thought I'd just copy this document I need."

"Oh, that machine is out of ink. Let me help you. We have another machine in here. I'll make a copy for you."

I wanted to kiss her, but just said, "Thank you very much." She went into another office with the sheet. Then I decided to press my luck. I went back to get some more of the report, and when she returned asked her to make some more copies for me. She gave me a slightly quizzical look, but silently went and did it. I thanked her profusely. "Do you want to leave a message?" she asked as I began to make my way out.

"No thanks, I'll just give him a call tomorrow," I said. Then I left, trying to look as if I wasn't fleeing the scene of a crime. I began to feel guilty. This information *ought* to be made public, I rationalized. I just hoped the secretary wouldn't get into trouble. So I did call the next day and asked a few innocuous questions. The official didn't even know I'd been there. No harm, no foul.

I was ready to write my big scoop. Back at the newsroom, I burst into the office of Pana Janviroj, the *Nation*'s editor, who had always supported our work on the environment desk. I explained what I'd uncovered, and he agreed it was big news. It could be the front-page lead, he said, if no other major stories came in. Excited, I phone around for a few last-minute comments, then sat down to write some crisp prose. I worked with the graphics department to design a diagram of the port showing where the high mercury and cyanide readings had been found. Finally, I waited as the copy editors went over the article, and kept an eye on the queue of other stories to see whether mine would indeed be the top one.

It was a proud moment for me when the story came out the next day—January 12, 1996—splashed in bold across the front page: "Deadly Toxins Found Off Eastern Seaboard: Mercury, Cyanide 'Contamination' Baffles Authorities." It was my first ever front-page lead.

But it was only the beginning of a long, uneven investigation. As I sat down to write a follow-up story, I received a call from Kraisak Choonhavan, the politician and son of a former prime minister, who had an interest in environmental issues. He left a message saying he was interested in learning more about the story and perhaps raising it in Parliament. But his attitude had changed by the time I called him back. "I've been asking around about it, and it seems you've come across a very sensitive issue," he said. "It's too hot. I'm afraid I can't touch it."

Further information came out in the following weeks. Sirithan, the PCD deputy leader, returned and confirmed that his agency was worried about the rising mercury levels in the waters off the coast of Rayong. "We have been watching with growing concern an increase in mercury levels in the Gulf of Thailand off Mab Ta Phud over the last year," he said. "The level is still under the limit, but we are worried about the rate of increase."[9] Coastal water quality tests around the port in 1995 had found average mercury levels of 0.09 micrograms per liter (ug/l), compared to an average of 0.07 ug/l the previous year. That brought it dangerously close to the standard of 0.1 ug/l. In several individual samples, the standard had been exceeded; the highest mercury level of 0.33 ug/l had been recorded in January 1995 in waters south of the industrial port. Because contamination seemed to be rising, it was unlikely to be a purely natural phenomenon, and Sirithan announced that his agency would investigate the source of the problem. That signified action would be taken and bolstered the credibility of my reports.

But the strangest thing of all was a story that ran on the front page of our main competitor, the *Bangkok Post,* on the same day as my big scoop. Datelined Rayong, home of Map Ta Phud, it reported that factory workers there were scared because of a prediction that had recently been made by a famous fortune-teller named *Moh* (Doctor) Yong. Such soothsayers are widely heeded in Thailand, and the good doctor's prognosis for the province was an ominous one: A terrible calamity was going to strike Rayong.[10]

THE OIL COMPANIES COME CLEAN

During the next few months, however, my investigation petered out. Unocal remained silent. So did the petrochemical firms and the industrial estate authorities in Rayong, which were now under close scrutiny. I made an investigative trip down to Songkhla, a pleasant southern port town that serves as a base for the offshore rigs, to try to uncover what Unocal did with its mercury-contaminated waste. Poking around the municipal dump revealed that someone had haphazardly left dozens of Unocal barrels lying around. But the barrels, which had contained drilling fluid, were mostly empty, and it eventually turned out that Unocal had donated them to the city to serve as garbage cans.

Back in Bangkok, I had a real breakthrough when I called a source who worked in the hazardous waste disposal business. I asked if she knew what Unocal did with its mercury-contaminated waste. Nothing I didn't already know, as it turned out, but she hinted at something else. I was missing something apparently, but what? "Oh, what always seems to happen when boys go playing in the mud," she suggested. What the heck did that mean?

And then it hit me. "Of course! Total! They have the same mercury problem as Unocal," I blurted out. I could almost hear my source smirking on the other end of the line. Silly me, why didn't I think of that before? But then I considered the implications. If this problem wasn't just unique to the Erawan field operated by Unocal, if Total had also found mercury in its gas from the Bongkot field farther south in the Gulf, perhaps mercury contamination was a far more widespread problem than initially feared. Perhaps it was endemic throughout the region. It was a chilling thought. After all, the countries that produce petroleum—Indonesia, Malaysia, Vietnam, Brunei, and soon Burma as well—were not terribly open. If their rigs were discharging mercury into their waste streams, the public might never know about it.

First things first: I had to confirm that Total had found mercury in its gas, and then determine whether the fish around its platforms showed signs of contamination. Finding another whistle-blower like Piamsak would be difficult, if not impossible. But an opportunity soon arose: Phillipe Persillon, a project manager for the French firm, was giving a dinner talk later that week for a club of petroleum engineers and industry executives. At the event, I looked around for people with Total on their nametags, and sat down next to François, a planning department chief for the company's Thai exploration and production arm. I told him who I was and tried to make small talk, recounting in French how I'd lived in Paris when I was a child. But he became guarded when I asked about the Bongkot field and whether Total had encountered any problems with mercury there. Over a meal of red snapper, he explained that mercury does collect at the installation, but he downplayed its seriousness. The equipment is cleaned out, and some of the waste put back into underground reservoirs through a process known as "deep-well injection," which Unocal also used, he said. "The flow of mercury is not constant," he added hastily, "and there is much less of it than at the Unocal operation, because we produce less gas." Fish surveys around the platform had not shown signs of contamination, François assured me, and the fish stocks in the area were so abundant that fishing boats tended to congregate there.

The room quieted down as Persillon, Total's manager for Bongkot, began his talk. He described how the project had progressed. Although Total was the operator, the Thai state-owned oil firm PTTEP was the largest shareholder, with a 40 percent stake, and was due to take over operations during the next few years. Total had a 30 percent share, British Gas a 20 percent share, and Norway's Statoil 10 percent. Bongkot was producing 350 million cubic feet of gas per day, but planned to increase production to 550 million cubic feet per day within a couple of years (which would mean the mercury quantities would increase, too). Persillon continued: The field also produced condensate, a liquid form of hydrocarbon that condenses out of the gas as it is brought up from high-pressure conditions deep underground. He explained that the gas is rich in carbon dioxide, which corroded the machinery, but he made no mention of mercury.

Immediately after the talk, I left my seat and went straight up to Persillon. I introduced myself as a journalist, and asked whether he'd encountered mercury at Bongkot. "Yes, it is a real problem. Our workers on the platform have to wear mercury detection badges [to monitor exposure to mercury vapors]," he said.[11] I began taking notes, trying to hide my excitement. "Were you surprised to find it there?" I asked.

"We didn't expect it," he said. "We didn't find any mercury in the test drilling. We only found it when we opened the pipes up for inspection." Apparently, the heavy metal liquefies out of its gaseous form as it's brought up to the surface. So it's found in the hydrocarbons and in the so-called "produced water" that comes up with the natural gas. "We're considering burying the water back in the well. The mercury is also in the gas, but what can you do?" he shrugged.

"How about the area around the platform?" I asked. "Has it been contaminated at all?"

"We're testing the water, but we haven't tested the fish yet," Persillon said as he turned to talk to another guest. I finished taking my notes, then turned around to find François behind me. He looked tense. I shook his hand, wished him a pleasant evening, and left. That was it. I had my story.

But the very next day, before I could even write it up, Konthi Kulachol, the PR director at Unocal, called to say that Unocal Thailand's managing director, Brian Marcotte, wanted to meet with me. The company had prepared an extensive briefing to explain how it has dealt with the mercury "issue of concern" (they didn't want it labeled a "problem"). Since I had reported on it first, Unocal would grant me the first briefing. After so many months of being stonewalled in my investigation, the doors had suddenly been flung open. On June 17, I was ushered into a Unocal conference room where a roundtable of company experts—aided by a profusion of documents, charts, graphs, and overheads—described where the mercury came from and how it was released into the environment. Marcotte, the American manager, led the discussion.

The mercury, he explained, originates in a layer of sandstone 9,000 to 10,000 feet deep, and probably enters the gas as the hydrocarbon is generated from the remains of ancient living matter. At least that seemed to be what happened in the reservoirs tapped by Unocal Thailand. Like Total, Unocal had been surprised back in 1985 when it discovered mercury accumulating in its equipment. But the presence of the heavy metal in oil and gas is not unique to Southeast Asia, he added, it's also common in the North Sea. (I later learned from other sources that high mercury concentrations are found in oil and gas produced in China, Venezuela, Algeria, and some fields in the Middle East. But apparently the highest levels are found in the Gulf of Thailand and Indonesia.)

As the gas is pumped up from deep underground, Marcotte went on, the free mercury condenses out of the gas. "It's relatively straightforward to take that out, and we then supply it to hospitals and industry," Marcotte said. "There's also a little bit of arsenic, but mercury is the one we want to pay the most attention to, since its concentration magnifies as it moves up the food chain." Nevertheless, the gas sent on by pipeline to Mab Ta Phud still contains mercury, and this is what fouled the PTT's gas separation plant when it opened in 1989.

A significant amount of mercury is also in the condensate produced from Unocal's fields. Most of the condensate is exported, but some mercury-contaminated material settles out while this relatively valuable form of hydrocarbon is kept in Unocal's storage tanks. Unocal long struggled about what to do with this sludge, since at the time there was no safe disposal facility in Thailand. The solution was to store it in Songkhla, although Marcotte revealed that 1,459 barrels had recently been injected back into disused wells drilled into the Gulf's seabed.

The biggest problem concerned the mercury found in the produced water that is pumped up along with the gas because it is discharged into the sea. Unocal treats the water, but Marcotte explained that it was quite difficult to separate the mercury out. "We have been pursuing this issue since 1985, and very aggressively since 1989," he insisted. "[But] there's just no [control] technology found to be acceptable, either here in Southeast Asia, in the North Sea or in North America." As a result, the average mercury content of the produced water dumped into the Gulf was around 200 parts per billion (ppb), although the fluctuating levels at times reached as high as 800 ppb.[12] That didn't break any laws, since Thailand doesn't have an offshore standard for mercury discharge, but the onshore standard for freshwater areas is 5 ppb because of the potential threat to drinking water. The standard for discharges in the North Sea, added the Unocal officials, is 300 ppb. Those numbers were interesting, but, since the volumes of water pumped up from underground could be huge, they didn't tell you how much mercury was actually being dumped into the Gulf of Thailand. But Marcotte continued: Altogether, he said, Unocal emits about 90 kilograms of mercury into the sea.[13]

"What about the fish?" I asked. "Are you willing to release the data that Piamsak has gathered concerning mercury contamination of the fish around the platforms?" Marcotte glanced at Konthi and Tawatchai Siripatrachai, the head of Unocal's environmental division, who looked uneasy. There was a silence, and then Marcotte began to praise Piamsak, explaining that Unocal had hired him back in 1992 as one of Thailand's foremost fisheries experts. Piamsak had decided to study "a complete spectrum of fish around the platforms, including those just passing through," he said. "These are the fish most affected [by Unocal's operations] and we have to concur. Is there an impact on the fish in the greater Gulf? Some comments have not been representative. All the impacts are very localized, to a small radius of less than 500 meters around the platform. By statute, fishing boats are excluded from this area, so it's very unlikely that the mercury would have any impact on commercial fisheries."

Marcotte was clearly hedging. He didn't want to criticize Piamsak, or discount his findings. But at the same time, he didn't want to acknowledge that Unocal's actions might have damaged Thailand's fisheries. That would anger the authorities, and potentially make the company liable in future lawsuits. But I argued that the public had a right to see the data and judge whether there was a serious health threat. I'd talked to Piamsak before visiting Unocal, and he had confirmed that the number of fish around the platforms found to have mercury levels higher than the acceptable standard had increased from 2 percent in 1992 to nearly 10 percent in his latest survey. "Dr. Piamsak was told that in the United States, if the number of fish with high mercury levels are found to exceed 10 percent, American authorities would shut down the platform," I said evenly. Marcotte looked upset and said there was no need for such drastic measures. But then Konthi stepped in, saying Unocal would consider the matter and get back to me.

During the following week, my stories on the Unocal revelations, the Total scoop, and the barrels found at the Songkhla dump appeared. The Total story included a map of the various gas fields in the Gulf of Thailand slated for development, and asked whether mercury contamination might be a region-wide problem. After all, production was due to expand in the Gulf, not just in the Bongkot field, but also further south in the Thai-Malaysia Joint Development Area. Meanwhile, Unocal had a concession to explore for oil in a Vietnamese part of the Gulf of Thailand, not far from its Thai operations. And various sources had told me they'd seen mercury removal units on rigs as far afield as Borneo and Sumatra.

Marcotte had acknowledged that if gas was found in the Vietnamese zone, it, too, might contain mercury since the geology there was similar to that elsewhere in the Gulf. David Watkins, Unocal's vice president of operations, explained that the volcanic activity underlying Southeast Asia could mean much of the gas produced in the region contains heavy metals and other contaminants. (It's not just petroleum that's affected: tailings from tin mines in southern Thailand have leached arsenic into ground water; and arsenic has contaminated water from thousands of tube wells in Bangladesh.) The Gulf of Thailand sits atop three geological basins: the Pattani Basin, the Khmer Basin, and the Malay Basin. Although separate, according to Watkins they share many of the same characteristics. They are all relatively hot and contain large amounts of carbon dioxide and heavy metals.

Unocal alone had eighty-seven rigs in the Gulf at the time, and a few weeks later took a group of journalists out to see what a massive, impressive operation it was to drill for petroleum. After we donned hard hats and special safety shoes for our tour, we were shown a hydrocyclone, a piece of equipment newly installed to purify the produced water. It had brought the average discharge of mercury down to 80 ppb. Marcotte (who was not on the tour) had told me that Unocal planned to install eight of these machines—at a cost of 14 million baht (more than half a million dollars) each—on various platforms in an effort to reduce the mercury loading. When I suggested that perhaps Unocal could afford to do it since the Thai subsidiary alone had earned profits of 6.5 billion baht on sales of 211 billion baht the previous year, he looked at me incredulously, rolled his eyes, and simply said, "It's a major investment."

Unocal also showed us the Songkhla storage facility where it keeps hundreds of barrels of mercury-contaminated sludge before they are sent down into tapped-out reservoirs. Tawatchai, Unocal's environmental chief, said the company would like to dispose of its produced water in the same way because it results in "zero discharge." But deep-well injection is not without controversy. The PCD's toxic waste officer, Prakan Boonchuaydee, complained he was not informed about the sludge disposal and questioned whether it was appropriate.

Environmentalists have also complained that the re-injected waste could leak from the wells and contaminate benthic fauna on the sea floor, or travel through the underground reservoirs and emerge through other wells. But Tawatchai said that it had received approval for its operations from the DMR, and added that the U.S. Environmental Protection Agency (EPA) accepted deep-well injection as a legitimate form of disposal. There would be no leaks, he added, explaining that the wells would be plugged with special cement that would last "at least a hundred years." The challenge Unocal faced is that the reservoirs available appeared to be too small to receive the huge volume of produced water that is continuously generated.[14]

There was no evidence to suggest that Unocal had lied to us—in fact, compared to many firms in Thailand, it had been relatively open about answering our questions—but the worrisome thing was that that no one seemed to be monitoring offshore operations. The company was operating way out in the Gulf with virtually no supervision. In theory, it could dump as much mercury as it liked into the sea and no one would know about it. Except for Piamsak, that is. And since Unocal had hired him, he was forbidden to reveal his data to the public. Merely speaking out in general terms had brought him dangerously close to a breach of contract.

The mercury story demonstrated a fundamental problem with protecting the environment in developing countries. Information, and the technology to gather it, is concentrated in the hands of corporations. In the developed world, at least, there are more independent monitors: academic experts, or NGOs such as Greenpeace, which are able to carry out tests for signs of toxic contamination. But such scientific expertise is not sufficiently widespread in Thailand and countries like it. Even government agencies—for example, Thailand's DMR and Pollution Control Department—often rely on the information provided by companies, or else have to hire outside consultants to run independent tests.

At any rate, governments are generally no more eager to reveal such information to the public than the corporations are. The DMR certainly was not going to reveal Unocal's fish survey data to me, and the public's right to access government information is extremely limited in most developing countries. Many times during my investigation I wished that Thailand had the same Toxic Release Inventory requirements as those enforced in the United States, where companies are ordered to reveal which hazardous materials they use and emit. The mere need to go public with this information has led numerous firms to drastically reduce the toxins they emit. In Thailand, on the other hand, I could not even persuade Unocal to reveal what was in the drilling fluids that were supposedly not present in the barrels lying around the Songkhla dump. They simply said it was a commercial secret.

SEX-CHANGING SNAILS AND OTHER FISHY BUSINESS

Despite Unocal's hospitality, the company had still not allowed Piamsak to reveal his fish data to the public. But I'd finally gained a better understanding of why it was so secretive—or rather, who. Apparently, Tawatchai, the company's environmental manager, disagreed with the way Piamsak had carried out his study, and distrusted his findings. It was hard to tell how bitter the dispute had become, because, for the record at least, the company respected Piamsak. But off the record, I was hearing more and more whispers questioning his methods, and even his integrity. "Bob," a foreign environmental consultant who worked with Unocal, had taken me aside a couple of times to explain the faults in Piamsak's surveying techniques.

Much of the debate was highly technical, and it didn't seem fair to criticize Piamsak when he wasn't present to defend himself. Nor did I want to be in a position of relaying each side's arguments back and forth. So during the Songkhla trip, Konthi and I agreed that we would try to arrange a meeting between Tawatchai and Piamsak, where the two men could sit down and hash out their differences. That, I hoped, would also finally enable me to publish the data on how much mercury was in the fish around Unocal's platforms.

In the meantime, I had begun the arduous task of trying to obtain information from secretive Fisheries Department researchers about levels of contamination in the Gulf. It turns out that the heavy metal found in greatest volumes was cadmium. Although less dangerous than mercury—the health standard for cadmium in seafood is 2 ppm, roughly four times higher than for mercury—cadmium is known to cause severe kidney damage to humans, and is associated with other diseases such as skeletal weakening, anemia, and heart disease. A 1994 study of moon scallops in the western Gulf found huge mean cadmium levels in the digestive gland (171.4 ppm) and kidney (20.4 ppm).[15] The authorities, concerned about maintaining the export markets for Thailand's huge fishing industry, did not reveal that Italian inspectors had rejected several batches of Thai octopus and cuttlefish because of their high cadmium levels. I uncovered the news only when a fisheries official I was interviewing stepped out of his office and I managed to rifle through the papers on his desk. The worrying thing was that no one was systematically testing the fish eaten by domestic consumers in Thailand. Nor did anyone know the source of the cadmium, which has various industrial uses and is also found in fertilizers.

The story was slightly different with tributyl tin (TBT). The source of this toxic compound is known. The tin-based chemical is an anti-fouling agent used in marine paints to keep barnacles and the like off ships and docks, and over time it washes off into the sea. But no one knows what happens to those who ingest it—or at least to humans. It has an alarming effect on certain species of shellfish, however: It causes them to change sex. When females of snail-like creatures known as gastropods become contaminated by TBT, they grow false penises and sperm ducts, causing their populations to plummet. In Europe, North America, and Japan, TBT has seriously affected oyster and mussel stocks. "TBT basically kills all the shellfish larvae in the area, including crab and lobster," explains Cornelius Swennen, a Dutch marine ecologist and shellfish expert.[16] Swennen and his collaborators in Thailand, Malaysia, and Singapore had found the greatest damage in heavily industrialized Chonburi and in the heavily trafficked straits of Malacca. Southeast Asian countries signatory to the Marpol Convention, which oversees marine pollution issues, are supposed to ban the use of TBT on small ships and set controls on leaching from larger ships. But authorities in the region—reportedly under pressure from the shipping and chemical industries—have yet to respond to the issue.

The mercury story picked up again when Total's boss agreed to talk to me. It was a far different meeting than my sophisticated briefing from Marcotte and his colleagues at Unocal. On September 9, 1996, the general manager of Total Exploration and Production Thailand, a soft-spoken Frenchman named Jean-Paul Azalbert, ushered me into his office and answered my questions for half an hour or so. He revealed that Total had discharged around 35 kilograms of mercury into the Gulf since it had begun operations three years previously—far less than Unocal because Total's operations are smaller. He also claimed that Total had known from the beginning that a mercury problem existed in the Gulf. The company's goal was to release no more than 100 ppb of the heavy metal in its wastewater, but "very occasionally" that figure was exceeded. Total also planned to send its produced water back underground. It had already "entombed" a dozen or two barrels of mercury-contaminated sludge. And it would begin a survey of fish in the area to check whether contamination was a problem.

The most surprising thing that Azalbert said was that Thai fishing boats frequent the area around the platforms, essentially confirming what his colleague François had told me.[17] Perhaps they thought this would be taken as a sign of healthy fish stocks. (Asked whether capturing fish in the vicinity of the platforms presents a danger, Azalbert replied, "It is difficult for me to comment because we have no idea of the level of mercury in the fish.") But somebody, perhaps the Thai authorities, must have scolded him about the statement, because he later denied saying it and accused me of misrepresenting him, even though I had been careful to tape the interview. Nevertheless, from that point on, Total refused to speak to me.

I finally got my meeting with Tawatchai and Piamsak in early October. We all gathered one afternoon on Piamsak's home turf, Chulalongkorn University's Aquatic Resources Research Institute. When I walked in, Konthi and Tawatchai had already arrived. They were sitting quietly around a table with Piamsak, a sturdy scientist with a black moustache and a calm demeanor. He might have been a whistle-blower in this case, but he was by no means an environmental activist. That, no doubt, was the reason Unocal had hired him in the first place. Throughout all my discussions with him, he'd never raised his voice, or even sounded excited. He always stated his argument simply, even stolidly.

No one was smiling as I entered, and everyone seemed tense. We immediately launched into a discussion of the data, which looked worrying. At the Erawan platform, the oldest rig and most contaminated site, fish sampled during 1994 and 1995 had an average mercury level of 0.28 ppm. That's well below the standard of 0.5 ppm, but well above the natural level of 0.1 ppm. What's more, nearly 12 percent of the fish collected there had mecury levels higher than the standard, and nearly three-fourths were contaminated above the natural level. Those numbers were all well above his results from 1993, when the fish tested at Erawan showed average mercury levels of 0.22 ppm, and only 3 percent of the fish were above the standard.[18] Tests carried out at Unocal's Platong and Funan platforms revealed lower but still elevated readings.

To assess these values, Piamsak carried out a control study at Bang Sarae, a site off the coast of Chonburi. He thought the samples there would represent a good example of background mercury levels in the Gulf. But the fish there were also contaminated at higher levels than normal—with an average level of 0.19 ppm, 5 percent of the specimens above the standard and nearly 60 percent above the typical background of 0.1 ppm—perhaps because Bang Sarae is only 40 kilometers west of Mab Ta Phud.

Piamsak's conclusions remained clear: "The mercury problem in the Gulf of Thailand seems to be expanding," he said. His studies had shown a significant increase in mercury levels since his previous examination of the fish at Unocal's platforms in 1993. The marine biologist was especially concerned with the high readings found in a species known as cobia (*plaa chon talae* in Thai), which are used to make fish balls, a popular treat.

Tawatchai didn't dispute the sampling procedures or the results of the individual tests, but took issue with the overall methodology. He argued that it was wrong for Piamsak to have combined results from 1994 and 1995, and from different sample sites at each platform. He brandished a report by an American firm called Battelle, which Unocal had hired to carry out an earlier study. It found that mercury levels in cobia and grouper had climbed between 1979 and 1993, but had declined in five other species, and concluded overall that no dramatic change had occurred during that period. And it complained that some important background data on the age, size, and health of fish from a study carried out by Piamsak in 1979 was not available.

Tawatchai also pointed out that the sample size for Piamsak's studies in 1994 and 1995, between forty and forty-five fish caught from each site, had been too small. Piamsak responded that he would have sampled more if Unocal had given him a larger budget and more resources. Besides, the 1993 Battelle study had collected about the same amount. "This is the standard number of fish caught in such surveys, even by agencies such as the U.S. Environmental Protection Agency," he explained. "But once an alarming trend is spotted, a bigger study is carried out."

Back and forth they went, Tawatchai at times becoming quite animated, his comments occasionally sharp, although the proceedings were always civil. Piamsak never lost his cool, and always talked evenly. I tried to focus on areas of agreement. Neither side felt the situation had become a crisis, although it could become one if serious action wasn't taken. Both wanted to see more studies conducted—Unocal had already hired an American firm called Tetra Tech to carry them out[19]—preferably with the government involved as a neutral arbiter. And they also agreed that a better control site had to be found because Bang Sarae appeared contaminated. Ultimately, however, some fundamental differences remained. Tawatchai could not accept the way different types of data had been combined, and Piamsak said the results do accurately reflect the mercury levels at the sites in question.

Despite the limitations of Piamsak's studies, the findings seemed significant enough to publish in the newspaper. There may have been problems with the methodology, but if they were so serious, why had Unocal continued to hire Piamsak? Why had they begun complaining only when he turned up results they didn't like? Certainly, from a public health standpoint, people deserved to know what had been discovered. I had to believe that readers of the *Nation* were mature enough to understand the results and what they meant without descending into mass hysteria. So publish them I did. A big table at the top of the front page and an article explained Tawatchai's and Piamsak's arguments.

Most people reacted to the numbers in the same way that I did: "So there really is a problem out there," one of my colleagues remarked. Unocal, meanwhile, seemed alarmed that the information had been made public, or at least eager to have its position clarified. The next day, Konthi sent in a lengthy letter to the editor, which we published in full.[20] He did not criticize my reports, nor did he contradict them, but merely stated that he wanted to add to them. He explained all the efforts the company had gone to in protecting the environment, and assured people that fish are safe to eat. Under the circumstances, it was quite diplomatic. I was surprised when Konthi suggested to me several months later that his job was on the line because of the mercury issue. After all, he wasn't responsible for the problems, only the fallout.

It was much harder to uncover the situation around Total's platforms at Bongkot because the company was stonewalling me. An industry quarterly called *PTIT Focus* reported that its first fish surveys had been carried out in

1996 (apparently only after the mercury problem at its operations were made public),[21] but did not provide the results. For that, I again had to rely on a couple of foreign consultants, "Joe" and "Nick." Like other consultants, they were environmental mercenaries, available for hire to whomever could afford them, and only willing to speak on the condition they were not named. And like other mercenaries, consultants come in all forms. Some are trustworthy, others not. So if you're finding it difficult to keep track of all these anonymous experts, imagine what it is like to try to piece together an accurate, reliable story based on their conflicting, whispered testimony. In the end, you have to rely on those with the most convincing scientific argument. The added uncertainty is the price we pay for a society in which information, and the right to access it, is jealously guarded.

I gradually came to trust Joe, who knew a great deal about the Bongkot fish surveys. But his accounts of Total's efforts were bewildering. The first samples were sent to a lab in Kuala Lumpur for analysis in 1996, he said, and showed mercury levels higher than the WHO standard, sometimes as much as 50 percent higher. Other fish samples were then purchased from the Songkhla market and the analysis allegedly produced "even higher figures."[22] However, an audit of the lab subsequently called those results into question. Apparently, the analysis had measured the mercury content of the entire fish, including the liver, where toxic substances are stored. Proper testing is supposed to examine only the wet muscle tissue. A second firm based in the United States that was then hired to study the remaining samples allegedly found mercury levels in the fish to be safely below the standard. To set the record straight, Total then gathered a team of Thai and international specialists to form the Marine Expert Mercury Environment Panel. The panel carried out its own surveys and published a paper in a U.S. journal reporting that the fish were not dangerous to eat.[23]

The entire process, as confided by Joe, didn't inspire a lot of confidence. Total never confirmed or denied whether this was indeed what happened. What's more, Nick had been out to the Bongkot platforms after the PTTEP took over the operations and was alarmed by what he saw. He has traveled around the world to work on mercury cleanups, and he gave a warning: "The Bongkot platform was the worst mercury situation I've ever seen," he said. "There was mercury rolling around on the decks in little silver balls." Nick also reported that mercury levels found in hydrocarbons from Indonesia and the Gulf of Thailand are far higher than elsewhere in the world; indeed, condensate from the Gulf may contain as much as 2 ppm of dissolved elemental mercury, whereas 10 ppb is typical for the rest of the world—that's a difference of two orders of magnitude.

The good news was that all the publicity appeared to have pushed Unocal and Total into disposing of its wastewater in a more responsible fashion. Tawatchai at Unocal reported that a new technology the company had tested, called PDT, could reduce mercury emission levels to around 10 ppb, a significant reduction from previous mean levels of 100 to 200 ppb. And Total announced plans to spend $4 million to carry out deep-well injection of its produced water.

THE EASTERN SEABOARD REMAINS MURKY

By the end of 1996, the picture of mercury contamination in the central Gulf had become clearer. The gas concessionaires had revealed how the heavy metal was entering the sea and had provided information about the quantities involved. Since they were the only source for such data, it was impossible to confirm whether they were telling the truth. But at least the companies and the government had promised to monitor and address the problem.

The situation was much murkier on the Eastern Seaboard, where the issue had first sprung up. The source of the mercury and cyanide in the sediment and water at Mab Ta Phud had never been found, or at least revealed publicly. It wasn't for lack of trying. The Pollution Control Department had inspected some of the more suspect factories in the estate, but came up with nothing. As Nick explained it, "All the drainage canals run together, so it's difficult to tell what effluent is coming from where." I attempted my own investigations. A firm named Thai Aromatics, which uses Gulf condensate as its raw material, had experienced some problems with its mercury removal unit, I discovered, but it denied that the matter had resulted in undue discharges.[24] TPI, one of Thailand's biggest and most notorious petrochemical firms, had been the subject of complaints by villagers in Rayong who accused it of polluting areas around its refinery. The Harbor Department subsequently threatened to file suit against the firm for discharging overheated water, but never followed through—and would never officially talk to me about the matter.[25]

The Eastern Seaboard situation was in some ways more serious than that in the central Gulf. By late 1997, Piamsak had overseen a new study at Bang Sarae; this showed that one in five fish contained mercury levels above the health standard.[26] "The thing which is worrying is that the mercury levels are higher now than when we measured ten years ago," he said. This contamination was much closer to the shore than the gas platforms, which meant that marine creatures in the area were more likely to be caught and eaten by local people. Shallower seabeds are also home to more of the methylating bacteria, which convert mercury into its more dangerous form, said Nick. And, he added, the amount of mercury entering the sea in Rayong was probably greater than that entering from the platforms. In fact, experts agree the amount of mercury entering the Gulf through natural means—rainfall, for instance—was also much greater than the amounts that came off the platforms. But this natural deposition is also more dispersed.

Nick was the source I trusted the most with this story. He was a mercury expert and, like other consultants, indirectly had an incentive to spread concern about the issues he worked on. But he was always willing to explain the situation carefully and calmly, and sometimes he downplayed the threat. He had been to the platforms in the Gulf and the factories at Mab Ta Phud, and his trained eye saw many lapses in the way mercury-contaminated waste was being handled. "A lot of the companies at Mab Ta Phud have the same sorts of problems. They are caught between a rock and a hard place. On the one hand, paying Genco to treat their waste is expensive. On the other, there is a lack of enforcement and no reporting requirements," he explained. "The air emissions are more locally hazardous. All sorts of fumes—benzene, formaldehyde—escape through the venting process.

"But the biggest problem is on the water side. They do a good job of keeping organic material out of the effluent, but their equipment is not built to handle metals like vanadium, selenium, and mercury," he said. The refineries use machines known as de-salters to clean crude oil with water. Salts are washed out of the oil and flushed into the waste treatment systems. But those systems typically aren't able to handle the metals, which are most likely discharged with the wastewater. Another major problem, he said, was sloppy maintenance. Mercury accumulates in the processing machines and at the bottom of the storage tanks where condensate is stored. "They have to shut down from time to time to clean out the equipment, and in the process try to corral the mercury," explained Nick. "But a lot of it just goes down the drain." There are similar problems at Laem Chabang, another industrial port up the coast, and in other industrial regions such as Samut Prakan, near Bangkok.

His explanation seemed the most plausible way to account for the elevated mercury levels in the Gulf. Not only was he an eyewitness, but the intermittent nature of these discharges would explain why it was so hard for authorities to locate the source of the contamination. Nick was particularly concerned about the fate of workers on the platforms and at Mab Ta Phud who were being exposed to mercury. But the gas producers aside, he would never give me the names of specific firms involved. The problems, he asserted, were too widespread to blame on a particular company.

The main reason I had to rely so much on anonymous consultants as my sources for the mercury story was that bureaucrats were inevitably cagey about the issue. The IEAT, the DMR, the PCD, the Department of Industrial Works (DIW), the Harbor Department, the Fisheries Department—there was a whole raft of agencies that wanted the problem solved, but quietly. Some of my most troublesome relations were with a PCD official named Pornsook Chongprasith. Capable and well-educated, Pornsook was also highly suspicious of the media, and of me in particular. Despite my repeated visits to her office, she was never forthcoming about which studies were being conducted, or what their results were.

So in November 1997, I was surprised to receive a phone call from Pornsook in which she told me about how several oil spills had occurred at the mooring point for ships supplying the refineries at Mab Ta Phud. Finally, I thought, she's decided to trust me. Assuming she wanted me to publicize the problem and thus pressure the companies involved to clean up their act, I called the refineries and published an article on the story. But the next day, Pornsook called back, and she was furious. She had given me the information in confidence, she explained, not to print. Even though I hadn't mentioned her name in the article, the companies would all know she had leaked the story. It was a simple case of miscommunication, but a damaging one. Pornsook, apparently convinced that I was out to embarrass her, was now frostier toward me than ever. My blunder had ruined any chance of our future collaboration.

Instead, I tried talking to her superior, the director of the PCD's water quality division, Yuwaree In-na. At first, she seemed more amenable to cooperation. When I visited her office in 1998, she explained that she didn't know much about the mercury situation because she had only just been promoted to her position, but offered to give me a report prepared by Pornsook and a colleague, Wimolpan Wilairatanadilok. Grateful for this veritable treasure trove of information, I quickly went home to study it.

The report was titled *Are Thai Waters Really Contaminated with Mercury?* which made the answer seem like a foregone conclusion. Sure enough, it concluded, "Generally speaking, elevated mercury concentrations are occasionally observed, but overall [are] still within [a] safe level." The report had some worryingly vague statements such as, "The results confirm that most total mercury in coastal waters do not exceed the National Coastal Water Quality Standard, except those of some locations and sampling periods." The report had not been published in a scientific journal, and so (like most of the data on this mercury issue) apparently was never subjected to peer review. Rather, it seemed to have been prepared for an ASEAN conference in Cebu. But it contained hard data—the only hard data I was likely to find at that point.

Most of the results were encouraging. For instance, the PCD had collected 119 tissue samples of fish, crab, shrimp, and scallops from all over the Gulf. Crabs tended to have slightly higher mercury levels than fish, but according to the report, "Overall, total mercury concentrations were very much lower than the standard of the [U.S.] Food and Drug Administration." It also warned that "mercury concentrations [at trawling] stations located in the vicinity of the oil and gas platforms tended to be higher than other areas."

But there were discrepancies between the report and other information I'd gathered. The PCD claimed to have found relatively low levels of mercury in the sediment samples taken far from the coast, with a maximum figure of 0.12 ppm. However, the natural gas producers admit that the sediment around their platforms is far more contaminated than that. Konthi cited a maximum figure of 50 ppm around the Unocal platforms, and the PTTEP's David Johns said that sediment mercury levels at Bongkot were as high as 40 ppm (those figures are over a hundred times higher than the amount of mercury originally discovered in the sediment off Mab Ta Phud). And Charlie Charuvastr, the PTTEP's director of external relations, claimed that the PCD had never entered the Bongkot area to carry out its survey, or at least had never asked permission to do so.[27]

In addition, the report contained some worrying new information. The PCD had also carried out mercury tests in rivers around the country. Generally, the results were neither surprising nor alarming. But for some reason, five out of the eleven rivers tested in northern Thailand showed mercury levels above the water quality standard of 2.0 micrograms per liter (ug/l). In fact, mercury concentrations throughout the north appeared elevated compared to those in other regions of the country: Out of the eleven northern watersheds tested, only two had mercury concentrations below 1.0 ug/l.[28]

The results might have been caused by systematic errors made by northern surveyors, but they could also stem from some real contamination by a localized source.

I contacted Pornsook and Yuwaree to ask whether they knew what the cause might be. They immediately made it clear that I was not even supposed to have seen the report in the first place. Yuwaree had apparently given it to me by mistake. When I asked about the elevated mercury levels in northern rivers, they seemed surprised and promised to look into the matter. But they never again responded to my requests for information.

That left me on my own to figure out why mercury was showing up in the north. Without access to scientific studies, I could only speculate. There is no oil or gas production in the rivers' watersheds, so a systematic error still seemed a likely cause. But another thought crept in. Thailand has a huge complex of power plants at Mae Moh, in the northern province of Lampang, that burns lignite, a dirty form of coal. Its vast emissions of sulfur dioxide have repeatedly caused public health problems among villagers in the surrounding valley. Perhaps the lignite also contains mercury, and it, too, is being spewed out.

I called officials at the Electricity Generating Authority of Thailand (EGAT), which runs the plant. But Akson Srisawat, who oversees operations at Mae Moh, assured me that there is no mercury in the lignite. Nor is any found in the fly ash left over from combustion, he said. So EGAT does not even bother to check Mae Moh's airborne emissions for mercury. "[Any] heavy metal trace elements will be captured in the electrostatic precipitator which collects fly ash dust," Aksorn said. Nick, on the other hand, claimed that coal always has at least some traces of mercury in it. "There is mercury in all coal," he said. "The only question is how much." And that is where the matter still stands today. EGAT apparently has not looked into the matter, nor has it been asked to do so. No explanation has ever been given for why rivers in the north of Thailand were found to have high levels of mercury.

Clearly, whether it was because of official obstinacy, public indifference, or my own mistakes, I was having trouble persuading others to focus on this issue with the attention it deserved. My last hope was to help produce a television show—part of the *Rayngan Si-khiow* series of environmental feature programs I worked on—about the mercury situation in the Gulf of Thailand. Perhaps we could draw in viewers by obtaining rare footage of work on the gas platforms. In 1998 and early 1999, I pushed the idea with our sponsors at the Department of Environmental Quality Promotion (DEQP), and finally received preliminary approval. We met several times with the PTTEP, which had taken over operations of the Bongkot field, and they agreed to take us out to the platforms on condition they could look at the script before we aired. From a journalistic standpoint, it was less than ideal. But I decided it wouldn't overly compromise our reporting, and realized that without the PTTEP's cooperation, we wouldn't obtain the footage we needed.

So in early 1999, we flew out to the Bongkot platform. The PTTEP showed us around the facilities, and we taped an interview with the director of operations. There was a slight kerfuffle about whether we would be allowed to air footage of the oily black smoke emanating from the platform's flare, but in general the trip went smoothly. Back in Bangkok, I wrote the script in English and sent it to the PTTEP. They asked for a couple of slight modifications that seemed reasonable, and it looked as if we were all set to go.

But then we ran into a brick wall. Our supervisors at the DEQP had sent the translated version of the script to Pornsook at the Pollution Control Department, and she expressed numerous objections to it. Some of her criticisms were legitimate. There were mistakes in the Thai translation that I had failed to pick up on. But when I suggested that they could easily be corrected, and that we could work together to air the show, they refused. The program was scrapped. As with our documentary on the filming of *The Beach,* our work had been wasted. All the time and money spent by the PTTEP in taking us out to Bongkot had gone for naught. It was embarrassing, and also ironic. We were stymied in the end not by the oil companies but by those charged with protecting the environment.

AN ALARM RINGS IN VIETNAM

Toxic problems do not end at national boundaries, of course. For all the obstacles to investigating the mercury issue in Thailand, it was far more difficult to find out what was going on in neighboring countries, which are all less open. But there were indications that mercury contamination was a regional problem. As Unocal officials in Thailand had pointed out, the geological conditions that lead to the presence of heavy metals in hydrocarbons are found throughout Southeast Asia.

In Malaysia, Nick claimed that mercury contaminates the gas produced off Terrenganu, but in smaller concentrations (about twenty times smaller, he estimated) than in Thailand.[29] The situation in Indonesia is more worrying. According to Nick, Indonesian hydrocarbons have some of the highest mercury content found anywhere. Mobil's Arun oil field, he noted, is equipped with the largest mercury removal units in the world. To make matters worse, because Arun is located in Aceh, a province in northern Sumatra rife with separatist violence, information there is especially difficult to obtain. Indonesia also produces gas off Borneo and in the South China Sea. Sources at Total and Unocal—both of which have operations in Indonesia—say the gas in some fields there contains mercury. Total's Azalbert also said he saw mercury extraction equipment in Indonesia. But within Indonesia, the biggest mercury issue concerns the run-off from gold mining operations in Borneo.[30]

It was in Vietnam, however, where real evidence of a mercury problem emerged. In 1997, a French-Thai team of researchers carried out a mercury survey in and around the Mekong delta. The results were alarming. They discovered "extremely high total mercury values" at a couple of sampling sites, according to the paper they presented at a conference in 1998. The highest levels were recorded in the waters around Vung Tau—a center for fishing, tourism, and offshore petroleum operations located east of Ho Chi Minh City—where the dissolved total mercury level was found to be forty times greater than the standard set for Thailand's coastal waters (the Vietnamese government, however, sets far looser standards). Samples of fresh water from the Bassac River, one of the many waterways that make up the delta, also revealed significant contamination by mercury. The highest level in the river was recorded at a site near the city of Can Tho, but mercury was also found farther downstream.[31]

The sources of the contamination remain a mystery. But around Vung Tau, attention immediately focused on the offshore oil producers, given Thailand's experience. "There is a probability that the [mercury] contamination could be caused by the produced water from the nearby oil extraction platform off Vung Tau," the paper noted. Vietsovpetro, a joint venture between Petro Vietnam and Russia's Petroleum Ministry, has been pumping out hydrocarbons from its Bach Ho field, located only 110 kilometers from Vung Tau, since 1984. What's more, Nguyen Duc Huynh, who runs Petro Vietnam's environmental and safety research center, confirmed that the company releases produced water from its Bach Ho and Rong fields into the sea after treating it. But Nguyen said he was "not anxious" about mercury contamination, because the gas and the produced water contain low levels of mercury. He also claimed that mercury levels in the sediment around the platforms were measured at around 90 ppb, which is barely above background levels and far lower (around five hundred times lower) than the levels reported in the sediment around some of Unocal's and Total's rigs in the Gulf of Thailand.

Whatever the source, the findings suggest that mercury contamination has expanded beyond the Gulf of Thailand and is turning into a regional phenomenon, posing a potential public health threat to many countries around Southeast Asia. Given the scanty and uncertain information available, it's difficult to say how serious a threat it is. Pornsook's report and the results provided by the Total and Unocal studies suggest the mercury situation in the Gulf of Thailand has stabilized. But Nick is less sanguine.

He describes the mercury problem in Southeast Asia as nothing less than a "ticking toxic time bomb." Not only do oil and gas in the region have higher mercury concentrations than anywhere else in the world, he points out, but production is expanding rapidly in many countries. In the Gulf of Thailand alone, Chevron, Arco, and Harrod's Energy have started up petroleum operations, and work in the Thai-Malaysia Joint Development Area is also progressing. Meanwhile, he adds, governments seem to be largely ignoring the problem. There has been no further research about the mercury situation in Vietnam, the Thai government seems to have lost interest in the matter, and Piamsak, who still makes fish surveys for the DMR, has reportedly been blacklisted by the petroleum industry and cannot do further research.

Discouraging whistle-blowers like Piamsak seems to be the norm when it comes to industrial issues. One of Thailand's bravest campaigners is Orapun Metadilogkul, a physician and award-winning occupational health expert at Rajvithi Hospital who has helped uncover numerous workplace safety scandals, and in return is frequently harassed by industrial authorities. This is the wrong way to go. Providing access to information about the chemicals being used and emitted would help bring the power of civil society to bear in fighting toxic contamination.

A good example is Indonesia's "Proper" program. Like America's Toxic Release Inventory in the United States, it uses publicity to pressure companies into improving their environmental behavior. Companies are ranked according to a five-tier system: Those that do more than required are given a gold rating and publicly applauded; those who fail to live up to their responsibilities are given a black rating and a six-month grace period before their names are publicized.[32] This "name-and-shame" program has reportedly worked well; many firms have rushed to improve their environmental performance, and the strategy is now being tried in the Philippines and Latin America.

More basically, developing countries determined to industrialize must do a better job of preparing to deal with hazardous waste. Of all the ways dreamed up to create a shortcut across the environmental Kuznets curve, building adequate industrial waste treatment facilities, preferably within industrial estates, is among the most obvious. But it is not easy. It requires that governments overcome lots of opposition—not just from factories unwilling to pay, but also from "Nimby" activists, and even from environmentalists who point out that such facilities produce toxic emissions of their own. Instead, say the greens, industry should do more to prevent the creation of hazardous waste in the first place, and they are right. But making industry pay to clean up its pollution would provide an incentive toward just that goal.

What Southeast Asian countries need to do, finally, is assess more systematically the extent of the threat posed by mercury. Perhaps some agency could fund an independent study that seeks to determine how much mercury is entering the environment, where it's coming from, and how seriously it is contaminating the food chain. Most important, we need to learn whether seafood consumers and petroleum workers in Southeast Asia are suffering from overexposure to mercury. That was an area of investigation in which I was never able to make much progress. Ideally, such a study should be regional in scope, and should examine other important toxic pollutants, too, such as arsenic. Not too far away, in Bangladesh, the arsenic contamination of tube wells has created a public health crisis that is affecting millions of people. I hope the mercury situation in Southeast Asia will never become as serious, but the best way to make sure of that is to begin taking the issue seriously.

9

GLOBAL ISSUES

NORTH VS. SOUTH, GREEN VS. BROWN

THE SKIES DARKEN OVER SOUTHEAST ASIA

In 1997, a cloud fell over Southeast Asia, in more ways than one. First came a currency crisis. On July 2, I left Bangkok for a trip to the United States. Since my salary was paid in baht, I needed to buy some dollars. But each bank I went to refused to change money, and none would say why. I finally came across one airport counter that was open and still exchanging at the pegged rate of 25 baht to the dollar. It wasn't until I landed in New York that I learned the reason for all the confusion: After months of fending off pressure from foreign exchange speculators, the Thai government had chosen that morning to float the baht.

The effective devaluation of Thailand's currency triggered a financial crisis that spread throughout the region; one from which, in 2002, Southeast Asia is only just recovering. The underlying causes have been widely discussed: ballooning current account deficits, rampant speculation (which in Thailand led to the bursting of a property bubble), poor banking oversight, crony capitalism, and some dubious International Monetary Fund (IMF) policies—they all played a role in worsening the downturn. Thailand's economy, which had previously grown by double digits, plummeted by 10 percent in 1998. Seven thousand Thai companies have since disappeared, and 55,000 bankruptcies are pending. With the collapse of eight banks and eighty finance firms, only four Thai-owned banks are left.[1] There was also a political fallout in the kingdom; reformers seized on public discontent with government mismanagement to gain the passage of a new Constitution, the most progressive ever passed in a country where such documents have a staying power roughly equal to that of a pop star.

But the crisis was greatest in Indonesia. It ravaged the country's economy. The loss of jobs and sudden increase in poverty destabilized the autocratic and deeply corrupt regime of President Suharto, eventually leading to its collapse after thirty years of power, and set the stage for ethnic strife that threatened to tear the country apart. Before any of those traumatic developments, however, there was an ominous portent . . . the haze.

"Haze" was the polite term for the huge cloud of smog that smothered the region beginning in September 1997. For weeks, noxious plumes of smoke from Indonesian forest fires, many of them lit illegally by plantation firms clearing land for their palm oil and pulp and paper operations, billowed out of Borneo and Sumatra to blanket Malaysia, Singapore, Brunei, most of Indonesia, southern Thailand, and the southern Philippines. A similar phenomenon had happened in previous years, but the scale of this event, exacerbated by an El Nino–inspired drought, literally took one's breath away. The smog hovered over the region on and off for months, the fires ravaging farmland, forests, and wildlife over an area of 9.76 million hectares (24 million acres),[2] an area roughly half the size of the U.K. The flames even spread to underground peat bogs, which, once ignited, can smolder for years.

But it was the economic and public health impacts of the "Suharto Smog" that captured the world's attention. The smoke from the fires not only was a danger in itself but also helped trap the industrial pollutants relentlessly spewed out by the region's cars and factories. Altogether, an estimated 70 million people were affected. In Indonesia, during the first month of the crisis alone, at least two people died as a direct result of the haze, and more than 35,000 others suffered respiratory problems.[3] In Malaysia, where at least 8,000 people were hospitalized and another 15,000 treated as outpatients, residents followed a rising Air Pollutants Index (API) as they would a plummeting stock market index: When the API reaches 250, people are told to stop working outdoors;[4] when it reaches 500, conditions are deemed seriously harmful. In Kuching, the capital of Sarawak, the index soared higher than 800, causing the government to declare a state of emergency (some even suggested that the state's 2 million residents should be evacuated). Everywhere in the region, schools and offices were closed. Flights were cancelled and shipping traffic disrupted. And, of course, there was the huge cost of putting out the fires.

The indirect effects of this unnatural disaster were equally serious. An Indonesian plane crash blamed on poor visibility caused by the haze cost the lives of 234 people. Another 28 people were presumed drowned when two ships collided in the smog-shrouded Straits of Malacca. And at least 270 people died in famine-wracked Irian Jaya because relief planes were unable to reach them.[5] No one knows what the long-term health impacts of the pollution crisis will be. "Inhaling pollutants like this is worse than just smoking [cigarettes], which is done intermittently," said Yeo Chor Tzin, a Singaporean physician and specialist in respiratory and lung diseases.[6] Some medical experts predicted that the pollution would eventually lead to higher lung cancer rates among the millions who were exposed.

The Indonesian government inevitably tried to duck its responsibility for the problems and pinned the blame on El Nino. "We are not late in anticipating the problem," said Azwar Anas, head of the country's National Disaster Management Coordinating Agency. "It's a natural disaster which no one could have prevented."[7] But meteorologists, including those in Indonesia, had predicted that the oncoming El Nino would be a strong one, and the drought in Indonesia correspondingly severe. "The Indonesian government says they knew nine months ago there was a drought coming," noted Gurmit Singh, a Malaysian environmentalist. "If they knew, then why did they not take action over the months to ensure there was no open burning by companies clearing the land for plantations? This is blatant negligence."[8]

Under the world's gaze, the smog ballooned into a huge symbol of Southeast Asia's environmental neglect and revealed how the region's governments had avoided cracking down on the well-connected firms and individuals responsible. Although the simultaneous financial and pollution crises appeared to be quite different from one another, in some ways their root causes were the same: an emphasis on rapid industrial growth instead of long-term sustainability; creaky systems of governance that failed to install proper regulatory regimes;[9] and, even more to the point, the reluctance of the region's leaders to work together in solving environmental problems that cross their borders. The crisis was a symbol, in other words, of the diplomatic failures of the Association of Southeast Asian Nations (ASEAN).

As a regional grouping of (now ten) Southeast Asian countries that regularly meet to discuss all kinds of issues, ASEAN would seem to be the ideal forum in which to come to grips with the region's many environmental problems. It had already agreed to attempt the challenging feat of establishing a free trade area, so it should also be able to cooperate on environmental matters. Besides air pollution, there are plenty of other transboundary issues—essentially, regional manifestations of the "tragedy of the commons"—that need to be addressed jointly, in particular marine pollution and the decline of fish stocks. But although ASEAN has set ambitious targets for reducing tariffs between member countries, it has made little progress in adopting a uniform system of environmental standards.

The problem lies in ASEAN's policy of "non-interference." Member states avoid publicly criticizing their neighbors about anything considered to be "internal affairs" (such as Burma's human rights record), and environmental issues seem to fall into that category. Even at the height of the haze, public officials were notably reluctant to censure Indonesia, much less threaten punishment. In part, this attitude reflects the generally low priority accorded to environmental concerns, especially compared to economic issues. Crossborder investment is not considered "interference" even though, for instance, officials in Kuala Lumpur quietly admitted that forty-five Malaysian plantation firms were responsible for some of the fires set in Indonesia.

But perhaps the most damning evidence of ASEAN's negligence was the fact that smog from runaway Indonesian forest fires had been a regional problem for years. The problem was preventable, but the grouping had never taken firm action against it. The worst previous incident had been in 1994, when Malaysia and Singapore were severely affected (and slash-and-burn farmers were blamed). The following year, ASEAN environment ministers met and ordered the ASEAN Institute of Forest Management to draw up a plan aimed at preventing the haze from recurring. The plan—designed to predict, monitor, and fight forest fires through the use of satellite technology and joint firefighting strategies—was unveiled in December 1996 in Kuala Lumpur. The Canadian International Development Agency even agreed to support it with US$1.48 million in funding, so long as ASEAN matched the offer.

But ASEAN waffled. In January 1997, ASEAN environment ministers met again, this time at one of the Thai Wah–owned resorts at Bang Thao Bay in Phuket. Rather than agreeing to fund the forest fire action plan, they simply released a statement that "expressed their appreciation [for] Indonesia's efforts on the issue of transboundary haze pollution."[10] Eight months later, tourists fled Phuket, Thailand's premier tourist destination, as the choking fog engulfed southern Southeast Asia. Ultimately, the total cost of the 1997 haze crisis was estimated at US$4.4 billion, according to the Singapore-based Environment and Economy Program for Southeast Asia. "This is more than the damages assessed for the Exxon Valdez oil spill and India's Bhopal chemical spill combined," said the program's director, David Glover. "The resources lost would have been more than enough to provide basic sanitation, water and sewage services for Indonesia's 120 million rural poor."[11]

To be fair, along with lacking the necessary foresight, ASEAN's environment ministers also probably lacked the power (and the budget) to take action against Indonesia's forest fires. We have already seen how the control of forests and other lucrative resources in Thailand is kept firmly in the hands of economic-minded ministries for which profit is the primary aim. The same is generally true throughout Southeast Asia and the developing world. Environmental agencies are charged mostly with trying to clean up the mess left behind.

When the pollution crisis broke out, ASEAN environment ministers happened to be meeting again, this time at another plush resort on the Indonesian island of Sulawesi. To escape the thick pollution there, they were promptly hustled over to Jakarta, where Suharto addressed the group and said sorry for the smog. He hastened to add, however, that "quite often the environmental issue—in addition to democratic and human rights issues—are wrongly used as a conditionality in international trade by certain parties."[12] It was a highly revealing statement: ASEAN strongly opposes linking environmental issues to trade, and that had clouded its judgment about how to handle a grave regional problem.

This hints at the broader reason why ASEAN governments refrain from commenting on their neighbors' environmental policies, much less cooperating with them: Like most countries, they zealously defend their sovereignty against international efforts to manage local resources for the good of the global environment. Malaysia, led by its strident Prime Minister Mahathir Mohammed, has been particularly outspoken in accusing foreign environmentalists of being "neocolonialists." But developing countries (the global South) in general have long feared that wealthy countries (the global North, essentially the European Union, the United States, Japan, Canada, Australia, and New Zealand) would force them to sacrifice their economic growth for the good of the global environment. At the 1972 UN Conference on the Human Environment held in Stockholm, North and South reached a famous compromise known as Principle 21, which acknowledged that states have both "the sovereign right to exploit their own resources pursuant to their own environmental policies, and the responsibility to ensure that activities within their jurisdiction or control do not cause damage to the environment of other States or of areas beyond the limits of national jurisdiction." But global environmental summits remain taut with the tension between these two conflicting principles.

In their determination to protect their sovereignty and pursue rapid development, ASEAN states often adopt a fiercely anti-green stance on the world stage, at times actively obstructing green agreements in various international fora. The main result of the ASEAN environment ministers' meeting in Jakarta, for instance, was a declaration rejecting a proposal requiring developing countries that have signed the Climate Change Convention to report on how they plan to help prevent global warming. Meanwhile, at World Trade Organization (WTO) deliberations held in 1996, Southeast Asian delegates to the committee on environment and trade refused to recognize the trade rules established by multilateral environmental agreements. Incredibly, the ASEAN environment ministers who gathered in Phuket in January 1997 did not even seem aware of this threat to their jurisdiction, no doubt because it had been engineered by the grouping's trade officials.

There was some hope that the sight of an apocalyptic cloud spreading across their region would spur ASEAN governments into greater environmental cooperation. In the years that followed, Thailand's young and dashing foreign minister, Surin Pitsuwan, urged the grouping to adopt a more open approach he called "flexible engagement." Certain issues that affected the region as a whole—not just the haze environmental problems, but also, for instance, terrorism, trafficking in women and children, and the spread of diseases—had to be discussed openly and addressed in a more cooperative fashion, he argued. "All these problems, because of the nature in which they are being transformed, cannot be regarded as domestic or exclusively internal like before."[13]

Although Thailand had the support of the Philippines in seeking greater collaboration on these issues, it's an uphill struggle. Malaysia, Indonesia, Vietnam, Laos, and Burma all still firmly believe that member countries should not tamper with the principle of non-interference, and even Surin acknowledged that this remained the foundation of ASEAN policy. So although Indonesia's forest fires, and thus the haze, have been brought under greater control in recent years, thanks to stricter enforcement in Indonesia and the adoption of a Regional Haze Action Plan by ASEAN,[14] forest fires remain an annual problem in Sumatra and Kalimantan, and there have been few signs of more general environmental cooperation. Like most governments, those in Southeast Asia turn green only when forced by a crisis.

When it comes to global environmental issues, ASEAN's stance is not much different from that of the rest of the South, particularly large developing countries such as China, India, Brazil, and South Africa. They all tend to accuse the North of hypocrisy and of holding ulterior motives in their international environmental dealings. In many instances, they have a point; the North in general has caused far more damage to the Earth's environment than the South (although that trend is changing). On the other hand, green activists note that many governments in the South seem willing to cooperate internationally, particularly in trade, if it's in their financial interest. So why the reluctance to cooperate environmentally, which also is supposed to be in everyone's interest?

The North-South schism over global issues is usually portrayed as an environment-development split, wealthy countries stressing the former and poor ones the latter. But this conventional wisdom is misleading. When you examine global issues such as climate change, the loss of biodiversity, and links between trade and the environment, it becomes evident that both sides still tend to place economic priorities over environmental ones. The two most important agreements signed at the 1992 Earth Summit in Rio de Janeiro—the Convention on Biological Diversity (CBD) and the Framework Convention on Climate Change (FCCC)—have partially morphed into commercial treaties. Some of the most bitter negotiations have been about how to establish a carbon trading mechanism and how to regulate trade in genetically modified organisms. So it's more accurate to conclude that disputes between North and South are based on *competing* economic interests and different environmental priorities.

It's often noted that developed countries tend to worry about global problems, and the developing world concentrates almost exclusively on local issues. But if you analyze the negotiations over the North's environmental aid for the South—that is, if you follow the money trail—the division might equally be characterized as a "green-brown" split. In debates about how to make development in the South more sustainable, the North is more concerned with so-called "green" environmental issues such as conserving forests and wildlife, whereas governments in the South would rather focus on "brown" issues such as pollution, urban environmental quality, and clean technology.

TUNA VS. DOLPHINS, SHRIMP VS. TURTLES, TRADE VS. ENVIRONMENT

Just before the onset of the haze crisis in 1997, ASEAN's agriculture ministers met in Bangkok and launched a seemingly laudable conservation initiative. Amid much fanfare, they signed a memorandum of understanding (MoU) urging the protection of the region's sea turtles. But only the most naïve of observers would conclude that these hard-bitten politicians had acted out of concern for the well-being of ocean-dwelling reptiles.

The MoU was actually a response to the embargo on imports of wild shrimp imposed the year before by the United States, which charged that shrimping fleets from ASEAN and other developing countries were slaughtering sea turtles. The impressive-sounding MoU, therefore, was mostly for show. Thailand's agriculture minister, Chucheep Harnsawat, admitted that it contained no concrete measures.[15] Plodprasop Suraswadi, the former head of Thailand's Fisheries Department, argued that by formally recognizing sea turtle conservation as a regional issue, countries would come under pressure to protect the creatures. The smog crisis, however, revealed just how worthless such "peer pressure" is. The agriculture ministers demonstrated their real concerns at the meeting by setting up a Shrimp Industry Task Force designed to do battle against foreign environmentalists who had forced the United States to impose the embargo and were threatening a campaign against shrimp farms.

But if there was a farcical element to the ASEAN announcement, that was only fitting because just about every step and counterstep between the United States and Asian governments over the issue was laden with sophistry. The U.S. government had never wanted to enact the trade ban against Asian countries in the first place, and carried it out half-heartedly. What's more, the embargo was not directed against shrimp farms, which are at least as destructive as shrimping fleets. Meanwhile, the retort from Asian governments that they already had laws to protect turtles and other marine creatures was laughable because everyone knows such measures are rarely enforced. And their indignant accusations that this was merely a disguised attempt at protectionism by American commercial interests were off base, as the origins of the embargo reveal. Where the sanctions had a real and far-ranging impact was at the WTO, since its handling of the dispute turned out to be a landmark case, discussed the world over, in how to reconcile trade with environmental issues.

It was not the first time that the United States and the developed world had argued about how to protect marine creatures. In 1987, Sam LaBudde, a fisheries biologist who worked for the San Francisco–based Earth Island Institute (EII), spurred major U.S. tuna firms into stopping the purchase of tuna caught by drift nets when he secretly videotaped the slaughter of dolphins off the coast of Mexico. To level the regulatory playing field, the United States then passed the Marine Mammal Protection Act, which sought to extend dolphin-friendly fishing provisions to imported tuna. But as economist Daniel Esty points out, although protection may be the ultimate good for environmentalists, for free traders it is the consummate evil.[16] The law was struck down by a nonbinding ruling from the General Agreement on Tariffs and Trade (GATT, the WTO's forebear) in 1991. Nevertheless, Mexico, which had filed the case with the GATT, reformed its fishing practices, and an unofficial U.S. boycott on tuna caught with drift nets effectively continued through consumer activism.

Thailand, which as of 1996 had the ninth largest fishing fleet in the world, reacted angrily to the boycott. Thai officials even suggested that ASEAN should retaliate against EII and Greenpeace for waging an "irrational" campaign to save dolphins.[17] But under growing pressure from American consumers, Thailand and other major tuna exporters soon realized they had to meet the demand for "dolphin-safe" products. With the help and supervision of EII's representatives in Thailand, led by a passionate Canadian environmentalist named Ian Baird, the country's tuna fleet adopted dolphin-safe nets. According to Baird, once the industry had agreed to make the switch, the process went quite smoothly.[18] At any rate, the controversy soon died out.

But in May 1996, a new controversy flared over the shrimp-turtle dispute, which had similar origins. Like dolphins, sea turtles surface periodically to breath air, and they drown if they are trapped in nets underwater. The nets used by shrimp trawlers are particularly damaging, and were considered by the United States to be the major reason that all seven species of sea turtles are endangered. Before 1989, the U.S. shrimping fleet alone was believed to be responsible for the deaths of about 55,000 turtles every year. Beginning that year, the United States required its own shrimping boats to install "turtle-escape devices" (TEDs), a kind of trap door that allows turtles caught in the nets to escape, and also passed a law extending these regulations to imported shrimp.

As a penalty, the law imposed an embargo on imports of wild shrimp (not farmed shrimp) from countries not certified as practicing turtle-safe shrimp trawling. At first, however, the United States enforced the law with only fourteen countries in the Caribbean and the Western Atlantic, with which it first carried out a series of multilateral negotiations. Only after a coalition of environmental groups—including the Sea Turtle Restoration Project, which was affiliated with Earth Island Institute[19]—won a lawsuit filed with the U.S. Court of International Trade was the U.S. government forced to extend the embargo to Thailand and fifty-five other countries. Although most of Thailand's shrimp exports come from aquaculture farms and therefore were not affected by the ban, it still had the potential to affect roughly from $60 to $80 million worth of the country's annual foreign sales.[20]

There were some important differences between the shrimp-turtle dispute and the earlier dophin-tuna boycott. Sea turtles, unlike the dolphins at risk from drift nets, are threatened with extinction. In Thailand, turtle populations had been so devastated—by the loss of their habitat, the collection of their eggs, and their drowning in trawler nets—that the Fisheries Department estimated that only about five hundred turtles of reproductive age remained along the country's entire coastline.[21] So the campaign to protect them was probably more important than the dolphin-safe tuna campaign. On the other hand, turtles are not as "charismatic" as dolphins, so it's harder to generate popular support for their protection.

Whereas the wild shrimp embargo was imposed by the U.S. government (after it lost a court case), the dolphin-safe tuna campaign was ultimately enforced by a consumer boycott. Free trade advocates in the North much prefer the latter, and rightly so, since it is more democratic and, as a market-based rather than a government-based mechanism, less likely to be manipulated by protectionists (although free trade advocates in the South see little distinction between the two methods, arguing that both are coercive). The problem is that consumer-based campaigns are only as effective as a cause's, or an animal's, popularity. It also often depends on how a product is marketed: Canned tuna was vulnerable to a consumer campaign because it is sold by brands; shrimp are not because they are usually sold fresh. Finally, there was a crucial difference in the certification processes for the two situations. Tuna fleets were monitored by NGOs such as the EII to make sure they were equipped with dolphin-friendly nets, but for shrimp trawling it was left to U.S. officials to assist and monitor the installation of turtle-excluder devices.

In Thailand, at least, the American inspectors charged with enforcing the trade action did not seem terribly keen on the job. They visited only one fishing port, Songkhla, before certifying Thailand as having a turtle-safe shrimping fleet in November 1996, just five months after the embargo began.[22] Perhaps they acted out of expediency, since it was questionable whether Thai shrimpers had actually installed the TEDs. According to the Southeast Asian Fisheries Development Center, TEDs were handed out to Thailand's 3,000 registered shrimping boats;[23] but many trawlers are not registered with the government, and there does not appear to be any monitoring of whether the boats actually use the devices. Since pre-modified TEDs allow a small proportion of the marine catch to escape trawling nets along with turtles, fishermen may be reluctant to use them. Peter Fugazzotto, an American sea turtle activist who worked to have the embargo imposed, acknowledged that there are serious monitoring and enforcement problems, not just in Thailand but all over the world, including the United States. But he claimed that "as a result of the [U.S.] law, sixteen nations [including Thailand] have improved their [turtle conservation] laws. Overall, I believe the law has done more to protect sea turtles from needlessly drowning in shrimp nets than anything else, especially in [the] absence of any international environmental treaties with teeth."[24]

Despite the quick certification, the Thai government and business community, along with their counterparts in other developing countries, were furious with the United States for what they saw as an attack on their sovereignty. Ampon Kittiampon, an economist and high-ranking official at the Office of Agricultural Economics, doubted that the embargo had genuine environmental motives: "I personally think that the [trade dispute] is a result of political and economic motivation."[25] Sirilaksana Khoman, dean of the faculty of economics at Thammasat University, claimed that the United States had no right to take measures on behalf of wildlife: "Environmental protection is the domestic business of any nation, so countries should not be able to force others to preserve their resources."[26] Environmentalists respond that protecting migratory species such as birds, dolphins, and turtles, some of which travel thousands of miles through international skies and waters, is not a purely domestic issue. Indeed, the question of whether the management of any endangered species is a national or a global matter—that is, whether countries have a right to exterminate creatures living within their borders—is a central issue surrounding the Convention on Biological Diversity.

In trade circles, the turtle and dolphin cases raised a crucial issue: Do countries have a right to regulate the way imported goods are made? Environmentalists believe the answer will go a long way in deciding the future health of the earth. If no linkage between trade and the environment is allowed, companies will seek out the cheapest "process and production methods" (PPMs in trade jargon), even if it means killing off endangered species or emitting toxic pollutants, and then sell them at a price advantage on the world market. According to a theory known as the Pollution Haven Hypothesis, dirty industries may flee countries with strong environmental regulations and enforcement (read: the North) to find sanctuary in countries with weak ones (read: the South). Not only would that wreak havoc on developing countries but environmentalists fear it could cause a "race to the bottom" or a "regulatory chill" that undermines green rules—rules that have been put in place only after decades of fierce internal struggle—in the developed world. If so, people in the North could be left unemployed while the South chokes on industrial waste.

Governments and the business community in the South, along with free trade advocates in the North, also recognize the issue as crucial; indeed, they are dead-set against allowing environmental, social, and labor concerns to be mixed in with the rules governing international commerce. Trade rules should not be discriminatory about the way something is made, they argue, because the benefits of trade come precisely as a result of differing production methods in different parts of the world. Realizing that this important principle was at stake, and upset by the unilateral nature of the U.S. embargo on wild shrimp, Thailand, Malaysia, India, and Pakistan filed a complaint with the WTO. And in its transformation from the GATT, the Geneva-based institution gained some formidable judicial powers: Its dispute resolution hearings were now legally binding. In effect, it had become a commercial court for the world.

If it all sounds confusing, Thailand's former deputy prime minister, Supachai Panitchpakdi, an outspoken critic of linking trade with environmental and labor rules, explains the issue much more directly: Linkage would rob the developing world of its main economic advantage—the availability of plentiful natural resources and cheap labor. And his views are important because in August 2002, he became director general of the WTO.

SUPACHAI: MAN OF THE SOUTH AT THE WTO

As an environmental reporter covering the first ever ministerial meeting of the WTO, held in Singapore in 1996, I was in a unique position to watch the debate about trade and the environment unfold. There were none of the protests or battles in the street that occurred three years later in Seattle. Environmental issues were discussed at the lively NGO forums held in a separate building, but were barely even addressed at the formal WTO conference. There was, however, a parallel dispute about whether to link trade and labor rules. Labor activists were quite vocal in calling for the trading community to recognize the universal right to, say, a healthy workplace, or collective bargaining.

The North-South split that was so evident at the 1992 Earth Summit in Rio de Janeiro was also present in Singapore, and the issue of linking trade and labor drew the most heat. The U.S. and EU governments publicly supported linkage. "Trade liberalization can occur only with domestic support," said the U.S. trade representative, Charlene Barshefsky. "That support, and support for the WTO, will surely erode if we cannot address the concerns of working people and demonstrate that trade is a path to tangible prosperity."[27] Governments and businessmen in the South, however, are even more firmly opposed to universal labor rules than they are to environmental ones. The last thing they want to see are stronger unions and more red tape.

They argue that such issues are better dealt with by institutions such as the International Labor Organization (ILO) and the UN Environment Program (UNEP), and they would be right if these agencies were stronger than they are and enjoyed the judicial powers of the WTO. The argument is also misleading because the WTO's bylaws do take some environmental factors into consideration; for example, the Agreement on Sanitary and Phytosanitary Measures allows member states to adopt measures that protect the lives and health of humans, animals, and plants, and Article XX of the WTO charter allows exceptions to WTO rules so long as they are aimed at protecting wildlife and conserving natural reserves. The problem, complain environmentalists, is that the trade lawyers on WTO panels almost always seem to rule against green measures. Fundamentally, note economists Kyle Bagwell and Robert Staiger, "the WTO is driven by exporter interests,"[28] and that is the reason the South is taking an increasingly strong stand at the WTO on all kinds of issues. The East Asian "economic miracle" was driven by export-led growth policies, still seen as the best way to catch up economically and technologically with the North.

More disconcerting, however, is that even NGO activists from the South who have devoted their lives to critiquing mainstream development seem distinctly uneasy at the prospect of linking labor and environmental issues to trade, if not downright opposed to the idea. The Indian activist Sunita Narain derides such measures as "environmental fascism."[29] In Singapore, the Penang-based Third World Network worried that allowing linkage "opens the door for other, competing lobbying groups with their own special interests." Martin Khor, the group's director, argued that there are too many problems and inequities with the current trading system to start looking at "new issues." Roberto Possio, a Uruguayan academic, complained that "the labor issue is just being promoted by developed countries for publicity purposes."[30] Walden Bello, a Filipino who runs the Bangkok-based Forum on the Global South, and Singapore's Simon Tay both complained that the agreements worked out at Rio had "unravelled": The North had failed to live up to its promises of funding sustainable development in the South. "I may disagree with my government and its policies," explains Malaysia's Chee Yoke Ling, another member of the Third World Network, "but I'd rather be ruled by them than by some transnational corporation based in Tokyo, or some faceless bureaucrat in Washington D.C."[31]

Some Asian grass-roots environmentalists, on the other hand, have a very different view of Western intervention. The Chinese anti-dam activist Dai Qing welcomed linking trade rights to environmental, labor, and human rights issues: "The charges of imperialism are just a fabrication that is used to reject human rights."[32] Wildlife Fund Thailand (WFT) quietly supported the U.S. trade sanctions aimed at helping turtles for the renewed attention it brought to the issue of conservation. Fugazzotto, the American turtle conservationist, explains, "The general paradox we found was that while many of the think-tank Asian groups opposed the embargo, on-the-ground sea turtle conservationists and activists supported it. Obviously some of the activists did not like the idea of the U.S. telling them what to do, but others saw the law as a way to pressure their countries to do the right thing for the turtles." In other words, there seems to be a difference of opinion within the environmental movement of the South. Groups such as Third World Network, which focus on global inequities, are against linking trade rules with social and environmental concerns. But the environmental and labor groups that deal with the particular issues at hand may welcome trade actions if they come with technical assistance.

Green groups in the North need to do a better job of consulting and networking with these environmental groups in the South if they are to advance their trade agenda. They may also need to take a hard look at the Pollution Haven Hypothesis, upon which they have based many of their arguments. It's a useful theory because it unites the environmental and labor causes at home; but trade economists say there is little real evidence to support it, and that Western corporations who move their dirty operations to some Third World country with lax environmental enforcement are fewer than commonly supposed.[33] The question is, even if companies don't move, are their supply chains moving overseas? Some highly polluting industries, such as power generation, find it difficult to migrate because they must remain close to their markets.[34] Also, environmental regulations, and the cost of meeting them, turn out to be a relatively minor consideration for firms when they are deciding where to invest.[35] On the other hand, this means that companies that argue against stronger environmental legislation and claim that the higher costs involved will force them to leave for other countries are often bluffing, and governments can be bolder in ignoring such threats. Because although there is little evidence of environmentally motivated industrial migration, the mere threat of it could be having an impact on environmental governance: "The evidence here is by no means as one-sided as many economists have come to believe," writes Yale's Daniel Esty. "Some recent empirical studies find races to the bottom."[36]

During the early years of the WTO, opponents to linking trade and the environment got their way. Seeing how effective the dolphin-safe tuna consumer campaign was, the WTO's committee on trade and environment (essentially a collection of trade bureaucrats) demanded strict regulations on the way eco-labeling is used. Most ominously of all, it has refused to recognize the jurisdiction of trade measures used by international environmental treaties. Of the more than two hundred multilateral environmental agreements (MEAs), the WTO has identified at least twenty-three that employ trade measures and therefore could come into conflict with the trade organization.[37] Some MEAs directly regulate trade in species or harmful materials, and are given teeth through their power to impose trade sanctions. But the WTO's combative stance means that a country can, in theory, break its commitments to an environmental treaty it has signed and then try to escape trade sanctions levied as punishment by appealing to the WTO.[38] More broadly, it portends a titanic struggle between the institutions of trade and those of the environment on the international stage.

For trade-related discussions, it's useful to distinguish between two types of environmental measures: Either countries may try to regulate products themselves or they may regulate the process and production methods (PPMs) that go into making goods. Disputes about the former are more common at the WTO, which ostensibly allows countries to impose safety, health, and environmental regulations on imported goods. But a U.S. measure that tried to set cleaner standards for imported petrol from South America, for example, was disallowed on the grounds that it was discriminatory. An EU initiative to block the import of hormone-treated beef was also deemed protectionist because the dispute resolution panel ruled that the risk assessments did not contain sufficient scientific evidence of the negative effects of ingesting growth hormones. Critics questioned whether trade lawyers were qualified to make such scientific judgments, and were alarmed when the WTO refused to embrace the precautionary principle, sometimes known as the "better safe than sorry" rule.

The shrimp-turtle complaint did not revolve around product standards but around PPMs, which the WTO eyes even more skeptically. Predictably, therefore, the dispute resolution panel once again ruled decidedly against the United States. The panelists—three trade experts from Hong Kong, Germany, and Brazil—dismissed the fact that Article XX of the WTO charter allows member states to adopt measures that protect wildlife and conserve natural reserves, and essentially ruled that such measures would "undermine the multilateral trading system."[39] The panel declared it could reject such measures merely for setting a bad example that could potentially undermine the system. Although it asked questions of a team of environmental experts, it refused to consider NGO submissions concerning the dispute. The United States was ordered to change its policies; and since the ruling was now legally binding, the plaintiffs could demand compensation or respond with trade actions of their own.

Even legal experts seemed surprised by this close-minded reading of the WTO charter. "The panel predetermined that measures having an environmental object and purpose could not be justified under Article XX," wrote Joel Trachtman, a professor of international law at Tufts University. The panelists' "uncompromising allegiance to the multilateral trading system . . . contradicts the clear intent of Article XX."[40] Environmentalists, meanwhile, were livid. Because the panel's finding followed on the heels of other anti-green verdicts, the system seemed rigged. "The WTO keeps moving the goalposts, creating new tests at every turn for any country wishing to prevent environmental damage associated with international trade," complained Charles Arden-Clarke of the World Wide Fund for Nature's (WWF) trade program. "What is needed is the reform of basic WTO rules."[41]

But slowly, things seem to be changing. The U.S. appealed the shrimp-turtle verdict, and this time the ruling was more ambiguous, asserting that countries may use trade rules to protect global resources, but that the measure in question had been applied too unilaterally and unevenly. In 1999, under its new director general, Mike Moore, the WTO released a report conceding for the first time that trade can harm the environment.[42] It correctly noted that subsidies promoting the profligate use of resources are bad both for trade and the environment. Later, a dispute resolution panel actually ruled in favor of a French measure that barred the import of chrysotile asbestos and concluded that there was sufficient scientific evidence to indicate the substance causes cancer.

This rapprochement may have been spurred on by the "Battle of Seattle" in December 1999. While delegates at the WTO ministerial meeting were locked in stalemate about how to proceed with a new round of trade talks, protestors representing all manner of causes—labor activists, green groups, human rights workers, consumer organizations—took to the streets. The television cameras focused on the vandalism of a few extremists, but the bigger story was that a grass-roots movement opposing the negative impacts of globalization was coalescing, and it would reappear again and again at global conclaves. It's hard to say how much impact the protests have had on trade negotiations. At the WTO meeting in Doha in November 2001, ministers called for the environmental review of ongoing trade negotiations, but they avoided calls for setting common environmental standards because, they argued, protectionists would manipulate them.

By reasoning in this way, they continued to respect the views of the South, and its spokesperson, Supachai, who will be a key figure in the coming years. Executives at the WTO claim that the agency—with a few hundred employees, most of them in Geneva—has little power; the power of the WTO resides with its member states. Nevertheless, the director general does have the authority to select dispute panelists if the parties involved can't agree on one, and his position also carries diplomatic weight. Supachai has plenty of political support in the South, which represents a majority of the countries in the WTO, even if they are not its most powerful members. That is how he gained his post, despite opposition from the United States and other powerful Western countries. An advocate of gradual liberalization and a strong supporter of China's entry into the WTO, he will no doubt be expected to stand up to demands from the North for increased recognition of environmental and social concerns at the WTO, among other issues.

Supachai is widely admired in Thai business circles, even among notoriously skeptical business journalists. The son of a civil engineer and a math teacher, he was considered a child prodigy and entered medical school at the age of fourteen, but was persuaded by his father to take a scholarship offered by the central bank of Thailand to study economics. Three years later, Supachai went to the Netherlands and became the youngest undergraduate to enroll at Erasmus University, where he studied econometrics and development economics. Upon returning to Thailand, he rose to become head of the Thai Military Bank, then followed the path of many Thai technocrats and joined the Democrat Party. He was elected to Parliament several times and served as deputy prime minister and commerce minister. A fan of chess rather than poker, Supachai is considered a good strategist, but his tenure wasn't smooth in the turbulent world of Thai politics; indeed, the feeling is that he will have to develop stronger political instincts if he is to be a successful WTO chief.[43]

So what are the prospects for linking trade with the environment? Based on the more recent dispute resolution verdicts, it appears the WTO will concede the right of countries to ban, regulate, or label imported goods that are environmentally harmful—so long as the measures aren't discriminatory and are backed up by science. On PPMs, the South may get its way in rejecting rules that seek to set common standards on the way goods are made so long as it affects purely domestic environmental issues. But there is still tremendous controversy about how to proceed if the resources at stake cross national boundaries—if they threaten the oceans, the atmosphere, or endangered migratory species, for instance. So far, the WTO, essentially a global Commerce Ministry run by trade officials, has tried to have it both ways. On the one hand, it says it's not equipped to establish fair social, labor, and environmental regulations, and should not be in the business of adjudicating them. On the other hand, it has not recognized the right of other institutions—the ILO, UNEP, the multilateral treaty secretariats—to do so with the help of trade measures. The WTO can't continue to monopolize authority on these issues.

Commerce can't be considered in isolation from the people and resources that make it run. The same pressures that drive all countries—from east, west, north, and south—to draft internal rules regarding labor and the environment are also making themselves felt in a world increasingly linked by trade. Wherever common markets have been created, both within and between countries, as with the EU and the North American Free Trade

Agreement (NAFTA), common environmental rules have followed. If ASEAN succeeds in setting up a free trade zone, it, too, will have to set some common standards. And if free trade is truly creating a global market, precedent suggests there will eventually be global environmental regulations. It's not clear whether the leaders of the South recognize this.

Supachai has a chance to make a huge impact on this debate. Will he seek merely to further the interests of the Southern leaders whose support won him his post? In theory, he could reach out to the North and use his clout in the South to broker a compromise on linkage that is acceptable to the world community—the "Nixon goes to China" approach. But he has a reputation as a man who sticks to his convictions, and so far he has continued backing policies the South favors. "For the really controversial issues of non-trade items, like environment and labor . . . cautiously. I don't think we should have them in the negotiations," Supachai said at the Asia-Pacific Economic Cooperation forum held in Brunei in 2000.[44] The South, angry because its interests have been poorly served by past trade agreements, seems to have momentum on its side in the new Doha Round, also called by some the "Development Round." What's more, it now has a powerful ally (on the linkage issues) in President George W. Bush, who also wants to keep labor and environmental issues out of trade rules. So there is no reason to think that Supachai will change his stance, although new trade deals will still have to be approved by a skeptical U.S. Congress and the more environmentally minded EU.

So far, Supachai's proposals have mirrored those of other free trade advocates: create new international institutions, or strengthen existing ones such as the ILO and UNEP, to look after these other global issues. In essence, set up global environment and labor ministries to go along with the work of the WTO and IMF as global commerce and treasury ministries. The plan could work, but it rests on some important assumptions. The WTO must finally respect the right of these agencies (or of the multilateral treaties that shape them) to have some say about trade and to use sanctions for enforcement. And the South, particularly its largest and most powerful countries, must genuinely engage with these institutions, not merely fob them off as talking shops while they maintain the trading access they covet so dearly. For if they stubbornly refuse to address the negative side effects of globalization, we will be back to square one, and growing grass-roots opposition could eventually undermine the multilateral trading system on which prosperity depends.

HUNTING FOR GREEN GOLD: BIOPROSPECTING OR BIOPIRACY?

All manner of miraculous compounds have been culled from forests. The anti-cancer drug taxol was isolated from moss that grows on the Pacific yew tree. Rubber is made from the sap of Brazil's hevea tree. Aspirin originated from a species of fungus. But a plant found in northern Thailand surely has the most seductive compounds of all.

The white *kwao krua* plant has the seemingly magical ability to enlarge a woman's breasts. The active ingredient turns out to be phytoestrogen, which is similar to the female hormone estrogen. When ingested in sufficient quantities, it can cause breasts to grow by as much as one inch in five days, according to Vichai Cherdchivasart, a plant expert at Chulalongkorn University.[45] And men may like the plant's cousin even more: the red *kwao krua* contains an elixir that Vichai claims is a sexual stimulant much like Viagra. No doubt it's already a big hit down on Pat Pong (Bangkok's red-light district).

Bioprospecting, the hunt for useful compounds amongst living organisms, was supposed to have an even more miraculous effect on the cause of conservation. The possibility of finding valuable natural products, "green gold" if you like, should in theory provide poor countries with a greater incentive to protect their forests, reefs, and wildlife, which can certainly use all the protection they can get; indeed, the decline in biodversity is considered one of the most serious environmental problems the world faces, and the most irreversible.

Some biologists claim that we are currently in the midst of an "extinction crisis," although there are heated debates about just how many species are dying out, and how fast they are going. Typical estimates claim that in the second half of the twentieth century, the earth lost 300,000 species. Biologists' best guess is that species are disappearing between 100 and 1,000 times faster than before humans entered the scene.[46] The fact is that the debate about biodiversity loss is marked by extreme ignorance. In Thailand, the Office of Environmental Policy and Planning (OEPP) reports that about 14,000 animals have been scientifically identified.[47] That number represents only 16 percent of the country's total estimated 86,000 animal species, which in turn is thought to amount to just 6 percent of the world's species—and these figures refer to animals only, not to plants or microorganisms.

Each extinction represents an incalculable loss to science, and possibly the loss of a vital strand in the web of life that sustains the world as we know it. For biodiversity also provides a crucial form of insurance. In the 1970s, for instance, Asia's rice paddies were attacked by a pathogen known as the grassy stunt virus, which defeated all farmers' and scientists' attempts to defend against it. The virus inflicted increasingly severe damage until finally a variety of rice was found that proved to be immune to it, and the immunity was bred into the crops. That variety happened to come from a valley in northern India, a valley that was subsequently flooded by a dam reservoir, demonstrating just how precarious our genetic reservoirs may be.

So it is that each specie may contain some valuable compound that can be extracted, commercialized, and sold at great profit. For all these reasons, the world community got together at the 1992 Earth Summit and signed the Convention on Biological Diversity (CBD), which was designed to protect this rich natural heritage for future generations. And the CBD is doing that with the help of funding from the Global Environmental Facility (GEF), a fund that uses money from the North to finance projects aimed at preventing climate change, loss of biodiversity, destruction of the ozone layer, and harm to international waters. But these conservation efforts have been largely overshadowed, and often diminished, by a fixation with economic issues related to biodiversity.

Sometimes, the act of bioprospecting can threaten individual species. Once the breast-enhancing properties of *kwao krua* became widely publicized, for instance, there was such a run on the plant that the government had to declare it a reserved species. Another example occurred in Kenya, where the entire adult population of a tree species known as *Maytenus buchananni* was harvested in an effort to test it for anti-cancer properties.[48] Admittedly, however, these are minor incidents compared to all the other factors—a loss of habitats, changing climates, toxic pollution, the spread of invasive or predatory alien species—blamed for causing extinction.

The most common fate for species that become commercially valuable is for them to be domesticated, or bred in captivity, even as they are hunted to extinction in the wild. That's what is currently happening to the giant catfish of the Mekong, and even to most species of tiger. Crocodiles are no longer found in the wild in Thailand, except when they escape from one of the ubiquitous crocodile farms. The danger is that the loss of wild populations will reduce a species' genetic diversity, causing it eventually to weaken and die out. But animals bred in captivity can, at great cost, be restored to the wild once proper protections are put in place, as has happened with wolves in the western United States and the golden lion tamarin in South America.

The possible fate of *kwao krua* reveals another major problem concerning bioprospecting. According to Vichai, Thais have known about the plant's titillating properties for decades. But a couple of Japanese pharmaceutical firms have sought to patent its active ingredient so that it can be developed into a commercial product. NGOs of the South have a term for this sort of behavior—"biopiracy"—and they claim that it occurs far more often than most people realize. Some scientists scouring the forests and reefs of tropical countries, which contain the vast majority of the world's biodiversity, exploit the knowledge of local healers and take away samples of anything that might be useful, sometimes without asking permission. Back home, if a natural compound seems to have commercial potential, a scientist, university, or corporation usually tries to patent it—or the means to extract and produce it—to gain monopoly rights over it.

Sometimes the organism in question might be rare or endangered. Sometimes it might have been known, protected, or even cultivated by local people for generations. But usually, even if the compound is turned into an immensely profitable product, the local people receive nothing. And since the companies and scientists who ultimately pocket those profits usually come from the North, researchers such as Cambridge's Bronwyn Parry question whether "the flow of material from the gene-rich developing world to the gene-poor industrialized world is a form of bio-colonialism."[49]

Actually, the history of exporting useful species goes back millennia, perhaps to the time when wheat was transplanted from the Horn of Africa to Mesopotamia. Transfers increased notably during the era of European colonialism. The potato originally came from the Andes but ended up becoming the staple crop of Ireland. And sometimes it may be that nobody minds. Just like modern-day intellectual property, genetic material is essentially "non-rival": One person can use it without diminishing the ability of others to do so. That's why many people do not consider the unlicensed use of such material to be thievery.

But at times history has been altered because useful plants carefully bred and protected by developing countries were stolen. In the nineteenth century, the newly independent Andean states lost their monopoly over quinine when British and Dutch collectors illicitly exported seeds from the chinchona tree. Growing their own plantations in colonial territories enabled the Europeans to manufacture sufficient amounts of the drug to fend off malaria when they set about conquering the tropics. Even more famously, the English explorer Henry Wickham, employed by the Royal Botanic Gardens in Kew, Surrey, managed to sneak a boatload of rubber tree seeds out of Brazil in 1876, thus enabling the British to grow their own plantations in Singapore, Malaysia, and Ceylon. By the time the automobile industry emerged in the early part of the twentieth century, increasing the demand for rubber to make tires, Brazil had lost its monopoly over the rubber trade.

Countries of the South have also benefited from species exchange, however. The rubber industry has been an invaluable source of income to many Southeast Asian countries, including Thailand, Cambodia, and Vietnam. Corn and potatoes, both of which come from the Americas, are grown extensively in the region. Altogether, an estimated 34 percent of the crops in mainland Southeast Asia are non-native, according to Boriboon Sumrit at Thailand's Department of Agriculture.[50] On the other hand, notes Sirina Teo of Singapore's Bioscience Centre, citrus plants originated in Southeast Asia.[51] Just as with goods and services, the trade in species provides net benefits to the world, but it imposes opportunity costs on the countries who owned the original DNA. And with the current patenting system, the South is losing control of its genetic resources to the hi-tech firms from the North.

The most infamous such event in Thailand concerns the *plao noi,* an unassuming plant found along the Thai-Burma border in Prachuab Khiri Khan province that contains a compound effective at healing peptic ulcers. Researchers from Japan discovered the plant's medicinal properties in 1970 and managed to isolate the active ingredient, plaunotol. The Japanese pharmaceutical company Sankyo then patented a process to extract and purify plaunotol, turning it into a drug known by the trade name Kelnac. Since the plant could not be grown in Japan's climate, Sankyo set up a plantation in Prachuap Khiri Khan to grow the raw material. The company has not publicly revealed how much money it has earned from sales of Kelnac, but when it began marketing the drug in 1986, it projected annual sales to be around 5 billion yen, or nearly US$25 million[52]—small potatoes to giant drug firms, perhaps, but a lot of money in Prachuab.

Similar stories have cropped up across the Third World. Two cancer drugs were developed from the rosy periwinkle found in Madagascar, one of the biologically richest but economically poorest countries on Earth. And yet Madagascar received no benefits. The female contraceptive pill, meanwhile, originated from a compound found in wild Mexican yams. Altogether, the World Conservation Union (IUCN) estimates that a quarter of the world's industrially produced drugs are based on natural compounds, and most of these modern pharmaceuticals are correlated to traditional medicinal use. But out of some 25,000 biotechnological patents granted worldwide between 1990 and 1995, writes the Argentine academic Carlos Correa, only about 6 percent originated outside of the United States, Japan, and the EU.[53]

Patents allow corporations to charge high prices, which draw particularly bitter complaints when the products in question are lifesaving drugs. It's an issue throughout the developing world, particularly in Africa, which faces an AIDS epidemic but whose people can't afford to buy the drug cocktails needed to stave off the disease. "When drug companies come here and collect samples, they say it is the collective heritage of mankind," complains Pennapa Sabcharoen, a physician and director of the National Institute for Thai Traditional Medicine. "Then they study it, develop it, claim intellectual property rights on it and come back to Thailand and make us pay for it." The pharmaceutical firms retort that patents and high prices are needed to recoup the massive research, testing, and development costs of producing safe and reliable drugs. And there's no doubt those costs are prohibitive. They are the main reason why companies from the South have found it so difficult to convert their own biological patrimony into profitable products.

Multinational corporations have found numerous ways to access the genetic resources of the South, often signing agreements with developing countries. A widely heralded example came when the U.S. pharmaceutical firm Merck purchased the right to prospect for useful compounds in Costa Rica by teaming up with a local research institute called INBio. Somewhat more dubious was a story I uncovered in Singapore. Glaxo Wellcome, Britain's largest drugs firm, managed to gain access to plants from all over Southeast Asia, and from as far afield as India and Latin America, through an agreement with the Singapore Botanical Gardens, whose collections go back to British colonial days. Access was gained through a S$60 million, ten-year joint venture with Singapore's National University and its Economic Development Board.[54] There is nothing illegal about this. The CBD does not apply retroactively to such collections, which are considered to be the "common heritage" of mankind, and thus free from benefit-sharing requirements. But apart from Singapore—whose scientists admitted that they need the help of a multinational pharmaceutical firm to develop drugs from raw genetic material—the countries whose plants are contained in the collection will see nothing from the deal.

An even more clever technique than scouring botanical collections is to take advantage of the traditional knowledge gained over the ages by indigenous people. Once again, however, there is a fine line between collaboration and exploitation when it comes to ethnobotany. In the strife-torn Mexican state of Chiapas, a consortium of local healers has opposed a project by researchers from the University of Georgia, who had planned to study the medical knowledge of traditional Mayan healers. In Thailand, similar outcries have been heard, first about an ethnobotanical study funded by a whiskey importer named Riche Monde, and later about the failure of a team of scientists from the University of Portsmouth to return some samples of medical organisms they had collected in Thailand (after a great hue and cry, the samples were sent back). In neither case was there proof that the researchers were up to no good, so the judgement as to whether these were examples of biopiracy or simply the figments of NGO paranoia depends largely on one's own inclinations.

The question of genetic property rights has even spread to humans. The United States has allowed patents to be filed on human genes, cells, and cell lines that may prove useful in fighting diseases. Applications have been filed for patents on cell lines taken from natives of the Solomon Islands and the Hagahai people of Papua New Guinea. And indigenous tribes are not the only ones being targeted. Metrodin HP, a hormone used for fertility treatments, was originally extracted from the urine of European nuns.[55]

There is now a growing movement in the South, aided by sympathetic groups in the North such as the Canada-based Rural Advancement Foundation International (RAFI), to challenge the alleged inequities of the industrial patenting system. The government of India, for instance, has successfully fought to revoke a couple of the more egregious patents—which are supposed to be based on novelty, usefulness, and non-obviousness—that were awarded on native compounds whose useful properties have been known for centuries: tumeric, which has medicinal qualities and was claimed by researchers at the University of Mississippi; and neem, used as an insecticide and claimed by the American chemicals firm W. R. Grace.[56] Dozens of other questionable patents are also being challenged.

NGOs from the South are particularly irate at what they consider to be a double standard. Although the intellectual property rights of multinational corporations from the technology-rich North are zealously defended, those of local villagers from the genetically-rich South are ignored. "Ironically, the companies that accuse the Third World of piracy and have created [intellectual property rules] to stop this piracy are themselves engaged in large-scale piracy of the biological wealth and intellectual heritage from the Third World," writes Vandana Shiva, an Indian scientist and activist.[57] These complaints are being heard. The issue of "access and benefits sharing" (ABS) is increasingly being recognized as a vital issue under the CBD.

THE BIODIVERSITY CONVENTION: WAYLAID BY TRADE

In Thailand, the issue of access and benefits sharing has become so controversial that, although the country has signed the UN Convention on Biological Diversity, it has yet to ratify it. On the surface, the delay has been the result of a bizarre and complicated constitutional dispute about whether the treaty needs to be approved by Parliament or just the Cabinet. But behind the legal debate lies a much more profound controversy about whether the treaty will benefit Thailand. NGO critics of the CBD that claim it will lead to a loss of sovereignty because foreign researchers will be able to exploit and profit from the kingdom's genetic resources, a concern shared by nationalists who have also weighed in against the treaty. Defenders of indigenous people and anti-globalization activists also argue that local people who have helped preserve biodiversity and sometimes bred species for their useful qualities, should receive a share of the profits—a principle known as "plant-breeders' rights."

These concerns haven't prevented other developing countries—even those normally outspoken in defending their environmental sovereignty, such as India, Malaysia, and Brazil—from ratifying the treaty; 183 states are now party to the convention, believing it has enough safeguards to protect their interests. Thailand's Office of Environmental Policy and Planning (OEPP) agrees, pointing out that the CBD leaves the issue of who owns genetic resources up to member states. But Thai NGOs worry about Article 15, which states that each member country "shall endeavor to create conditions to facilitate access to genetic resources for environmentally sound uses by other Contracting Parties." They want a law protecting local rights over genetic property before the treaty is ratified. "We're not trying to stop commercial development and we're not challenging the principle of intellectual property rights," says Pennapa, who runs a government institute on traditional medicines, but shares the views of Thai NGOs. "But we don't want to be taken advantage of any more. We pay a lot for drugs and we need to have some rules, some self-defense."[58] The OEPP counters that Article 15 also states, "Access, where granted, shall be on mutually agreed terms and shall be subject to prior informed consent," and its officials are upset that, by failing to become party to the treaty, Thailand has denied itself access to conservation funding from the GEF.

The biggest wallflower of all to the CBD, however, is the United States, where powerful drug and pharmaceutical firms oppose ratification because the convention's objectives include not just "the conservation of biological diversity [and] the sustainable use of its components," but also the "fair and equitable sharing of the benefits arising out of the utilization of genetic resources." The United States nevertheless still manages to throw its weight around during treaty negotiations, while the Thai observers generally watch quietly from the sidelines. Essentially, these two countries represent the extremes of opinion about the issue of access and benefit sharing: The United States zealously protects the interests of its hi-tech corporations, while Thailand seems unsure whether the treaty will allow it to maintain control of its biological wealth. In both countries, the treaty has essentially been waylaid by property rights issues related to international trade.

Actually, that may be true for the convention as a whole, because scientists increasingly complain that all the concern about access and benefit sharing has damaged the cause of conservation. Due to increasing concern about biopiracy, researchers—foreign and domestic, commercial and academic—have found it increasingly difficult to carry out biodiversity fieldwork in tropical countries. It can sometimes now take more than a year to obtain the necessary permits to study remote areas. Even the basic but vital taxonomical work of describing and recording previously unidentified species has been set back. The problem seems especially serious in Latin America, where some researchers have even been chased out by suspicious locals, but it has also hampered study in Southeast Asia. Navjot Sodhi, a biologist at the National University of Singapore, finally had to cancel his survey of bird species in Sarawak because of all the red tape and official harassment he had to endure after rumors spread that an AIDS drug had been found in the area. "I couldn't take the nonsense any more, and we pulled out," he said. "I was willing to sign anything saying that we were not doing any bioprospecting."[59] But as journalist Andrew Revkin reported, there was nothing to sign. Perversely, restricting access probably hampers the legitimate researchers more than the biopirates, who may carry on their work without seeking permission, or may simply move to another country.

If the dispute about bioprospecting seems familiar, that's because it touches on similar issues to the dispute about the establishment of wilderness parks (discussed in Chapter 5). Once again, the South's environmental democracy movement claims that local people are sacrificing their resources to further the cause of conservation. And with bioprospecting, it is even clearer that the economic benefits are going to the urban-industrial sector—to the corporations of the North, no less. More ecologically minded groups again worry that this focus on local rights is coming at the expense of conservation, and generally support the CBD. In Thailand, however, the OEPP does not have as much clout as the Royal Forest Department (RFD), and has not been able to convince the government that ratifying the treaty is in the country's interests.

Access and benefits sharing is not the only tangential issue to steal the limelight at the CBD. The North has paid far more attention to negotiations about the convention's Biosafety Protocol, which deals with another trade-related matter: how to regulate commerce in genetically modified organisms (GMOs). The split here is between the "Miami Group" of agro-industrial exporters—including the United States, Canada, Argentina, Australia, New Zealand, Chile, and Uruguay—and a group of countries more skeptical about the benefits of GMOs, led by Europe with its deep-seated fears about food safety. The environmental ramifications are also huge. Once released into the environment, hardy GMOs could spread like weeds and push out indigenous species. For instance, pollen from bt corn—a genetically engineered species popular with farmers because it produces its own pesticide—has been shown to harm the larvae of monarch butterflies. "Irreversability distinguishes the release of GMOs from other technologies," argues Greenpeace's Jan van Aken. "Once they are in the environment, they cannot be recalled."[60] Similar concerns are aired within the hi-tech industry; Bill Joy, former chief technology officer of Sun Microsystems, warns that we are now entering a dangerous new era in which technology can reproduce itself.[60a] That offers benefits, as well, say biotech's defenders, by potentially reducing the need for chemical inputs.

It took years of intense negotiations—talks collapsed at the Cartagena conference in 1999—before agreement on the Biosafety Protocol was finally reached at a meeting in Montreal in 2000. Facing down opposition from the Miami Group, most developing countries called for strict regulation of trade in genetically modified crops and seeds to protect their environment, farmers, and biological diversity. European delegates, at the behest of consumer groups, pushed for labeling transgenic crops. The treaty had "a lot of holes," concluded the Third World Network's Chee Yoke Ling, "[but] it's historic in the sense that international law is recognizing that GMOs are distinct and have to be regulated separately." It requires the labeling and segregation of all living GMOs (seeds, fish, and trees, for instance) traded between countries—since they can spread if released into the environment—but not of genetically modified (GM) crops or other products.

Most important, perhaps, the protocol repeatedly and explicitly enshrines the precautionary principle, allowing states to take protective action even if there is no scientific certainty that a certain GM product is dangerous. Such provisions stand in marked contrast to the view of the WTO. "Thus, a potential for conflict clearly exists . . . it is only a question of time until a dispute before the WTO is initiated," writes Eric Neumayer, author of a probing book titled *Greening Trade and Investment,* which contends that the WTO's Sanitary and Phytosanitary Agreement should be reformed to embrace the precautionary principle.[61] The framers of the protocol seem well aware of the coming confrontation: "We ensured that the Biosafety Protocol is independent of and equal to the WTO, and not subject to their free market rulings," said Jo Dufay, a Canadian activist.[62]

That could set an important precedent for future negotiations concerning access and benefits sharing, because the biggest threat to the South's control of its biological resources probably comes not from the CBD, as Thai NGOs suspect, but from the WTO's Trade-Related Intellectual Property Rights regime (TRIPs). In the words of Third World Network's Martin Khor, Article 27(3)(b)* of the TRIPS agreement "will be the key to the next century, which will be dominated by biotechnology and genetic engineering."[63] It covers the right to patent life forms and according to Khor's interpretation requires member states to protect plant varieties to the satisfaction of countries such as the United States. All developing countries were supposed to have such protections in place by 2006, but following complaints from the South, the WTO meeting in Doha agreed to postpone that deadline. Kenya has spoken on behalf of many developing countries in calling for an end to the patenting of living organisms because of fears that farmers will rush to plant monocultures of high-yield standardized crops, thus concentrating market power in the hands of the Northern companies that own the

---

*It reads: "Members may also exclude from patentability plants and animals other than micro-organisms, and essentially biological processes for the production of plants and animals other than non-biological and micro-biological processes. However, members shall provide for the protection of plant varieties either by patents or by an effective *sui generis* system or by any combination thereof."

seeds. It's not only farmers' well-being at stake, say the NGOs, but the biodiversity of cultivated species.

The more you study the issues surrounding biotechnology the more you realize that the divisions over them are actually between those who hope to capitalize on hi-tech advances and those leery about how their food, resources, and livelihoods will be affected. Often, that schism follows the North-South line, but not always. European consumers and health and farmers' groups have pressured their governments into carefully controlling the introduction of GM food. In Latin America's southern cone, on the other hand, agro-industry has set the agenda, causing Chile, Argentina, and Uruguay to join the Miami Group in pushing GM crops; and hi-tech Singapore has sought to become part of the bioprospect*ors* instead of the bioprospect*ees*. This reveals an interesting dynamic within and between developing countries. Virtually everyone in the South agrees that the global governance system—whether it concerns trade, the environment, or some other issue—is inequitable and favors the North. The response of Southern NGOs is to challenge that system and try to reform it. Rhetorically, Southern governments may support such efforts, particularly in international confabs. But when they have the opportunity, their broader strategy is to try to increase their technological capability so that they can become part of the North, even if it comes at the expense of their neighbors and their environment.

The key question now before the CBD is where the ABS debate will go. Regarding access, there is a growing consensus that a two-track system of permits is needed: one for bioprospectors that protects the genetic property rights of host countries; and another less onerous one for purely academic researchers. But creating this system, and distinguishing between the two types of researchers won't be easy. "Academics have been kind of naïve to the question of ownership of genetic material," says Eric Mathur of Diversa, a San Diego–based firm that carries out bioprospecting. "They think that under the guise of academia they can do whatever they want. But if their work results in any kind of invention—and most come serendipitously—you can be sure their institution will want to own it and make money from it."[64]

An effective permitting system should help governments of the South protect their genetic property rights. But shouldn't some of the benefits also go to the people living in the area where useful organisms are found, particularly if they helped to breed it or sacrificed to protect it? The statement of a

U.S. trade official expresses the cynical view of benefit sharing: "That anybody thinks they should get a share of the profits because they happen to be squatting on the forest where the resources are is laughable."[65]

Actually, it's not, because if you believe in the power of incentives, it could help save the forest. Establishing genetic property rights should give governments and people in the South another much-needed motive to protect areas with large biodiversity, much as intellectual property rights help spur innovation among scientists and corporations. After all, Thailand's RFD likes to claim that the activities of local people—poaching, encroachment, setting fires—are now the primary threat to protecting nature reserves, so you'd want to make sure that local people see some of the benefits if they are convinced to help conservation efforts. Just as NGOs of the North say that trade and the environment are clearly inter-related, those of the South say the same about conserving biodiversity and sharing the benefits.

There are difficulties with this approach. First among them is that all the hype surrounding bioprospecting has created inflated expectations. "[Bioprospecting has] never really panned out and was totally oversold," says George Amato, director of the Bronx Zoo's conservation genetics program.[66] Although finding a useful compound can be immensely profitable, researchers estimate that only about one in 100,000 has potential for development,[67] suggesting that the marginal value of genetic resources is actually too small to provide an incentive for conservation. But while that may be true if the search is random, other studies indicate that information having the potential to help isolate useful organisms is valuable enough to command payment.[68] In other words, there is a strong argument for at least rewarding local people who provide ethnobotanical knowledge.

The danger in this debate, however, is that we forget the bigger picture. All the publicity about biosafety and ABS has stolen the limelight in the press (and in this book) from the conservation work that needs to be done, and skewed the focus of the Convention on Biological Diversity. "It is much more about sharing the profits from genetic resources than it is about conserving biodiversity, about science," says Stanford Law School's John Barton, an expert on the treaty.[69] Benefit sharing is important, but it must be done in a way that supports conservation efforts, for there are other important reasons to protect biodiversity. Are we simply to ignore the vast majority of species that can't be turned into commercial products? Only at our own peril.

CLIMATE CHANGE: THE KYOTO COMPROMISE THAT WASN'T

At a conference on the oil industry and the environment held in Vietnam in 1996, Thai environmentalist Pisit na Patalung raised an intriguing question: "Nature usually has a purpose in doing things, so what is its purpose in keeping all this oil underground?" he asked. "How are we disrupting the natural system by taking it out?" The engineers and businessmen at the conference didn't know what to make of such a question. They hemmed and hawed, and even snickered a little bit.

But a climate expert could have answered him immediately: Deposits of fossil fuels such as oil, coal, and gas, which are the remnants of ancient life forms, help store carbon. By digging them up and burning them, humans have inadvertently sent huge quantities of greenhouse gases—the bulk of which are carbon-based—into the atmosphere. A sizable portion is absorbed by the oceans, and some is taken up by vegetation to grow; but a major chunk remains in the atmosphere for centuries, where it has the effect of retaining heat that would otherwise radiate into space. In pre-industrial times, atmospheric concentrations of carbon dioxide, the main heat-trapping gas, averaged about 280 parts per million (ppm). It is now around 370 ppm, and increasing by roughly 1.5 ppm every year. The Intergovernmental Panel on Climate Change (IPCC), a huge team of scientists brought together by the UN, has been studying the effects of this accumulation; in the mid-1990s, it finally concluded that "the balance of evidence" suggests that humans are having a discernible impact on the Earth's climate.

The debate goes on about what the impacts of climate change will be, and there are still skeptics—particularly in the Cabinet of President George W. Bush—who want to deny that it's a problem at all. But the latest IPCC assessment released in 2001[70] contained some alarming forecasts. Scientists now have "observational evidence" that the Earth is warming at a faster rate than at any other time in the last 10,000 years. Average temperatures are expected to rise between 1.4 and 5.8 degrees Celsius by the end of this century. Impacts will likely include water shortages, the spread of tropical diseases, increased droughts, floods, landslides, and sea storm surges. Sea levels have risen by from 10 to 20 centimeters (cms) since 1900 and are predicted to rise anywhere from 9 to 88 cms during the next century. The IPCC's then-chairman, Robert Watson, said he expected "tens of millions" of people to be displaced by rising seas.[71] Some countries, such as the Maldives, a low-lying archipelago in the Indian Ocean, may be wiped out. And there are potential catastrophes: the sudden collapse of a gigantic polar ice sheet, for instance, or the switching off of the Atlantic Conveyor, a warm-water current that helps keep northern Europe warmer than its latitudinal cousins.

Not everyone will be affected equally. Some regions—particularly in the industrialized North, largely responsible for climate change—may benefit from increased crop yields, longer growing seasons, and shorter winters. It's the poor who are most vulnerable, even though they are least responsible, because they don't have the resources to adapt. "Those people most at risk from global warming are those that make their living from the land and the seas in the tropics and subtropics of Africa, Asia, and Latin America," Watson warned. He called it "a tragic irony" that sub-Saharan Africa, with the lowest level of emissions of greenhouse gases, could be one of the most affected regions if the earth's climate warms significantly.

This potential disconnect between those largely responsible for climate change and those most likely to suffer from it makes preventive action all the more difficult. In Southeast Asia, although there is general awareness about the dangers of global warming, the issue receives little attention. With the help of GEF funding, governments took the important first step of measuring how much greenhouse gas their factories and farms were emitting, and assessing the status of their "carbon sinks"—essentially forests, wetlands, and tree plantations—which can store carbon that might otherwise enter the atmosphere as greenhouse gases. But beyond that, policy measures are haphazard at best. Indonesia, an equatorial nation made up of thousands of islands, and thus one of the countries most vulnerable to climate change, generally puts junior officials in charge of the issue, signifying that it was of minor importance. The same is true in Thailand, although the situation there is somewhat different. The Thais have for security reasons already committed themselves to energy efficiency and have received $12 million in GEF funding for demand-side management programs.[72] On the other hand, the Philippines, another archipelagic nation, reportedly has a well-coordinated policy team working on climate change issues; and, by all accounts, so does China.

The reason climate change receives little attention despite the threat it poses is that most people in the South feel, justifiably, it is up to the North to take the first steps to solve the problem. Also, the presence of much more immediate problems—environmental, economic, political, you name it—makes it difficult to focus on a problem that is so huge and that will play out for such a long time span. People may die in a few decades from storm surges and reduced fresh water supply, but in the developing world many are dying today from floods, famines, and diseases caused by unsanitary conditions. Perhaps that explains why I tended to write more about climate change than most of my colleagues in the Thai press.

In doing so, I realized that the issue saw such little play for another important reason, one the environmental movement barely mentions because they fear the cure more than the disease, so to speak. If rapidly industrializing countries such as Thailand seek to release fewer greenhouse gases from generating electricity, they wouldn't halt development but rather rush to build more hydroelectric dams and probably go nuclear—technologies that generate far more opposition in developing countries than fossil fuels. Some visionaries, among them Amory Lovins, founder of the Rocky Mountain Institute, believe that efficiency gains could allow countries such as Thailand to expand their economies without having to increase their electricity supply. But technocrats are skeptical of such claims.

Certainly within industrialized countries, the reduction in greenhouse gas (GHG) emissions required to stabilize concentrations in the atmosphere is so huge, explains Richard Bradley, a policy analyst at the U.S. Department of Energy, that efficiency can help only at the margins. "Ultimately, to reach the levels we're talking about—roughly 550 ppm of carbon dioxide—we need fuel switching. We need a transport system that runs on hydrogen."[73] Humans have been creative at finding new fuels. We moved from wood to coal to oil and now to natural gas, becoming cleaner along the way, each new fuel containing more hydrogen and less carbon. Unfortunately, we've never ditched the old fuels.[74] On a planetary scale, even wood is being used now more than ever. That's why there is no Kuznets curve for GHG emissions; as countries become richer, they just keep on growing. The new technologies needed to mitigate climate change will therefore require a massive mobilization of research and development. So for practical and moral reasons, the North needs to take charge.

That's what was expected at Kyoto, a conference of parties to the Framework Convention on Climate Change. By negotiating binding targets for reducing GHG emissions, the North was to demonstrate its seriousness on the issue, signaling corporations that it was time to seek energy alternatives. The EU was ready to accept such limits; it was in a good position to meet targets because England and Germany had recently phased out their coal industries. Japan was nervous about it, but as host was expected to agree. The United States and its CANZ allies (Canada, Australia and New Zealand) were more recalcitrant, but it was hoped that Al Gore, America's environmentally minded vice president, would rally the green troops.

The United States, quite reasonably, wanted a "cap and trade" system similar to the one for sulfur dioxide and that has worked so well and at such little cost domestically. Essentially, a cap is set on the total amount of pollution allowed, and then polluters buy and sell the right to emit sulfur dioxide (or carbon dioxide, or anything) within that limit. Creating a market in emission permits or credits not only provides companies with flexibility and an incentive to reduce pollution but also takes advantage of the greater efficiency some firms have over others. Bradley, an economist, estimates that a perfect global carbon trading system would cut mitigation costs by about 70 percent.

Another key issue at Kyoto was how best to channel money from the North to projects in the South that would help prevent climate change. The North in general and the United States in particular were eager to see these Joint Implementation (JI) projects included in the Kyoto Protocol because they are more cost effective than similar programs in their own countries. In the years leading up to the Kyoto summit, the South was cautious toward JI, fearing it was a ploy by the North to avoid making difficult reductions in their own emissions. "We see JI as a way to shift responsibility [from the North to the South]," said Saksit Tridech, one of Thailand's top environmental negotiators.[75] Such stridency was in part a negotiating ploy because the South generally welcomes projects that involve technology transfer. Thailand has undertaken several JI projects in cooperation with Japan, all of which relate to brown issues: One seeks to improve energy efficiency in a Thai industrial plant,[76] and the other two are smaller traffic management schemes.[77]

The real controversy concerned whether JI should include projects to help developing countries conserve carbon sinks. The North favored these provisions, arguing they would help provide incentives to conserve and replant forests, particularly in the South. The Washington-based World Resources Institute noted that "trees grow faster in the tropical regions . . . [and] deforestation has occurred in the developing world more recently, suggesting a greater need and potential benefit for successful forestry-related projects."[78] The South, however, feared such projects could eventually constrain how it uses its own resources. "Brazil led negotiations for the G-77 in Kyoto," said Bradley, "and it did not want to commit its forest to a global service." Malaysian and Filipino negotiators expressed similar misgivings.

Critics have even labeled JI projects that protect or plant sinks "carbon colonialism," arguing they will hinder development in poorer countries because land reserved for forests can't be used for other purposes.

But what the United States wanted from developing countries most of all at Kyoto was "meaningful participation," some kind of commitment from the big emitters of the South that they, too, would eventually accept binding targets. American businessmen feared that accepting restrictions for themselves while allowing developing countries the freedom to emit GHGs indefinitely would allow countries such as China and India to become free riders, able to lure industries that wished to burn all the fossil fuels they wanted. U.S. diplomats, meanwhile, worried they would have no leverage in future efforts get commitments from the South. And there was a strong environmental argument against leaving these countries out. The United States may be the top emitter of greenhouse gases, but China is number two and India number five. By 2010, the South is due to match the North in emissions,[79] and when you take sinks into account—developing countries are currently cutting down their forests while those in the North are generally growing—the South (with far more people) may already be contributing just as much as the North to the greenhouse effect.

In response, the G-77 negotiators at Kyoto, representing a vast Southern coalition, pointed out that it was the accumulation of carbon-based gases during the last two hundred years that was driving climate change, and the industrialized world was responsible for the bulk of that. The United States, with only 4 percent of the world's population, currently emits a quarter of the earth's greenhouse pollution. What's more, a previous agreement known as the Berlin Mandate, had assured developing countries that they would not have to make commitments at Kyoto. Why the United States agreed to that in Berlin, and then tried to go back on the deal in Kyoto, remains unclear.*

In Kyoto, attempts for a compromise centered on Article 10 of the draft protocol. Under a procedure known as "evolution," developing coun-

---

*A diplomatic source claims that in the Berlin talks, the United States missed a chance to get "meaningful participation" from the South. China offered to make some kind of commitment if the United States agreed to accept the word *target* in discussions over emission reductions. But the United States was so afraid of binding limits at the time that it backed off and sought a less restrictive "policies and measures" approach. Within a few months, the United States realized that setting targets was inevitable, but by that time it was too late to get commitments from the South.

tries could join the emission limits regime voluntarily, without stating a specific time frame. The article had some support in the South. Argentina declared it would eventually be interested in signing on. And there were other nuanced views from the G-77, including those of Kasem Snidvongs, Thailand's top environment official, who recognized that the South included countries with vastly different means and interests. The alliance of small island states, whose physical existence is threatened by rising sea levels, wanted to see the biggest mandatory cuts possible. Saudi Arabia and other oil producers preferred there to be no mandatory cuts. And industrializing giants such as China, India, and Brazil want to make sure their fossil fuel–driven development remained unchecked. A Thai proposal boldly suggested differentiating responsibilities among countries in the South,[80] in effect accepting Thailand's own eventual responsibility to help prevent climate change.

Until the end, it looked as if Article 10 might sneak into the Kyoto Protocol. Thailand and some other developing countries seemed willing to accept it because it contained no deadlines. In a press conference held on the last evening of negotiations, Gore suggested that a "version of Article 10" could play a crucial role in a compromise between developed and developing countries.[81] But China, India, and Brazil all continued to demand that the provisions for evolution be deleted. In the end, armed with the Berlin Mandate, they got their way. The EU's agreement to accept cuts without commitments from the South let the G-77 unite in support of that stance. To be fair, even if the evolution provisions had been included, the U.S. Senate probably still wouldn't have ratified the Kyoto Protocol. Noted Senator John Kerry, a supporter of the treaty, "I think it's going to be very difficult to ratify without more participation from key developing countries. That may take a long time."[82]

In return for agreeing to cut its greenhouse gas emissions to 7 percent below 1990 levels, the United States won a lot of what it wanted in the Kyoto Protocol. The treaty recognizes the role of sinks, despite difficulties in determining how much carbon a forest or plantation actually stores and how to monitor them. The Clean Development Mechanism was set up to help carry out JI projects. And after years of negotiations to overcome fears that the United States would use emissions trading to avoid making difficult cuts at home, a cap-and-trade system was established to create a more liquid market in pollution permits. Talks broke down at the 2000 conference in the Hague, before the system was finally ironed out in Marrakech the following year. By then, however, Bush the younger had already rejected the Kyoto Protocol as too costly and impractical. The Kyoto process continues—in fact, the U.S withdrawal has made it easier for the remaining countries to carry on and meet their targets—but it is now considerably weakened because the number one, two, and five emitters of greenhouse gases are not contributing.

In hindsight, there were many key points in the process where it is tantalizing to speculate on what might have happened if different decisions had been made. What if the United States had taken up the alleged Chinese commitment offer in Berlin, or if there had been a compromise in Kyoto that included provisions for evolution? What if the Europeans had agreed to a fuller trading system in The Hague, when Clinton was still president? What if Gore had received a few more votes in Florida and won the 2000 U.S. presidential election instead of Bush? On such turning points does the fate of the Earth depend.

CARVING UP THE COMMONS

In the 1960s, the British atmospheric chemist James Lovelock teamed up with scientists working at the National Aeronautics and Space Administration (NASA) and the Jet Propulsion Laboratory to study whether there was life on Mars. Lovelock and his colleagues eventually predicted that none would be found because the Martian atmosphere was in a chemically dead equilibrium. It was more than 95 percent carbon dioxide (but too thin to have much of a greenhouse effect), with a little bit of nitrogen and some trace amounts of oxygen and water vapor. Venus's atmosphere was thicker, but also about 95 percent carbon dioxide—it suffered, in effect, from a runaway greenhouse effect. What was it, asked Lovelock, that kept the Earth's atmosphere, made up of 77 percent nitrogen and 21 percent oxygen, in a *dis*equilibrium that seemed so suitable for life?

The answer: life. Three billion years ago, the earliest microorganisms began extracting carbon dioxide from the atmosphere and releasing oxygen. Gradually, concentrations of nitrogen and oxygen built up until the atmosphere could support other forms of life powered by aerobic combustion—including humans. Lovelock came to view the earth as a living system that regulates its own environment even as the organisms within it continue to evolve. It was not just life that made Earth hospitable, but its interaction with the air, rocks, soil, and water around us. "Let the image of a giant redwood tree enter your mind," wrote Lovelock. "The tree undoubtedly is alive, yet 99 percent of it is dead. The great tree is an ancient spire of dead wood, made of lignin and cellulose by the ancestors of the thin layer of living cells which constitute its bark. How like the Earth, and more so when we realize that many of the atoms of the rocks far down into the magma were once part of the ancestral life of which we all have come."[83] Legend has it that the name for the theory came from William Golding, Lovelock's neighbor and author of *Lord of the Flies,* during a stroll in the Wiltshire countryside. He suggested it be named after the Greek goddess who drew the living world forth from Chaos: Gaia.

The Gaia Hypothesis seemed to herald an extension of the Copernican and Darwinian revolutions: Humans were not at the center of the universe, nor were they even the culmination of evolution on Earth; they were just one of many organisms feverishly molding a superorganism as they go about their business. But how could such a superorganism, composed of many finely-tuned component species, come to be? Lynn Margulis, a microbiologist from the University of Massachusetts at Amherst, proposed a theory explaining how life could evolve under a Gaian model. Known as "endosymbiosis," it postulates that species arise from the cooperative evolution of constituent organisms acting in a mutually beneficial way, much as the zooxanthellae and coral help each other to thrive. "Darwin's grand vision was not wrong, only incomplete," Margulis suggested. The driving force behind evolution, in other words, could be cooperation between organisms instead of, or in addition to, competition among individuals.

The theories of Lovelock and Margulis have sparked heated scientific and spiritual debate, but at the very least they offer an interesting model with which to pursue the management of the global environment. For that is the task with which humans are now confronted. We need to recognize our symbiotic dependence on each other, and on the earth systems around us. Global environmental governance is such an ambitious goal that we've divided it into separate issues. Reconciling trade and the environment, conserving biodiversity while sharing its benefits and mitigating climate change are merely some of the larger components. We also must figure out how to manage the oceans and their denizens, how to protect migratory species, how to regulate trade in endangered species and hazardous waste, how to control acid rain and persistent organic pollutants—and each must evolve with the others into a working whole.

Establishing a Global Environment Organization with binding judicial powers similar to those of the WTO would help this effort. Even without it, however, protecting the environment can come without protectionism. The WTO needs to improve its transparency and explicitly recognize the right of MEAs to use trade measures. In return, process and production standards that affect the global environment should be set through treaties rather than through the imposition of unilateral sanctions, which should be a last resort. Forcing governments to enact environmental rules, like forcing farmers to respect conservation edicts, can be counterproductive because it stirs up so much resentment. Surely the North can find other ways to urge better environmental governance in developing countries, many of which already have decent laws on their books? As for benefits sharing under the Conservation on Biological Diversity, it seems only fair to recognize the contributions of local breeders and healers.

Climate change seems a far more daunting challenge. The problem is so huge, and encompasses so many aspects of modern life that it's unclear whether we can mobilize the will and technology to prevent global warming. With three out of the top five greenhouse gas producers refusing to cut their emissions, the Kyoto process cannot be considered a success. Some diplomats believe the negotiation simply grew too complicated. It would have been more effective, they say, to bring in the top dozen or so emitters from the North and South, hash out an agreement among them, and then let others join in. It also would have made more sense to iron out the process first, before negotiating specific targets for each country. With an emissions trading system now in place, the debate will ultimately come down to how to allocate emission quotas to each member state. The South would like emission rights to be dispersed democratically, according to the size of a country's population; the North would prefer a formula based on how much countries are already emitting. The most likely compromise is an "indexed GDP" system: Under an all-encompassing cap, emission rights would be granted according to a country's share of the gross global product and would be adjusted according to the growth of each country's economy.

In the short term, developing countries can focus on "no-regret" options: efficiency and conservation projects that help reduce greenhouse gas emissions while also helping the economy and/or local environment. But assuming they continue to industrialize, they will need access to technologies that not only boost efficiency but also enable fuel switching and carbon trapping (which, according to Butler, seems feasible for smoke stacks, but not for tail pipes). Those technologies will almost certainly be developed and controlled by the North, so a way must be found to transfer this technology to the South.

It's the same old problem of seeking a "shortcut" across the Kuznets curve, a means to "leapfrog" over the dirty phase of development: Who is to pay for all this leapfrogging? Governments in the North claim they don't own the technology; the patents are in the hands of corporations. More to the point, they are reluctant to provide the aid that could finance technology transfer. Even when environmental aid projects are approved, many seem aimed primarily at promoting the donor country's technology (and consultants), and may be inappropriate for the recipient. A better approach is to improve the capacity of developing countries to carry out their own research and development (R&D) by urging educational reforms, funding scholarships, and promoting hi-tech investment. The International Rice Research Institute, a Consultative Group on International Agricultural Research (CGIAR) center that helped produce the green revolution, is often held up as a model of North-South cooperation. And projects under the Clean Development Mechanism can be thought of as hi-tech trade: The South can gain access to green technology and the North will earn carbon credits in return.

This is part of a much broader trend with immense implications for the global environment. To halt tragedies of the global commons, resources are being divvied up and handed out as property rights with the goal of creating markets for trading in public goods (fish, wildlife, and other common resources) and bads (carbon dioxide, sulfur dioxide, and other pollutants). The hope is that the market can succeed in protecting the environment where open access regimes have failed. The success of the U.S. sulfur dioxide emissions trading system demonstrated that market-based mechanisms can yield environmental gains at lower cost and with greater flexibility than traditional "command-and-control" regulations. Emission permits, carbon credits, and genetic property rights are just a part of this trend. New Zealand is handing out individual transferable quotas (ITQs) to fishermen—essentially, tradable permits to catch a certain amount of fish. Meanwhile, the Law of the Sea has granted coastal countries exclusive economic zones offshore where they control fisheries, mining, and petroleum development.

But there are dangers to this approach. Market-based mechanisms have to be carefully planned. Simply granting property rights to a resource does not ensure that it's used sustainably, particularly if it's poorly monitored. The mercury problem in the Gulf of Thailand and the forest fires in Indonesia are a testament to that, as is the havoc wrought by mining companies around the world. As the principles of adaptive management suggest, even a well-organized measure such as the sulfur dioxide cap-and-trade system may need continual adjustment; and although the cap-and-trade system has greatly reduced pollution, distributional problems mean that some "hot spots" of acid deposition remain.[84] ITQ systems have been subject to complaints that fishery quotas end up concentrated in a few hands. And participants in trading systems set up under the Kyoto Protocol may find they don't always get what they want. "The majority of [Clean Development Mechanism] projects are now geared towards reforestation rather than technological upgrading of industry," explains Tim Forsyth, a British academic. "This is unpopular with a lot of governments [in the South] who see technology transfer as a necessary requirement."[85]

Global negotiations have revealed a striking difference of environmental priorities between developed and developing countries: The South wants brown aid that involves technology transfer; the North prefers to help with green issues, particularly conservation. At the Earth Summit, the North pushed for a global convention on forests, but the South rejected it. The United States also took the drastic step of carrying out trade actions to protect marine species. The largest allocation of GEF funds by the UN Development Program goes to protecting biodiversity in developing countries.[86] The second largest amount goes to preventing climate change, which has green and brown components. But the North was particularly eager to include forestry schemes for carbon sinks as part of the Kyoto Protocol, in contrast to developing countries, led by Brazil, which were reluctant to include it.

The South's wariness of green aid is not from lack of concern for forests and wildlife, but because it is reluctant to commit these green resources toward a global agenda. Forestry agencies love receiving aid, but not the strings that come attached to it. NGOs suspect that foreign donors are more concerned with charismatic mega-fauna than with poor people's livelihoods, and fear that villagers may suffer for it. That helps explain why Thailand has refused to ratify the biodiversity convention, and why its joint climate change projects have focused on technology transfer and shied away from forestry schemes.

An even more recent example came in March 2002 when Thailand turned down an offer by the United States to reduce part of the kingdom's debt in exchange for setting up a Tropical Forest Conservation Fund that would help protect mangroves.[87] The same environmental democracy groups that have railed against the Convention on Biological Diversity attacked the debt-for-nature swap as a U.S. ploy to gain access to Thai biological resources or collect carbon credits should the United States ever join the Kyoto Protocol.[88] When the United States failed to provide the necessary guarantees, the Thai government backed off.

In short, the South jealously guards its sovereignty over its resources, which helps explain why carving up the global commons and handing out property rights to resources has proved so popular. All in all, on the eve of the World Summit of Sustainable Development, to be held in Johannesburg, the sovereignty of the South has held up rather well. At Kyoto, developing countries were able to withstand pressure from the United States to "meaningfully participate" in the reduction of greenhouse gas emissions. They can be selective, as Thailand has been, in deciding which treaties to ratify. Their sovereignty did suffer some small blows from the U.S. trade actions, but they were able to seek recourse through the WTO, and now their representative, Supachai Panitchpakdi, is in charge of the organization. The North does have more subtle ways to influence the South's environmental policy, but overall developing countries have been able to reject overt pressures. Arguably, their sovereignty has been strengthened by the codification of their right to exploit their own resources into international law.

The danger, of course, is that the South's emphasis on sovereignty has hampered cooperation in solving global environmental problems. To be fair, the country that has been most zealous in defending its national interests, and is the biggest obstacle to working out multilateral environmental agreements, is the United States. As an American, I am pained to admit that. But what other conclusion can be drawn? The United States has backed out of the Kyoto Protocol without offering an alternative. It has failed to ratify the Convention on Biological Diversity or the Biosafety Protocol despite flexing its muscles during negotiations. It has rejected other MEAs, as well, such as the Basel Convention, and is resolutely opposed to establishing a global environmental organization with powers akin to the WTO. Its contribution to overseas environmental aid is pathetic. Even when the United States government imposed trade sanctions to protect turtles and dolphins, it was partly trying to protect its own fishing fleet. Environmentalism is a powerful force within the United States, but on the international stage the U.S. government continues to put short-term economic interests ahead of long-term environmental interests.

The global problems we face, however, cannot all be laid at the feet of one nation. They stem from our most basic human urges. We are becoming so numerous, and so expert at exploiting resources, that, to paraphrase the Buddha, there is virtually no place on Earth that has not felt man's impact. If we are to maintain this hospitable environment, our climate and oceans and soil, we face the choice either of living within the limits of our resources or of becoming Earth systems engineers: managing the carbon cycle, farming the oceans, genetically modifying other species—and maybe ourselves. It's not at all clear that we have the capacity to manage such a huge, complex system. "To what extent is our collective intelligence also a part of Gaia?" asked Lovelock. "Do we as a species constitute a Gaian nervous system and a brain which can consciously anticipate environmental changes?" If we fail to manage the global environment, we risk knocking the Earth into a far less hospitable equilibrium. Gaia, if there is such a thing, will go on. But the quality of our lives and the stability of our societies could lie in the balance.

10

CIVIL-IZING SOCIETY

THREE BLOODY DAYS IN MAY

COUP DU JOUR

February 23, 1991, didn't even start off like a typical Saturday. Iraq had invaded Kuwait six months earlier, the U.S.-led coalition had been bombing Iraqi troops for weeks, and now Saddam had been given an ultimatum to get out of Dodge by noon, or else. The world was on tenterhooks, waiting for ground troops to attack and the Gulf War to begin in earnest. Southeast Asia seemed peaceful by comparison. In Thailand, some political bickering had generated rumblings from the military, but that was business as usual.

Saturday was my busiest day at the *Nation.* At the time, I was still the science and technology editor, responsible for putting out a weekly eight-page section, and this was the day I edited stories and Scot Donaldson laid them out. Scot was a capable and easygoing sub-editor from California, and I was thankful that he had replaced a good-for-nothing layout man named Francis who'd made my first six months at the newspaper harrowing. On this particular Saturday, however—my twenty-sixth birthday—I was distracted and kept an eye on the wires to see whether the ground war was breaking out in the Middle East. I'd been a backpacking tourist in Iraq just a couple of years earlier, and had actually celebrated my twenty-fourth birthday in Baghdad. It was disconcerting to see places I'd visited pop up now in the news as refugee camps or chemical weapons testing sites.[1]

Yes, the newsroom had a buzz that day, but it wasn't all about the Middle East. Around noon we began to hear some rumors. Nick Keyes, a copy editor on the news desk, wandered over and suggested in a conspiratorial whisper that there were rumors of a coup in Thailand. "Didn't Hang say there might be a coup last week?" I asked, referring to the hyper-energetic politics editor who banged at his keyboard all day and never seemed to leave the newsroom.

"Yeah, but Hang says that every week." Nick had a point, and there was nothing on the wires about a coup, so we went back to work. But half an hour later, the editor Thepchai Yong walked into the office looking pensive. Something was up. Thepchai never came into the office on a Saturday. It was his one day off. That was part of the reason he'd hired me the year before. Lacking an editor for the *Technology* section at the time, he had been forced to come in himself on Saturdays to make sure the section was properly laid out. He was desperate to find someone to take over, and although it was unusual for the *Nation* to hire a foreigner to run one of its sections, there weren't many local journalists who could match my science background.

Now the news reports started coming in. Yes, there had been a military takeover, by all accounts a peaceable one. Prime Minister Chatichai Choonhavan, who had been feuding with the generals in recent weeks, had been scheduled to fly up to Chiang Mai in an air force plane that morning. The air force commander, Kaset Rojananil, had gone to the Bangkok airport, apparently to see him off, but in fact to put him in detention. At around 11:00 A.M., shortly after the plane with Chatichai and his aides began taxiing toward the runway, five men in safari suits whipped out their pistols and ordered the prime minister's startled security guards to stand down. Chatichai was then whisked off the plane and taken to a hidden location at the airport.

As the details of the coup started coming in, the newsroom began to fill up. We were still in our cramped old office near Sukhumvit Soi 42, a tiny room jammed with desks and computers, wires dangling in every direction, and a sign urging "Please Don't Feed The Newshounds!" (Everyone ignored the warning and ate constantly around the computers.) Then a special announcement began to air on television, and we all crowded around the television set, standing on chairs and tippy-toes for a glimpse.

A line of uniformed generals appeared on screen, seated behind a long table and looking grim. In the middle sat Supreme Commander General Sunthorn Kongsompong, who began reading a formal announcement of the military takeover. A National Peacekeeping Council (NPKC) was being formed, he declared. Sunthorn was its nominal head, but we all knew the real power lay with the army chief, General Suchinda Kraprayoon. The Constitution was abrogated, Parliament and the Cabinet were dissolved, and the political parties were ordered to suspend all activities. Democracy would be restored "as soon as possible." Then, about a third of the way through his speech, Sunthorn stumbled. Flustered, he glanced away from the camera and said in a half-whisper, *"Rong mai"* (try again), apparently not realizing that he was being filmed live. In the newsroom, our tension was released in a roar of laughter.

It was typical Thailand. Even on the sternest of occasions, their human side seems to flash through. The scene grew stranger still a few minutes later, after we'd all returned to our desks, when a bevy of tall, strikingly beautiful women suddenly swept in through the front doors of the newsroom, the leading lady sporting a gorgeous gown, a diamond-studded tiara, and a fake smile. It was Miss Thailand and a group of her runners-up in the recently completed beauty pageant—an event of such importance in Thailand that it is roughly on a par with the Oscars in the United States. They were on a scheduled visit to the newspaper and hadn't realized that their legally elected government had been overthrown that morning. So poor Thepchai was forced to put aside the most important news event to hit the country in years and instead meet and greet the young ladies. The rest of us just looked at each other: What would happen next?

The coup, however, was no laughing matter. It was a serious setback for Thailand, and for democracy in the region. But it was far from unprecedented. At least seventeen coups have been attempted since the country became a constitutional monarchy in 1932, ten of which resulted in a change of leadership. There hadn't been a successful coup in fifteen years, however, and many thought they'd been relegated to the past. "I didn't expect this to happen to our country again," said Surakiart Sathirathai, one of Chatichai's advisers, echoing the views of many. "It's a pity that the country has come this far and that we may have to start from zero again."[2]

Ominously for the press, the NPKC imposed the toughest censorship measures since the bad old days of dictatorship in the 1970s. All news stories "affecting national peace and order" were to be cleared by the authorities before being published. In a bold front-page editorial the following day, Suthichai Yoon, the *Nation*'s founder and publisher asserted that "power belongs to the people, nobody else." He condemned the coup and warned that in attempting to censor the press the NPKC was simply damaging its own credibility. The stage was set for a war of wills between the military and the forces of democracy in Thailand.

Among the general populace, however, the reaction to the coup was essentially a shrug. People were embarrassed and frustrated that the political squabbling had led to a military takeover, but there were no protests to speak of, and hardly any show of force was required. Although the Thais say that you can tell a good coup from a bad coup by noting whether the tanks on the streets stop at the traffic lights, this time there were barely any tanks on the roads at all.

The reason for resignation was clear: People were fed up with the rapacious corruption of the democratically elected government. Such was the avarice exhibited by the administration that it had been dubbed "the buffet Cabinet." The NPKC listed government corruption as the number one reason for the coup. Given the military's own reputation for sordid dealings, such rationalizations were generally taken with a pillar of salt. Political conflicts and naked ambition on the part of a group of military college classmates were more fundamental factors. But the Cabinet's brand of patronage politics had become so odious that few came to its defense. The Chatichai government's shameless greed had rotted away its own legitimacy.

SOMETHING'S ROTTEN IN THE TROPICS

Time and again, we've seen how corruption has exacerbated environmental damage in Southeast Asia, or prevented the effective management of natural resources. As Pisit na Patalung, the head of Wildlife Fund Thailand, often points out, the one thing the Thai government seems to have perfected is a system of corruption. "Thai officials operate like a finely tuned orchestra," he says. "Each knows what his role is, what notes he has to play, without even the need for a conductor." It's a phenomenon endemic throughout the developing world. Corruption exists in wealthy countries, too, of course; wherever you go, moneyed interests find a way to gain undue influence. But if you define corruption as the abuse of one's position for personal gain, then virtually anyone who has lived in both North and South would agree that it exists on a whole different scale in the developing world. It is so prevalent that a country's level of corruption—or more broadly, respect for the rule of law—might be considered as much an indicator of development as incomes, educational levels, and mortality rates.

Estimated corruption rankings seem to bear this out. Of the twenty-five "cleanest" countries on the 2001 Corruption Perception Index tabulated by Transparency International,[3] a global anti-corruption NGO, only Chile at #18 can be considered a developing country (Singapore, tied for #4, and Hong Kong at #14 are also near the top of the list, but they are probably too wealthy to be considered part of the South anymore). Then come other developing countries that (like Chile) are more economically successful, such as Botswana (#26), Malaysia (#36), and Uruguay (#44). Thailand (#61) is in the lower half of the rankings, but among developing countries it is about average. Still further down are the Philippines (#65), Vietnam (#75), and down near the bottom is Indonesia at #88. Burma, Cambodia, Laos, and Brunei are not listed. Clearly, there seems to be a relationship between a country's wealth and its level of corruption, although it's hard to say which is cause and which effect.

That the Third World is corrupt is hardly news, but the impact of corruption on the environment is given surprisingly little mention. At international conferences, such as the Earth Summit, the subject is virtually taboo for fear of offending dignitaries. Even in academic circles, discussion usually focuses on the broader issue of "environmental governance," which is hampered by other factors, as well, including a lack of manpower, small budgets, weak laws, poor coordination, and plain old negligence. Besides, resources such as clean air and watersheds are just some of the many public goods—from security to street lighting—whose management requires some kind of public intervention, whether through regulation, setting standards, handing out property rights, or preventing free riders. Corruption's impact on the environment simply doesn't receive the attention that, say, a corrupt police force receives.

But resources are particularly vulnerable to corruption because money *does* grow on trees, or rather *as* trees. In the modern economy, forests and fish stocks are seen as "natural capital," and the temptation to convert them into regular capital, particularly in poor countries, is almost irresistible. It's an easy way to turn power into cash. Meanwhile, the benefits of clean air, living forests, or healthy reefs, the advantages that come with managing them for the long term, are discounted since they're spread over generations and among groups having little power.

Corruption damages the environment by granting access to resources either in contravention to existing rules or at a price below that which society has attributed to them, explains Achim Steiner of Transparency International.[4] The damage can come in the form of illegally extracting a resource such as timber, getting away with toxic dumping, or preventing effective environmental management, adaptive or otherwise. But most damaging may be how corruption eats away at the trust and cooperation needed to manage a resource wisely. People don't want to sacrifice their own time or money by, say, getting smog checks for their cars or cleaning up wastewater emissions, because they figure that others get away with being free riders, through bribery if necessary, creating a runaway tragedy of the commons. "Corruption essentially perverts societies' choices and values attributed to . . . resources," says Steiner. "[It] generally results in a use of environmental resources not sanctioned by society. Otherwise there would be no need for bribery or other means of perverting the decision making process and ultimately public choice. The end result is a misallocation of resources."

This sheds a different light on the environmental values of people in the developing world. It is typically claimed that they're willing to sacrifice the environment for higher growth, and that is partly true. But many such countries (including Thailand, the Philippines, and Indonesia) have good environmental laws, often modeled after those in the West. And when polled, most respondents claim they want a cleaner environment.[5] If you accept that these laws and poll results represent people's actual desires, you'd have to conclude that environmental protection is a high priority but that the government is simply unable to provide it effectively.

The truth is probably more complicated than that. Yes, people want a clean environment, but they don't want to make the sacrifices themselves. Similarly, everyone in Thailand complains about police corruption, but if they're pulled over for a parking ticket, they're usually relieved to slip the officer a small bill rather than go through the hassle of paying a larger fine. Admittedly, it may be necessary to differentiate between different kinds of corruption. We may excuse a bribe paid for survival—to flee persecution from a military junta, for instance, or to poach game for the sake of hungry children—more readily than we accept an attempt by some large corporation to get around environmental regulations. But in the end anyone who's been party to such activity is partly responsible for the rotten system that's created. Then inertia takes over: A lack of faith in honest administration yields a dirty cycle of corruption. Finding a way to break that cycle, or perhaps work around it, requires that we examine its social and political roots.

One thing you notice in Asia, and indeed in most of the developing world, is that everything is more *personal.* It's certainly true in business: Collaborative ventures are often based more on personal ties than on legal contracts. The lack of bureaucracy makes business more efficient, but also more volatile and subject to abuse. In Thailand, most political parties (and many NGOs) are also highly personalized. They serve as the vehicle for a particular leader or group rather than a set of policies.

A bit of pop sociology holds that the key to understanding Asia is the saying, "I'll scratch your back if you scratch mine," and in a simplistic way that does capture the Asian approach to process. Some would add by way of contrast that the "golden rule" in the West is to "do unto others as you would have them do unto you." Although that may be a normative principle in Judeo-Christian culture, people in the West seem no more or less selfless than people elsewhere. What's instructive about those two sayings is how much more abstract it is to "do unto others" compared to the more personal image of "scratching each other's backs."

The difference reflects the formalistic nature of Western society, built around laws and institutions that strive for impartiality and revolve around the individual. There is a level of trust in these institutions—the courts, the police, and, yes, even the media—that people in industrialized countries take for granted but that is severely lacking in the developing world. Personal ties among families and clans have been weakened in the West, but cooperation among larger groups who share common interests is easier. That cooperation increases social mobility and helps the formation of, for instance, large environmental groups funded by mass membership.

By contrast, familial and patron-client bonds within the "communitarian" cultures of the global South remain strong. I've often noticed how my Asian friends are much more group-oriented than Westerners, and Westerners don't mind being alone nearly as much as Asians. "In white middle-class America people think of themselves as a bundle of traits, preferences, and desires," explains Shinobu Kitiyama, a psychologist at the University of Oregon. "But in Asian culture people conceive themselves in terms of a dense web of social relations."[6] Businesses are run as family enterprises, as opposed to the West, where nepotism is more frowned upon. Think back to the description of urban architectural styles in Bangkok: People build walls around their family compounds, which often include homes for several generations of relatives, as opposed to the atomized housing patterns of the West, where children typically move out once they've reached adulthood. Or recall the story in the Introduction of how the European visitor was aghast at the garbage littering the Bangkok slum. Wealthy countries are better at managing public spaces, but people in the developing world are more concerned with personal space.

So although the global South is communitarian, these communities aren't necessarily formed along lines that allow them to band together to protect their common environment. The cities of the global North have consumer groups, neighborhood coalitions, park enthusiasts, mass transit advocates, and bicycling campaigners that fight for their causes. But these organizations take time to form as the forces of urbanization slowly crush traditional bonds and new interest-oriented groups emerge. In Southeast Asia, the development of civil society simply hasn't kept pace with the region's spectacular economic growth. In the countryside, traditional groups such as irrigation and fishing co-ops still exist, tenuously, and can sometimes rally with the help of NGOs to fight for better, more localized natural resource management. The revival of these communities is one of the great successes of the environmental democracy movement. But rural society is being continually weakened by the spread of modernity and the pull of the cities.

The other type of relationship that still ties people together in the South is that between patron and client, ubiquitous not just in business, but also in politics, which is often run as a business to the detriment of clean, effective governance. The root of Thailand's political problems is that people, particularly in the countryside, sell their votes. Sometimes, canvassers for candidates will go around and secretly hand out 500-baht bills on behalf of parliamentary candidates. Since vote buying is technically illegal, they find clever ways to purchase political allegiance, perhaps by treating voters to a lavish banquet, for instance, or offering them one shoe before an election and the other afterwards if the right candidate wins. To recoup all the money they've invested to gain office, members of Parliament subsequently need to profit from their position through kickbacks and other forms of corruption. As a U.S. diplomat stationed in Thailand in the 1990s explained to Thomas Friedman: "We've just opened a dozen or so new embassies in the former Soviet Union, and our job there is to explain to people that there is something called 'a market.' Our job in Thailand is to explain to people that there is something other than the market."[7] Thailand is an extreme example of how the patron-client relationship has created a society rife with corruption, but it's not just in the global South where that occurs: Italy and some of its southern European neighbors also fit that description.

If Thailand is considered the "Italy of Asia," Singapore can claim the mantle of being Asia's Switzerland. Clean, efficient, peaceful, and well-run despite the presence of multiple ethnic groups, Singapore is more centralized than its European cousin, but its success can at least partially be attributed to a government remarkably free of corruption. In 1960, soon after it gained independence from Malaya, it passed a Prevention of Corruption Act that was made even more stringent in later years. An Auditor-General's Office and the Public Accounts Committee in Parliament were established to make government more accountable. A Corrupt Practices Investigation Bureau was given extensive powers to investigate officials, including their bank accounts. Thailand's Counter Corruption Commission (CCC) has similar powers, but as our land title investigations at the *Nation* revealed, the kingdom's prosecutors rarely acted on the commission's findings (although in recent years it has been strengthened thanks to the new Constitution). So although Singapore officials convicted of corruption must pay fines and forfeit all the money they have accrued, officials in Thailand are rarely even charged with malfeasance. If their transgressions become too obvious, they are "relegated to an inactive post"—they remain in the not-so-civil service but, in effect, lose the opportunity to take more kickbacks.

Again, it seems that the difference between well-governed and poorly governed states in Southeast Asia lies not just in their laws, but in the way those laws are enforced. So perhaps the most important ingredient in Singapore is what David Seth Jones, an academic at the National University of Singapore, calls a "strict anti-corruption ethos" that places an "emphasis on clean and honest government."[8] To Singapore's longtime leader, Lee Kuan Yew, and his Malaysian counterpart Prime Minister Mahathir Mohamed, this ethos is part of the "Asian values" system that helped these countries achieve rapid economic development. There is no doubt that traditional ideals of hard work and the priority placed on education—along with open economic policies—have helped them succeed, but skeptics claim their praise of obedience and defense of "guided democracy" is largely self-serving. The lack of a large and poor rural population in Singapore makes the city-state easier to administer, they point out. "Asian values is a red herring," concludes Pana Janviroj, editor of the *Nation.* "Dealings are more personal here. But how is that a value?"[9] Ian Buruma, an author of many books on Asia, argues that Mahathir's and Lee's version of Asian values "vulgarizes" traditional Confucian values and more closely resembles the British colonial ethos they grew up with. "Mahathir and Lee were greatly influenced by

British education," he argues. "They admired imperialism, and think the democracies in the West have now 'gone soft.'"[10]

The authoritarian brand of democracy advocated by Lee and Mahathir does seem to provide some environmental benefits, however. Although Malaysia and Singapore have not been shy about exploiting their resources, or those of other countries, their environmental governance seems better than that of neighbors such as Thailand and Indonesia. Their cities are pleasanter to live in, and their resources generally better managed. Malaysia has cut down large swathes of its forests, and Malaysian logging firms are responsible for ravaging forests from Cambodia to Latin America, but the country's forest service is less corrupt than Thailand's. And Thailand's system of environmental governance also improved following the coup. That was not because the Thai generals were green, but because they appointed a highly regarded industrialist and former diplomat named Anand Panyarachun to run an interim government until new elections were held in 1992.

Anand turned out to be a blessing for the kingdom. After naming a group of honest and capable technocrats to the Cabinet, they proceeded to pass sweeping reforms of virtually every aspect of Thai administration, and improving environmental management was one of their major goals. Among the new laws they passed was a landmark Environmental Protection Act that created three new environmental agencies—in charge of policy and planning, pollution control, and environmental promotion—and placed them in a newly named Ministry of Science, Technology, and Environment. Provisions in the Factory Act and Hazardous Substances Act were updated. And all sorts of green policies got a boost, from cleaner car exhaust standards to energy efficiency projects to zoo improvements. Catalytic converters became mandatory on all new cars. A huge tract of wetlands south of Bangkok named Bang Krachao was protected to serve as the city's "green lungs." Even the golf course in Khao Yai National Park was removed.

So does this that mean less democracy is the answer to the South's environmental ills? No, because although political scientists as far back as Plato have recognized the advantages of enlightened dictatorship, it's hard to find dictators who are enlightened. The majority of despots—including Suharto, Ferdinand Marcos, and the SLORC in Burma—decimate their country's environment.

The question of whether democracy is good for the environment needs to be rephrased: What type of environmental issues does democracy help? It clearly hampers the building of big, high-impact development projects such as dams and power plants. It allows the formation of civic pressure groups and watchdogs, and this gradually improves air, water, and general environmental quality. Democracy can also delay and obstruct sewage and hazardous waste treatment plants through what critics dub the Nimby (Not in My Back Yard) phenomenon, but this has the long-term beneficial effect of spurring waste reduction. On the other hand, given the high deforestation rates of democracies such as Thailand and the Philippines, and the fact that Cambodia and Indonesia both experienced sudden increases in deforestation as they loosened up politically, democracy doesn't seem good for conservation whether of forests or of biodiversity, at least in the short term, although these trends may be due mostly to attendant economic liberalization and political decentralization. Apparently it's only in the long run, as democracies become wealthier and stabilize, that the urge to conserve gains some clout.

THE POOR PAY MORE

A week to the day after the coup, on March 2, 1991, Scot and I were back in front of our computers laying out the *Technology* section when we heard a huge WHUMP! The building swayed, and the windows actually seemed to buckle inwards before snapping back. Everyone looked around, startled, our nerves already on edge from the unstable political situation. "What the hell was that?!" someone called out. A hubbub of voices followed, then a shout, "To the roof!" We dashed up the dingy stairway and burst outside. There, perhaps a mile to the southwest, a mushroom cloud was billowing into the sky, a wavy pillar of oily black smoke streaked with orange flames. There must have been a huge explosion in the general vicinity of Klong Toey port. Was it an act of political sabotage, we wondered, or an accident?

The latter, it turned out. Barrels of chemicals stored haphazardly at a port go-down somehow ignited, creating a huge fireball. The port is also home to one of Bangkok's more notorious slums, and a fire subsequently tore through hundreds of shanty houses. Four people died during the incident, and in the weeks that followed, scores of squatters became sick. Checkups on 350 slum dwellers revealed that over 200 suffered itching rashes and another 100 unspecified illnesses.[11] The fear was they had been contaminated by toxic fallout, but testing was complicated by the fact that the Port Authority of Thailand at first refused to reveal what chemicals had been stored at the site. Or perhaps they didn't know, since over the years, barrels with mysterious contents had been left at the port by midnight dumpers. Eventually, the authorities released a list of twenty-three chemicals, including formaldehyde, a preservative, and methyl bromide, a pesticide. A survey of 6,000 slum dwellers displaced by the fire found that 215 had elevated levels of methyl bromide in their bloodstream.[12]

It was yet another vivid example of how poor people are most vulnerable to environmental damage because they lack either the mobility to move away from environmental hazards or the skills to adapt. Virtually every issue examined in this book has provided similar examples. It is the world's poor who will find it hardest to adjust to climate change. They live in flood-prone zones and near toxic waste dumps. They can't afford to buy clean drinking water. In the countryside, they form the majority of people displaced by dams. In the city, everyone suffers from traffic and air pollution, but the poor have to take the bus or work on the street. Small-scale fishing communities are most at risk from mercury contamination in the Gulf of Thailand if residents regularly eat their catches. Fishermen suffer when their stocks are depleted and farmers when their soil is eroded and their watersheds dry up. On the other hand, they are also asked to sacrifice when conservation areas or marine reserves are created. So yes, the poor get the short end of the stick with just about everything. But although few claim that people should sacrifice their health or their education for the sake of development, the refrain that environmental concerns must make way for rapid growth is a common one. And that often just doesn't make sense.

The open market policies that have fueled industrialization and globalization in the South have increased overall wealth, but have also created greater inequities. Through a process that Peter Bell calls "maldevelopment," the combined incomes of the richest fifth of Thailand's population rose from 49 percent of total income in the mid-1970s to 57.5 percent by 1994.[13] Meanwhile, the incomes of the poorest fifth declined from 6 to 4 percent. The uneven impacts of environmental problems has surely contributed to that growing gap—although it's difficult to quantify by how much—and makes the inequities even worse than such statistics suggest.

Consider the case of Usa Rojphongkasem, an eighteen-year-old resident of the Koh Lao slum in Klong Toey and student at Ramkamhaeng University. Usa was by all accounts healthy and happy when the explosion occurred. Following the blast, she raced into the smoke to help her relatives evacuate their belongings before they went up in flames. After the incident, she began to suffer increasingly from headaches, nausea, and exhaustion, and eventually she was forced to drop out of school. Then she began to feel pain in her muscles and a rash broke out all over her body. Hospital tests showed high levels of methyl bromide and formaldehyde in her bloodstream. In 1993, she was diagnosed with brain cancer and underwent surgery. She is alive today, but approximately fifty of her neighbors have died of mysterious illnesses since the explosion occurred.[14] Usa is now blind and deaf on her right side, and her vision and hearing capacity are reduced on her left side. The hair on the right side of her head has fallen out, revealing a long thin scar from her surgery. The Port Authority provided compensation to victims who lost their houses or suffered injuries from the fire, but refused to pay those who became sick after the fact because they lacked evidence that their illnesses resulted from the accident.

In the North, the environmental justice movement has emerged to fight for communities like Usa's, and for other green causes on behalf of minorities and the poor. Occasionally that requires fighting against other environmentalists—most of whom represent middle-class interests—for instance if a waste treatment plant is being built in a low-rent neighborhood, or if indigenous people have their hunting rights circumscribed. The difference in the South is that there are a lot more poor people. So the environmental justice movement—or rather its Southern equivalent, what I've called the environmental democracy movement—form the mainstream greens. And they have fewer avenues for advocacy open to them.

In particular, environmentalists in the South rarely take their cases to court. There are exceptions, of course. India has a well-developed legal system that has been used to protect the environment at times, and the courts in emerging democracies such as Thailand and the Philippines are slowly becoming more progressive. But a host of obstacles remain to filing environmental lawsuits in Asia, the most basic being a general antipathy to legal confrontation. Chinese-Thais have a saying: "It's better to eat dog shit than to go to court." Besides, victims of environmental negligence often can't afford to pay legal frees, although assistance is available from groups such as the Law Society of Thailand (LST), which does pro bono work for the poor.

Even if you make it to court, the Thai legal system is not very accommodating for environmental cases. There is a lack of environmental awareness in the justice system, says LST attorney Theerasak Chiukhuntod,[15] which makes it difficult to convince the courts that the damage inflicted is meaningful. Imagine the tribulations of Erin Brockovich compounded by an unsympathetic judge. Filing suits against the Thai government is especially challenging because there is a long-established precedent to respect the decisions of the executive branch, and protect it from litigious harassment, says Sooboon Vuthiwong, a criminal court judge.[16] Plaintiffs must provide proof of "special damage" before a court will even consider a case against the government, and that has almost never happened. That means it's virtually impossible to prevent dubious projects from going ahead because there is no way to prove that a future event will cause damage. Finally, victims of environmental damage face an uphill struggle in court because they are considered civil cases, so the burden of proof falls entirely on them.

In Usa's case, however, there seemed to be strong medical evidence that she had suffered as a result of the Klong Toey explosion. So in 1996, Theerasak and the LST filed a lawsuit against the Port Authority on behalf of Usa, her mother, and another victim.[17] It was a bold move if only because an environmental lawsuit had never succeeded in the Thai courts. Even when the government sought to file suit against a sugar refiner accused of dumping waste into the Nam Phong River in 1992, causing a massive fish dieback, prosecutors dropped the case for lack of evidence. As Supreme Court Judge Vicha Mahakun acknowledges, when it comes to pollution cases, "the judiciary system in Thailand is quite outdated."[18]

In the absence of effective legal recourse, the media's role as an environmental watchdog in developing countries becomes even more important. In Asia, a growing number of countries and territories—including Thailand, the Philippines, Japan, South Korea, Taiwan, Hong Kong, Cambodia, Indonesia, and most of South Asia—have a free press, even if their television outlets are under tighter control. Because people fighting to protect their environment often have nowhere else to turn to, green issues are often played out in the media and ultimately decided by public opinion. Newspapers become a court of last resort—a role for which they're not really suited. The media is often seen as a kind of public service, and thus is supposed to cover events that affect the public in general. But news is a business, too, one that caters to customers who can afford to pay for information. So in developing countries, coverage is skewed toward the interests of the urban middle and upper class, with scantier reports about the poorer rural sector where most people still live. And although many Asian newspapers—including the *Nation* and the *Bangkok Post*—do an excellent job of covering environmental issues, some periodicals suffer from corruption and irresponsible journalism, particularly in countries where the press has only recently been set free.

So it is a good omen that in 2001, a decade after the Klong Toey blaze, a Thai court ruled in favor of Usa and ordered the Port Authority to pay her damages, in large part thanks to the hard work of Theerasak and Orapun Metadilogkul, the physician and environmental health crusader who helped gather evidence for the case. Environmentalists believe the verdict sets an important precedent for the Thai judicial system; certainly it should spur victims of other industrial accidents to file suit against responsible parties and, in the long run, make industry more accountable. Rather than serving as a court to decide important public issues, the media is much better suited to fulfilling its traditional role as a watchdog. Bangkok's *Nation* in particular—founded by Suthichai in 1971 as a homegrown alternative to the foreign-owned *Post,* which had set a standard for accuracy and reliability that the vernacular papers had failed to match—has sought to serve as the progressive voice of Thailand's rising middle class. So we were well situated to cover one of the most momentous events in Thai history: the democracy uprising of 1992, arguably the nation's, and the *Nation*'s, finest hour.

HOT FOR DEMOCRACY

A year after the military coup, Thailand was in turmoil. A new Constitution, the country's fifteenth, had been passed and it was notable for its lack of faith in democracy. Although the 360-member lower house of Parliament would be selected through national polls, neither the prime minister nor his Cabinet members were required to be elected officials. And the Senate, an unelected body whose 270 members were appointed by the prime minister, was given strong powers. The military, which stood to benefit from these measures, claimed they were necessary to prevent vote buying from corrupting the government. Elections were then held on March 22, 1992, and the generals said they would not interfere. But it all seemed part of an elaborate political stage play, and the five pro-military parties that subsequently formed a ruling coalition had clearly read the script. After the first nominee for prime minister was dismissed for alleged involvement with narcotics, they nominated Suchinda, the unelected military strongman behind the coup. The general graciously accepted, and then formally took power.

All during his first month in office, however, protests against his rule grew. A former member of Parliament named Chalad Vorachart, who had gone on hunger strike to protest military rule, became the opposition's conscience. But another general named Chamlong Srimuang, who had been elected Bangkok's governor, became its leader. A founder of the Palang

Dharma (Power of Dharma) party and a follower of a strict Buddhist sect, Chamlong cut quite a different figure from the typically venal Thai politicians. A vegetarian who crops his hair in a flat-topped crew cut and was usually seen wearing a blue denim farmer's shirt, he was known as "Mr. Clean," although his puritanical zeal and ambitious nature made many people uneasy. As an elected official, however, he was able to enunciate a principle around which the democratic opposition, and eventually the bulk of the middle class, could rally: Thailand's prime minister should be democratically elected, and Suchinda was therefore an illegitimate ruler.

The rallies grew more tense in early May after Chamlong announced that he, too, was going on a hunger strike, supposedly "to the death." It was an exciting time to be a journalist, and my colleagues and I—now ensconced in our shiny new office at Bang Na on the eastern edge of Bangkok—would regularly make the trip over to the western side of town where the demonstrations were held. They ultimately gathered at Sanam Luang, the Royal Field, a vast oval of public land outside the Grand Palace that had been the epicenter of Thailand's previous political upheavals. In 1973, students from nearby Thammasat University helped usher in the country's first truly democratic government by demonstrating on the site. In 1976, the democratic experiment ended when further student-led protests were brutally suppressed in a military backlash that sent activists fleeing into the forests and intellectuals into exile. Now, in 1992, it was the height of the hot season, and as protestors camped out under the searing sun, their emotions at times threatened to boil over.

But these protests were different from those in the 1970s. They were far more broad-based, and included members of the now much larger middle class. Traditionally considered apathetic, if not fatalistic, something in the current situation had sparked their political passion. A democratic flame had been lit, and it seemed to feed on the middle class's aspirations for development. On May 8, I went to the commons with two fellow *Nation* reporters, Oh and Kunsiri, to interview the "yuppie protestors," who dressed in ties and clutched cell phones. It was hard to tell where their economic concerns ended and political concerns started. Suchinda was "bad for business" because his heavy-handed style didn't allow room for compromise. He had tunnel vision, they claimed, and he was tied to vested interests. Many were angry with his arrogance, his conviction that he deserved to rule. "We follow the world," said one protestor. "Thailand relies on export markets and it will affect us if we have dictatorship." Another was even more direct: "If we stay like this, with political conflict decided by force, then we will never develop."[19]

We followed the crowds as they filed out of Sanam Luang and headed toward Government House, but first we stopped at the venerable Royal Hotel on the northeast corner of the commons. There's nothing particularly attractive about its gray concrete façade or the heavy postwar décor of its spacious lobby, but there's a historic feel to the place because of its role in previous political uprisings, having served as both a den of intrigue and a base camp for journalists. We faxed our report from the hotel, then headed out to Ratchadamnoen Klang Avenue, a stately boulevard lined by low-rise Art Moderne buildings and elegant lamp posts adorned with statues of mythical *kinaree.* Designed to host royal processions and military parades, it had now been taken over by people marching for the right to elect their leaders.

After walking for several blocks, we arrived at the Democracy Monument, a potent symbol of political aspirations framed by four massive winged finials soaring out of a swirling circle of traffic. Built in 1940 to commemorate the bloodless coup that had replaced absolute monarchy with a constitutional form, I had always found the Democracy Monument to be stolid, unlovely, and vaguely fascist-looking. But on this day, draped with banners and people and the spirit of democracy, it looked positively beautiful. The crowd was prevented from moving much farther by police stationed at Paan Fah (Heavenly Gate) Bridge. But the mood around the monument remained more festive than fervent. Some chatted, others debated. Vendors selling food and drinks did a roaring trade, and people long starved of a cause drank in the idealism.

The rallies were suspended for a week to see whether the Constitution would be amended to make it more democratic; but when that effort failed, they started up again with renewed vigor, culminating in a massive demonstration planned at Sanam Luang for Sunday, May 17. I ran into Paul Handley, a correspondent for the *Far Eastern Economic Review,* and we wandered behind the stage that had been set up and climbed atop a generator truck to take pictures. A buzzing mass of people lay before us, the spires of Wat Phra Keow and the Grand Palace in the background. Kites flew high above, bobbing on the uncertain hopes of the throng below. The scene still had the peaceful feel of a festival. Families were having picnics. Besides selling food, vendors hawked videotapes of how the protests had evolved; you could buy the abridged version or the five-volume set.

As night fell, the Royal Hotel became a hive of activity. Many of the rooms had been booked by journalists, some from local newspapers, others international correspondents who had jetted in to Bangkok in expectation of a major conflict. I ran into Yord and Lin, fellow reporters from the *Nation,* and together we went back to Sanam Luang, where the masses glowed under lamplight. Ad Carabao, a folk rock star, was on stage playing politically charged tunes. Chamlong had yet to address the crowd, but he strode by us on three occasions, making a tour of the commons surrounded by a Praetorian coterie of youths, their hands locked together to keep an open space around him as he waved at us. Each time he passed, a charge went through the crowd. Finally, at around 8:30 P.M., he went on stage and began his harangue. It was a moving speech, outlining again why Suchinda had to resign, why the time for military interference in politics had passed. He had the crowd behind him, and we all wondered where it would lead. And then we found out—he called on the crowd, now more than 100,000, to march toward Government House.

Yord and I looked at each other, and each silently thought, "Uh oh." The night had turned humid and the crowd was riled up. There was no way the police were going to let an angry mob march on the seat of government. This was clearly a recipe for confrontation. Yord and Lin decided they were going home. I was torn, but decided this wasn't my fight. I caught the bus with Yord. A few hours later, back on Sukhumvit Soi 42/1, I stepped into a bar with a television set to see whether anything had happened. Despite all the premonitions, I was shocked to see the special bulletins. The protestors had run into the police at Paan Fah Bridge again, and this time the confrontation had exploded into violence. Police were firing water cannons into the crowd. The demonstrators threw rocks back. Some had allegedly ransacked a police station. And finally, there were pictures of soldiers shooting. The government declared a state of emergency.

At the *Nation*'s editorial office, it was the moment of truth, or rather the moment *for* truth. It was unclear whether newspapers were allowed to report all that was going on. The press censorship laws were vague, but clearly could be used to close publications deemed a threat to national security. The *Bangkok Post,* we later learned, had called the police to find out what they were allowed to run. When told to censor themselves, they decided to comply and leave columns of white space where the stories would normally run—innovative, perhaps, but not very brave. "I couldn't believe that when I saw it," recalls Tulsathit Taptim, the *Nation*'s news editor. "It never occurred to us to censor ourselves, or ask permission. We just went ahead and printed our stories." Suthichai, we were all convinced, *wanted* to be closed down. It would have made him famous the world over. The *Nation* was one of three local newspapers to report events freely. Local television, firmly under military control, was providing biased reports. CNN coverage was blocked in Thailand by communications authorities. News on the BBC World Service was available only to the small minority who had access to satellite television. Nantiya, a fellow environment reporter who was in Denmark at the time, had to call home and tell her parents the situation since they were completely in the dark.

So when the *Nation* hit the newsstands the next day, it provided a vital link to current events. A bold headline screamed "Protestors Battle Police," and underneath it stories described how the violence had broken out. Most striking of all was a brutal photo of three policemen flailing away with nightsticks on a protestor cowering on the ground in fear. "When our photographer Nararat Disyabutr saw the scene, he only had time to take one picture, no reserves. He just took it and ran," explained Tulsathit. Raj Gopalaram, the *Nation*'s veteran chief subeditor, quickly devised a simple but dramatic layout. The result was a front page that became an emblem for the uprising, and eventually won the *Nation* several international press freedom awards—a front page that made, and marked, history.

THE "BLACK MAY" MASSACRE

The next morning, the whole city felt electrified. There were so many questions to be answered. How many people had been hurt or killed? Why hadn't the police used tear gas instead of live ammo? What was going on now? I was supposed to be working on a series of stories about the Earth Summit in Rio, which was only a couple of weeks away. But there was only one story in Thailand now, and I was drafted to help edit articles about the evolving massacre.

For that's what the uprising was turning into. The violence had broken out the night before at around eleven, when protestors tried to storm through the police barricade at Paan Fah Bridge. This was when the yuppies and other members of the middle class probably went home. From then on, the fight was taken up predominantly by angry young men of the working class.[20] Following the exchange of water cannon and rocks, police waded into the crowd and started clubbing people indiscriminately. That's when Nararat apparently took his photo (the fate of the victim being beaten in the picture was unknown at the time, but we later reported that the man had died). A seesaw battle ensued. Chamlong Srimuang, shouting from atop a van through a loudspeaker, called for the attacks to stop, but he was too far away from the front lines to be heeded. Around midnight, the situation spun completely out of control. Molotov cocktails were lobbed at the police and a couple of fire trucks were set alight. The authorities accused protestors of torching the Nang Lerng police station, but Chamlong blamed a "third party" for instigating much of this violence, providing the military with an excuse to take drastic action.

That they did. At around 12:30 A.M., Suchinda and his interior minister announced a state of emergency in Bangkok and surrounding provinces that allowed the use of all measures to quell the "riots." Army troops had been brought in from the provinces, and in the early morning darkness of May 18, they opened fire with semi-automatic rifles. Some shot into the air as a warning, but others aimed directly into the protestors. Obtaining an accurate figure on the number of casualties was impossible, but reports from hospitals suggested that 150 to 250 people were wounded during the night, and anywhere from 3 to 15 were killed. Just before dawn, when protestors were singing the royal anthem and waving their hands to show they were unarmed, troops started shooting once again.

When daylight broke, an uneasy truce took over the streets. Chamlong met with a military officer and agreed that the protestors would remain seated if the soldiers would stop firing. Government agencies and schools were closed, and people were urged to stay at home. As the day wore on, troops tightened their grip on the area around Democracy Monument. At about 3:00 P.M., while firing thousands of rounds into the air, they arrested Chamlong. The soldiers had difficulty finding him at first because he was lying on the ground with the rest of the protestors, his hands covering his head to ward off the incessant sound of gunfire. Once located, he was hauled to his feet, handcuffed, and taken away. His crew cut and hangdog expression made him look like a sad and tired little boy. His wife, Siriluck, sat up and raised her arms in desperation, but was pushed aside. Watching the events unfold on television back in the newsroom, we couldn't help but be moved.

The soldiers then rounded up hundreds of other protestors, ordered them to take off their shirts, tied their hands behind their backs, and herded them into military trucks. But the soldiers were unable to secure the avenue. Onlookers fled down the side streets, then regrouped in front of the Public

Relations Department at dusk, defying the troops to open fire and shouting epithets about Suchinda ("Hiya Su!" they shouted, cursing him as a "monitor lizard," one of the vilest insults imaginable in Thai). The crowds swelled, joined by gangs of motorcyclists. Some enterprising youths commandeered city buses and drove them to the front lines to act as barricades amid heavy gunfire. One bus rode slowly into the troops as bullets riddled the driver's body. By 10:30 P.M., pitched battles raged in front of the Public Relations Department, resulting in perhaps a score of deaths and dozens more wounded. The marble and brass lobby of the Royal Hotel was turned into a blood-spattered triage unit as volunteer doctors and nurses frantically tended the wounded. Finally, the Public Relations Department and the neighboring Tax Revenue Department—two fitting symbols of the government—went up in flames.

It seemed like a repeat of Tiananmen Square, where the Chinese military had brutally repressed another democracy uprising just three years previously, or Rangoon, where the year before that Burmese troops had meted out similarly harsh measures on protestors opposed to the military dictatorship of Ne Win. But Thailand's May Massacre, or *Prusapha Tamin* (Black May), as it came to be known, would end differently from those incidents. The closest parallels were instead with the People's Power revolt of 1986 against the Marcos dictatorship in the Philippines, or the uprising in Kwangju in 1982, which eventually helped bring democracy to South Korea. Bangkokians had concluded that it was now or never for democracy. It was as if someone had thrown a switch and the populace had suddenly become politicized, even radicalized.

The change in attitudes was typified by my friend Giep. Young and well educated, she had always been uninterested in politics. When Chamlong began his hunger strike, she claimed it was "no use." And when the mass protests began, she refused to attend because her father had warned her against it. But on Monday night, she went down to Sanam Luang to see for herself what was going on. Witnessing the military's behavior firsthand made her so angry that she, too, was determined to fight for democracy. In her mind, as in so many others', the tide had turned.

The next day, I made my own trip to the protests before heading to the office. For better or worse, however, by the time I got to Ratchadamnoen, only the remnants of battle were visible: people milling around, firefighters still dousing the flames at the Public Relations Department, debris scatted all over the avenue, the hulks of charred and abandoned buses, and even an oil tanker that had been brought into the fray. Soldiers were everywhere, of course, and they had blocked access to Sanam Luang and the Royal Hotel. I had arrived too late.

Scot, however, had visited the area in time to be in the thick of things. Something of a night owl, he had gone down to Ratchadamnoen with Chris Burslem, a *Nation* colleague, around 4:00 A.M. Tuesday morning to find hundreds of soldiers asleep on the street, cradling their weapons, even as the sounds of gunfire came from the Public Relations Department building 500 meters up the road. About a half hour later, the army began its final push to clear the streets of protestors. "The image was one of awesome firepower: M-16s, belt-firing M-60s, M-89 grenade launchers, jeeps mounted with heavy machine guns, armored personnel carriers with rapid-fire cannons," Scot recounted.

> At one point I even saw a soldier with a clumsy-looking bazooka strapped to his back. Neither truncheon nor riot shield could be seen. These troops brought to the streets of Bangkok were a fully armed division headed into combat.
>
> The columns marched forward . . . I was cut off by the last man in the formation. Five of us were left standing on the outside with M-16s pointed over our heads. . . . A squad of soldiers surrounded us and forced us to the ground. My shirt was pulled over my head and I was pushed flat on my chest. An officer began yelling out questions. . . . I raised my head to tell him I was a journalist, only to have it pushed back down by the surrounding soldiers who stood boot-to-boot. All I could see were those shining boots, their toes menacingly close to my face.
>
> Two of the four [Thai] photographers with me were blindfolded and bound. As soldiers went through their equipment, exposing each and every roll of film, the sergeant was bearing down on one of the photographers. I couldn't comprehend the meaning of his interrogation, but I knew the photographer's answers were not the correct ones when soldiers began kicking him in the back. The blindfolded photographer grimaced, but remained silent. By then the sky was lighting up with neon streaks of tracer bullets and the chattering roar of hundreds of automatic weapons could be heard. There was no apparent pattern to the gunfire; it was just an orgy of shooting. Fixed machine guns on the roofs above sent tracers off into the empty blackness toward Khao Sarn Road. Only the sergeant's barking could be heard over the white noise of the machine gun fire, which lasted well into the dawn.[21]

What Scot didn't realize was that the troops were mounting an offensive on the protestors in and around the Royal Hotel, its lobby now a makeshift surgery unit and morgue. Rushing into the building, soldiers shouted orders for everyone to lie face down on the floor, beating and kicking anyone who didn't do it quickly enough. Then they went room to room hauling out anyone who looked suspicious. Kamol Sukin, a future colleague at the *Nation,* was in the hotel as a reporter for *Phujadkaan* newspaper, which also flaunted its defiance of the authorities by reporting events openly. "When the soldiers came in, we raced out the back entrance to the hotel and into the alley. Luckily, people there showed us the way to go or we wouldn't have escaped," he said. "Troops had surrounded our editorial office [in nearby Banglamphu], but we were able to send our reports directly to the printing plant across the river." After securing the hotel, the soldiers rounded up thousands of other protestors along the avenue. About 1,500 were brought to the front of the hotel where they squatted, tied up and shirtless—including a couple of *farang,* Scot and Chris.

Scot recounted how they got there:

> We [prisoners] were marched off towards the Royal Hotel, with occasional shoves for those who walked too slowly. When we arrived at the hotel, we saw hundreds of protestors sitting cross-legged outside on the hot sidewalk, their hands tied with T-shirts and heads bowed. A soldier took my shirt off and fastened my hands behind my back . . . and forced me down onto the sidewalk. [Chris], who had come along to act as an intermediary, stepped in, but he, too, was tied up and forced on the sidewalk. Several protestors turned their heads towards us in bewilderment. We were the only *farang* in sight. A teenager sitting in front of me tried to signal his solidarity with us by nodding his head in my direction. But a soldier quickly came and pushed his head back down.
>
> After watching the military parade around all morning as if evil invaders had just overrun the country's borders the absurdity of the whole thing was more than I could fathom. This wasn't a war. These weren't enemies. The army had brought all it had to bear against these people—their own people. One of the photographers, who was sitting next to me, slumped and began to cry. Again, a new group of soldiers approached me and asked to see my ID. I pointed to my pocket. One of the soldiers drew out my wallet, then after much fumbling, retrieved my newspaper identification card. 'I'm sorry, I'm sorry,' he said, as he untied my hands. 'You can go now.' I

put my shirt on and tried as best I could to wish the photographers well. As I turned to leave, I looked back at the teenager who, from behind his back, flashed me the victory sign.

THE KING PUTS OUT THE FIRE

As foreigners, Scot and Chris had gotten off lightly. The Thais, meanwhile, had to undergo a terrifying ordeal: They were herded onto trucks to be taken to an unknown destination. Rumors raced around the city about soldiers stuffing dead protestors into sacks and feeding them to the crocodiles, or (more credibly) of secretive cremations being carried out at a temple near the airport. But most demonstrators, at least 2,500, were simply taken to the Bang Khen Police Private School, where they were kept under guard.

The presence of foreign camera crews undoubtedly increased the pressure on Suchinda and his allies. Scenes of military brutality, particularly from the forced entry into the Royal Hotel, raced around the globe. Equally striking were images of ordinary Thais, usually so calm, venting their indignant rage—a young student in glasses trying to explain in English that "we just want demo-cracy!"; a thin forty-something woman shrieking (in English for the camera's benefit) at soldiers about to fire on demonstrators, "Don't shoot! If you are patriot, don't shoot!"

By Tuesday afternoon, the protestors had been routed from the Ratchadamnoen area. Rallies began forming around Ramkhamhaeng University, a hotbed of student activism. An uneasy truce settled over the rest of the city on Tuesday evening and into Wednesday. Streets were eerily empty, and passengers on the bus to work maintained a tense silence. But clearly a denouement was approaching on Wednesday night. There were reports that army units loyal to former Prime Minister Prem Tinsulanonda were making their way to Bangkok to fight against those loyal to Suchinda. The government slapped a curfew on the city; no one was to be on the streets from nine in the evening until early the next morning. And then we got word that the government had ordered the *Nation*—along with two other papers, *Phujadkaan* and *Naew Na*—closed down.

We'd been expecting a closure order all week. Now, Suthichai called us in for a meeting and delivered a rousing speech. We had not received the closure order yet, but it was sure to come soon, and it might arrive with a heavy military escort. Officially, our license to print the *Nation* would be suspended, but we had licenses to print other publications, including weeklies such as *Business Review* and *Nation Junior,* a light-hearted magazine for teenagers who wanted to learn English. So we pondered printing the newspaper under a different name, such as *Nation Junior Daily.* But Suthichai favored not running the paper at all; it would make more of a statement, he argued. In fact, the closure order never officially arrived, and we later learned that the police had decided to hold off delivering it because they weren't sure which side was going to win.

The winning side would be determined by the one man with the moral authority to decide the matter—the king. Rumors about his intervention turned out to be correct. On Wednesday night, he summoned Suchinda and Chamlong to Chitrlada Palace. In an amazing display of traditional respect, aired on national television around midnight, the two principal combatants crawled in before His Majesty, gave him a deep and courteous *wai,* and then listened silently as he lectured them for half an hour. The scene seemed even more surreal because the audio pickup was so poor that we couldn't make out what the king was saying! We just sat, glued to the television, watching the drama unfold, unable to understand what was going on. Only his final few phrases proved audible. "*Chuay kan khit,*" he commanded. "Consult with each other [to work out a solution]."

We were stunned. That was it? Work with each other? After all the bloodshed and the mayhem?! The thousands of protestors who had been detained were going to be released, but the government had officially announced that forty people had been killed in the uprising and more than six hundred wounded (to this day, many think the number of fatalities was much higher). What good was a compromise going to do them? I've never seen the *Nation*'s newsroom so depressed. We wanted justice, and a definitive end to military interference in domestic politics. It didn't look as if we were going to get either. One reporter began singing, "How can we go on?" One of the more hotheaded news editors began shouting furiously. But most of us, tired and resigned, started winding down, thankful at least that the violence seemed to be over. The curfew was still in effect, so a bunch of people gathered in Thepchai's office and cracked open a bottle of whiskey. I pushed two chairs together and crashed for the night.

The situation turned out better than it had initially appeared that night. Although we hadn't heard it, the king had called on the government to pass the constitutional amendments that required a prime minister to be elected. Suchinda resigned, and Anand and his team of technocrats took over again until another round of elections were held in October. No one ever accepted responsibility for the bloodshed, and no one was ever formally punished for it. That confirmed an awful but valuable lesson for all of us: Justice is not a condition of nature. If people want it, they have to fight for it, but it is by no means guaranteed. And if it's achieved, that as much as anything else is a sign that a society is "developed."

Nevertheless, the uprising was a victory for the people of Thailand. Democracy, albeit a flawed and incomplete version of it, was restored, hopefully for good. Set amidst the debate over Asian values, Thais showed that although it may be possible to have economic growth combined with authoritarian rule, they felt strongly that democracy goes hand in hand with development. More than anything else, Thais want to be developed, in every sense of that word. And if the determination they displayed in May 1992 is any guide, they will get their wish, even if it came at the cost of another generation's innocence lost.

TROPICS AND POLITICS

Although Thailand's political revolt had a unique form of resolution, the quest for democracy was evident in uprisings throughout Asia in the 1980s and 1990s. The king's intervention does not explain why the people of Bangkok, like those of Manila in 1986 and Jakarta in 1998, were able to regroup and ultimately throw out the generals. What made these results so different from the uprisings in Rangoon and Beijing, which were stamped out? Part of the reason is that, unlike in China, the protestors in Thailand, Indonesia, and the Philippines all had an organized opposition to provide legitimate leadership (Burma had similar opposition leadership in the form of Aung San Suu Kyi and the National League for Democracy, but it was brutally suppressed following its electoral victory in 1990).

Equally important is that these countries all have far more open societies. The Suchinda government's decision to clamp down on television and radio broadcasts was akin to erecting a house of straw amidst a media storm. It proved counterproductive, stirring up even greater public resentment. The continued outspokenness of the written press was the main outlet for information on the latest events, but there were plenty of alternative channels of communications. The proliferation of VCRs turned out to be a key; indeed, video entrepreneurs moved amazingly quickly to sell footage of the street battles only a day or two after they had happened. *Samizdat* spread through fax machines. Mobile phones—previously reviled by the politically correct for the disturbances they create in public places—proved crucial in helping to rally the protestors. And all this happened before anyone had even heard of the Internet.

Nor should one overlook the conventional explanation for why Korea, Taiwan, Thailand, the Philippines, and, tentatively, Indonesia, have been able to make the transition to democracy: They all had a more extensive middle class from which to draw support than, say, Burma. The political scientist Barrington Moore[22] has theorized that peasant-based societies tend to be characterized by either revolutionary fervor or military rule—an analysis that describes the less-developed countries of Southeast Asia, and used to fit Thailand. But the situation changes with the emergence of a thriving middle class, whose members understand that political development must coincide with and support economic development.[23] The transition, however, is rarely smooth, and is bloodier in some countries than in others.

The rage felt by the democracy protestors of Bangkok over having their voices and rights ignored was a familiar feeling to environmentalists, who naturally saw in the uprising many similarities to their own movement. "The middle class in Thailand wants a better environment," said Dhira Phantumvanit, a well-respected economist who sought to merge the goals of development and environmental protection. "They are the same people who went out on the streets of Bangkok." Witoon Permpongsacharoen, the anti-dam crusader and founder of the Project for Ecological Recovery (PER), pointed out that the two movements have fed off each other. "When you look at the development of the people's movement, you must give some credit to the anti-dam protests of the 1980s. The struggle to block construction of the Nam Choan Dam was the first push to bring power back to the people after the1976 uprising. It also showed that protest movements can be broad-based. Previously, protests were considered to be something only for students, or for those with left-wing ideology. But environmental issues are not divided by ideology; everyone can agree with them."[24]

Perhaps, but the aftermath of the democracy uprising in 1992 also show how environmental issues are perceived, and acted upon, differently by the urban middle class and rural villagers. In Bangkok, the person best able to tap into the environmental aspirations of the middle class was Bhichit Rattakul when he was elected governor of Bangkok on a green platform in 1996. Meanwhile, Dhira, who argued that NGOs should do a better job of engaging the middle class and raising funds from them, subsequently set up the Thailand Environment Institute, a policy-oriented think tank. But the institute mostly relies on research contracts and working with business, for instance to establish ISO14000 environmental standards.

In the provinces, the impact of greater freedom emboldened villagers to protest much more forcefully over resource and development issues in the following years. Most dramatically, NGOs teamed up with fishing and farming communities on the Moon River to attempt to block construction of the Pak Moon Dam; that protest led to a violent clash in Ubon Ratchathani. Continued opposition has exposed the project's flaws and forced the dam's gates open for a while. The strength of the environmental democracy movement has ballooned to the point where thousands of protests occur across the country every year. Meanwhile, the Assembly of the Poor's gatherings at Government House are now an annual event.

When we think about environmentalism in the West, we mostly think of the middle-class variety. It's commonly understood as part of a hierarchy of needs, as something to clean up once we've reached a certain level of development, which is why we get that maddening Kuznets curve. But in the developing world, there is a different kind of environmentalism among the poor, at least in the countryside. They care about their resources, but to them protecting the environment is about protecting their livelihoods. Demands for greater control over rural resources can be viewed as part of a power struggle, a way to prevent those resources from being appropriated to the national or global level; or it can be seen as local people re-inventing themselves as environmentalists, perhaps reviving more sustainable traditions along the way.

The continuing urban-rural divide, so stark that the sides appear to be different nations, is one of the reasons politics remains so unstable in developing countries. Until recently, Thai politicians seemed incapable of bridging that gap. But the current prime minister, Thaksin Shinawatra, tried to change that pattern, at least initially, which helps explain his electoral popularity, according to academics Chris Baker and Pasuk Phongpaichit. Thaksin is a telecoms businessman whose main appeal is to the urban middle class, but he also made a direct appeal to rural voters, promising to spread the wealth through a debt moratorium and a cheap health care scheme. He sought out ideas for agrarian reform, and even included some in his platform, notes Baker. On his first day in office, he had lunch with the Assembly of the Poor, which promptly closed down its long-running protest outside Government House. Relations have since cooled, "but what made Thaksin's approach revolutionary was that no one had tried it before," says Baker.

"Thai Rak Thai [Thaksin's party] is still an urban party, but he realized that an accommodation has to be reached with the rural sector."[25]

Still, while Bangkok's middle class elected Bhichit, environmental democracy hasn't made many inroads into rural electoral politics. The uncomfortable truth is that after getting rid of military rule, Thailand is now governed by provincial tycoons elected through political machines in their home provinces. The business interests and the corporations make the rules, not the generals. Governance has slightly improved, but the system remains corrupt. The Democrat Party's Chuan Leekpai, elected prime minister when democracy was restored in late 1992, had a clean image thanks to his modest tastes, but was unwilling or unable to rein in his colleagues.

In the long run, the most important legacy of the May 1992 events could be that they helped pave the way for a much more democratic constitution passed five years later. Thongchai Winichakul, a Thai academic at the University of Wisconsin, argues that the democracy uprising does not need a monument to commemorate it because the new "People's Constitution" serves as that monument.[26] It contains numerous measures near and dear to the hearts of environmentalists, including provisions for the freedom of information and greater public participation in managing natural resources. Progress in carrying out these and other reform measures is uneven, but there is hope that the electoral process, and governance in general, will slowly be cleaned up.

Ultimately, that may be the best hope for the environment, as well. A better balance is needed between the three types of institutions that make up society: government, business, and the civil sector. In developing countries, Thailand among them, the business sector performs its role ably once it's allowed to transform resources and skills into wealth through open market policies. The civil sector, meanwhile, clearly needs strengthening. Civic groups can help create a sense of public trust necessary for people to join in protecting the environment. These groups on their own are not equipped to manage the environment, much less all the other facets of modern-day society. But by keeping a close eye on the government and lobbying for reform, NGOs, the media, and other watchdog groups can balance corporate influence and bring constant pressure to bear in favor of cleaner, more effective governance.

Indeed, the sector that has largely failed in the South is government. Somehow it needs to be weaned away from the corrupting influence of vested interests to establish a responsible, long-term outlook that can propel society toward a better quality of life and that tries to make development truly sustainable. And ideally, it should attempt this before sinking into the environmental abyss at the bottom of the Kuznets curve. An excellent start would be to use "green GDP" indicators, rather than simply chase high economic growth rates, and to replace labor taxes with taxes on pollution. But numerous other approaches have also been proposed.

NGOs in the South support political decentralization to increase the authority of local communities in managing their own resources. It could work, if a careful balance is kept between the central and local authorities. But it could also lead to administrative gridlock. And if it is not accompanied by greater local democracy, it could simply present local officials with more opportunities for corruption. Phaichitr Uathavikul, Thailand's environment minister under the Anand administration, felt that holding elections for provincial administrators was too dangerous because of the dominance by local mafias. "Thailand today is like America in the 1920s," he explained. "It would be like letting Al Capone take over."

Another option favored by free market economists is to use market-based mechanisms to manage public goods and bads. This would entail providing incentives for conservation, for instance by establishing a fair and sustainable bioprospecting regime, certifying eco-tourism efforts according to proper standards, and paying rural inhabitants to sequester carbon and protect watersheds, endangered species, and ecosystems. Sound environmental management could also be rewarded by awarding property rights over resources to the private sector, either as deeds for individuals or long-term concessions to companies. These new overseers would then have the incentive to manage the resources sustainably and to use them efficiently. The fishery ITQ (individual transferable quotas) system used in New Zealand follows this approach, and the United States—along with using emissions trading to regulate air pollutants—has experimented with handing out tradable water property rights. But bear in mind that Thailand has run into controversy in merely awarding deeds as part of land reform; the conflicts in deciding who would receive the rights over a more fungible resource such as water could be even greater. And the interests of private owners don't always match those of the public. So to look after societal goals, some kind of regulation—such as cheap public access to nature sites and the conservation of rare species that have no economic value—would still be needed. Nevertheless, public-private partnerships clearly have a great deal of potential.

Another approach could be to partially globalize natural resource management by creating more multilateral environmental agreements or extending the reach of present ones. Ironically, governments in the South tend to respect the measures in these treaties more than their own laws, perhaps because they fear the economic penalties that would come with failing to uphold their obligations. But MEAs (multilateral environmental agreements) with binding provisions are viewed with tremendous suspicion by the United States and the global South because they fear they will give up more in lost sovereignty than they will gain. So reaching agreement on such treaties has proved difficult.

Finally, the problem of corruption could be tackled directly. Singapore has done so by combining internal means—for instance, raising the salaries of civil servants—with external accountability. But perhaps an alternative approach could try to draw these informal markets out into the open. For instance, Thailand's National Parks Division (NPD) is beset with corruption because high-ranking officials and park chiefs typically have to pay for their posts, and so force their subordinates to do the same. Rather than try to break this corrupt cycle, some have suggested simply taking a World Bank loan and buying out all these "contracts" so that the officials would no longer have to recoup their investments through corrupt practices.

A similar approach could be used to address complaints made by conservation officials that they are saddled with unfunded mandates. Asked to protect areas without the necessary budget or manpower, they could sell some resources—for instance, by allowing hunters to shoot some game—to support the broader goal of conservation. Mightn't it be better to let them do this openly, and thereby help ensure the money actually goes toward conservation and not into the officials' pockets? In practice, this would mean relaxing the restrictions on these areas and allow hunting, the collection of forest products, or community forestry. But perhaps that is simply bowing to local realities, and it may improve the area's integrity in the long run.

In the end, development is marked not only by a rising level of wealth or technological prowess but also by a respect for public institutions and the rule of law. To achieve that, the middle class in Thailand and other developing countries must reach some kind of accommodation with the rural majority that still decides who is elected into office; that way, the rural sector improves its standard of living and everyone gets cleaner government along with a better quality of life. The alternative is to continue business as usual and hope the country keeps growing to the point where the middle class forms a majority.

Ultimately, a similar accommodation may be required on the world stage. It's hard to see how the North can protect the global environment and achieve ecological security without coming to terms with the aspirations of the South, including a desire to be developed. This is not the threat that many in the North seem to fear it might be. In fact, it is a tremendous opportunity to tap the creative, entrepreneurial, and intellectual potential of the majority of mankind.

But it also represents a formidable challenge. For in striving to fulfill that potential, there is a risk that the resulting political and environmental shocks could significantly alter the world as we know it—for the worse. The good news is that, at a basic level, people the world over not only recognize the need for democracy and a good environment, they are willing to risk their lives to achieve those goals. By and large, the green urge is universal and transcends national or cultural boundaries. That is a powerful sign of hope.

EPILOGUE: JOHANNESBURG
A CIRCUS OF SUMMITS

A CIRCUS OF SUMMITS

Romani Ben Zekri has a different view of garbage. Where others see refuse, he sees a resource. Romani is one of Cairo's thousands of *zebaleen:* garbage collectors who every day gather 3,000 tons of the city's trash, sort through it and recycle about 80 percent of it, generating jobs and income for roughly 40,000 people.[1] Many developing countries, including Thailand, have similar sanitary entrepreneurs, although they are rarely so efficient. It's a wonderful example of how protecting the environment can go hand in hand with providing jobs—an example of sustainable development at work.

But Romani and his colleagues have a problem. Even though they dispose of a third of Cairo's garbage daily at no cost to the government, the authorities decided to sign a contract with some multinational waste management firms to haul away the city's trash. Not only does this threaten the livelihood of the resourceful *zebaleen,* but the contract stipulates that only 20 percent of the garbage needs to be recycled, so the decision could prove to be more damaging to the environment, as well.

Romani told his story to the world, or at least to anyone who would listen, in August 2002. He was one of some 20,000 NGO representatives who attended the World Summit for Sustainable Development held in Johannesburg. The summit was a disappointment to many because government negotiators managed to come up with only a few desultory agreements. Even the most lauded achievement—a vow to halve the number of people without access to proper sanitation facilities—was not legally binding. It was a far cry from the historic treaties on climate change and biodiversity, signed a decade earlier at the famed Earth Summit in Rio de Janeiro.

On the other hand, the much-maligned meeting in Jo'burg was not the complete disaster many had feared. This was largely due to the feverish networking carried out on the fringes of the official summit: at the Global Civil Society Forum held across town, the Children's Earth Summit in Soweto, the Indigenous Peoples Summit in Kimberley, the Ubuntu Village and Water Dome exhibition centers, at the scientific discussions held in the IUCN Environment Center, the gatherings of landless people, the forum on environmental justice, the meetings of legislators and local governments, the international business conventions, and so on. As noted by Tony La Vina, former undersecretary of the environment for the Philippines, Jo'burg was a convening of many summits.[2]

All this networking produced some promising results. Take the example of community-based forestry. Although still the subject of struggle and controversy in Thailand, this movement has grown to the point where it is now being exported from the developing world and gaining a foothold in North America, where proponents hope that it can one day form the basis for a compromise in the long-running battle between environmentalists and logging interests. In Johannesburg, community forestry federations and other grass-roots groups gathered to create the Global Forum on Community-Based Resource Management. Will it amount to anything? Nobody knows. But its potential—to serve as a voice for the burgeoning environmental democracy movement around the world—is tantalizing.

For it is becoming increasingly clear that, barring some kind of crisis that forces authorities to act, the impetus will have to come from the bottom up if advances are to be made toward sustainable development. Governments and corporations must be pushed into protecting the environment by civil society. That was illustrated at the summit by a group rarely heard from, a global gathering of 130 chief justices and senior judges who issued some of the most insightful statements in Johannesburg. They noted that most countries already have decent environmental laws, but they often lack the will to apply them. They called on the judiciary the world over to "fearlessly implement and enforce applicable international and national laws," and even proposed the establishment of an International Environmental Court, although such an institution would no doubt face stiff resistance from many national governments. And they emphasized that the lack of access to the law was often linked to the lack of public access to information—problems that clearly do not afflict Southeast Asia alone.

Whereas Rio was an environmental summit, attended primarily by environmentalists and held at a time when the concept of sustainable development was newly formed, Jo'burg was to a much greater extent about development. All manner of groups—from disabled Africans to spiritual Hawaiians to protesting lesbians—showed up to put forward their vision of how to make the world a greener, cleaner, fairer, and wealthier place. That made the summit less coherent and more chaotic, but was also arguably a mark of progress, a sign that the message of sustainable development is spreading to places it had barely touched before. NGOs from countries such as China and Vietnam, formerly represented solely by government officials, were present in unprecedented numbers. The Arab Environmental Caucus, previously obedient to officials, was taken over by activists and issued a noisy declaration.

And those *zebaleen*? They kicked up such a fuss in Johannesburg, and received so much attention, that Egyptian authorities promised to negotiate with them. They also went to Soweto and set up an exchange program with local garbage collectors, who proved eager to visit Cairo and learn about the *zebaleen*'s innovative recycling methods. Perhaps one day Romani and his colleagues can make a similar trip to Thailand, where they will no doubt view the litter-strewn slums not so much as an eyesore but as an opportunity.

NOTES

INTRODUCTION

1. James Fahn and Sara Colm, "Aid That Hurts," *Nation* (Bangkok), 13 January 1993.

2. Thanks to Ted Bardacke, former *Financial Times* correspondent in Bangkok, for this observation.

3. David Kinsley, *Ecology and Religion* (Englewood Cliffs, N.J.: Prentice Hall, 1995), 87.

CHAPTER 1

1. "The Brown Revolution," *Economist,* 11 May 2002.

2. Philip Blenkinsop, *The Cars That Ate Bangkok* (Hong Kong: White Lotus, 1997).

3. Edward McBride, "A New Order: A Survey of Thailand," *Economist,* 2 March 2002. A major reason for the popularity of pickup trucks is that they are subject only to a 3 percent excise tax, but passenger cars are taxed up to 60 percent. More than half the cars sold in Thailand are pickups, many of which are essentially converted into station wagons and used as passenger cars.

4. "Vehicle Sales Plummet," *Nation* (Bangkok), 18 November 1997.

5. "Ford Conducts Survey on Drivers in Bangkok Jam," *Nation* (Bangkok), 9 March 1998; and Joshua Kurlantzick, "Livable City or Urban Nightmare?" *Bangkok Post,* 22 August 1999.

6. "Racing Against Time in Bangkok's Winding Traffic Snarl," *Jakarta Post,* 10 August 1997.

7. Seth Mydans, "Bangkok Opens Skytrain, But Will It Ease Car Traffic?" *New York Times,* 6 December 1999.

8. Walakkamon Eamwiwatkit, "Official Silence Over City's Noise Pollution Deafening," *Nation* (Bangkok), 30 May 1996. See also John Hail, untitled UPI report, 30 August 1995, and "City Noise Could Cause Hearing Loss," *Bangkok Post,* 17 January 1998. Workers in noisy factories are also vulnerable, as are residents who live close to loud construction sites. Government officials, meanwhile, seem to be largely deaf to the problem: The fine for construction companies found to be disturbing the peace is a mere 100 baht, or roughly $4 at the old exchange rate.

9. The estimate was calculated by the research center of Thai Farmers Bank. See "Bangkok Think Tank Finds Silver Lining in Traffic Gridlocks," *Deutsche Presse-Agentur,* 10 August 1995.

10. Ibid.

11. A description of the incident is recounted by Nick Cumming-Bruce, "Thai Police Discover 'Reprisal' Bomb in Abandoned Hire Truck," *Guardian* (London), 19 March 1994.

12. "Racing Against Time in Bangkok's Winding Traffic Snarl," *Jakarta Post,* 10 August 1997; and Seth Mydans, "Bangkok Opens Skytrain, But Will It Ease Car Traffic?" *New York Times,* 6 December 1999.

13. "Bangkok Jams to Get Worse As Projects to Ease Congestion Kick Off," *Straits Times* (Singapore), 15 January 1997.

14. For an interesting comparison of car ownership growth rates in Asian cities, see Marc Levinson, "To Travel Hopefully: A Survey of Commuting," *Economist,* 5 September 1998.

15. "London to Take Delivery of Low Emission Buses," *Environment News Service,* 30 March 2001.

16. James Fahn, "Officials 'Stall' As Bangkok Chokes," *Nation* (Bangkok), 14 September 1992.

17. In 1993, roughly 40 percent of Bangkokians owned cars, compared to 27 percent of Tokyo's residents and just 5 percent of Hong Kong's residents. See Levinson, "To Travel Hopefully."

18. "Hopewell Plan Indicates Govt Haste," *Nation* (Bangkok), 3 September 1997.

19. For insights on Bangkok's planning history, see Tsuji Shingo, *Bladerunning: The Dialetics of a Restless Urban Landscape: Transformation of the Bangkok Metropolis in the Late Twentieth Century* (master's thesis, Urban Planning, Columbia University, 2000). Shingo also points out that an exception to the lack of planning rules lies in Rattanakosin, a historic area in western Bangkok that has been designated a conservation district.

20. Peter Janssen, "It's 'From Dust to Dust' for Bangkok Residents," *Deutsche Presse-Agentur,* 7 May 1996.

21. Judith Scherff, "Environmental Ills Threaten Fitness Fans," *Los Angeles Times Syndicate,* published in the *Nation* (Bangkok), 22 October 1993.

22. Anil Noel Netto, "Gasping for a Solution to Urban Air Woes," *Inter Press Service,* 8 May 1997.

23. Victor Mallet, "Third World City, First World Smog," *Financial Times* (London), 25 March 1992.

24. Netto, "Gasping for a Solution."

25. Daniel Litvin, "Dirt Poor," *Economist,* 21 March 1998. This is also the source for the data below on the health costs of air pollution.

26. Another World Bank report, which confirmed that in the late 1990s, China lost from 3.5 percent to 7.7 percent of its potential economic output as a result of pollution's impacts on workers' health, apparently convinced the country's leadership to take the problem seriously. See "A Great Leap Forward," *Economist,* 11 May 2002.

27. Walakkamon Eamwiwatkit, "Alarm Sounded Over Air Pollution," *Nation* (Bangkok), 18 April 1996.

28. Litvin, "Dirt Poor."

29. Kunda Dixit, "Choking on Urbanization," *Inter Press Service,* 20 April 1994.

30. Netto, "Gasping for a Solution."

31. Gene Kramer, "People Urged to Get the Lead Out of Their Gas Tanks—and Lungs," *Associated Press,* 18 May 1996.

32. Dixit, "Choking."

33. Netto, "Gasping for a Solution."

34. Kramer, "People Urged to Get the Lead Out."

35. James Fahn, "The Fighting Khunying Speaks Her Mind," *Nation* (Bangkok), 4 November 1994.

36. James Fahn, "It's No Stroll in the Park," *Nation* (Bangkok), 17 May 1996.

37. Thongchai, who is an avid cyclist and head of the Bicycling Club of Thailand, also disagrees with the notion that Thailand's climate is ill suited for nonmotorised transport. "Actually, it's harder to walk or ride a bicycle in a cold climate where there's snow and sleet," he explains. "I found it much harder to ride a bicycle in the United States. Here, if you get wet with perspiration, you just take a shower."

38. James Fahn, "Who Is That Masked Man?" *Nation* (Bangkok), 7 January 1994. This article contains an interview with Bhichit from a couple of years before his election.

39. Pasuk Phongpaichit and Chris Baker, *Thailand's Boom and Bust* (Chiang Mai: Silkworm Books, 1998), 161.
40. "Poll Finds Youth Put Environment Before Economy," *Nation* (Bangkok), 11 September 1996.
41. This curve is named after Simon Kuznets, a Nobel Prize–winning economist who created many of the techniques for measuring economic analysis that are still in use today. Intriguingly, his first assessments carried out for the U.S. Bureau of Economic Analysis contained measurements of environmental degradation, which he felt were important to include in the data, but these subsequently proved controversial and were dropped. Environmentalists are still trying to have similar green indicators included in GDP measurements today.
42. For instance, Scott Barnett, a London Business School researcher who presented as part of the Global Environmental and Trade Study at the 1996 WTO Summit in Singapore noted, "There is nothing inevitable about the relationship [between growth and the environment]." What he calls the "inverted-U model" may not hold for developing countries, and some types of environmental degradation, such as the loss of biodiversity, may end up being irreversible. See James Fahn, "Trade Liberalization Not So Green As Painted," *Nation* (Bangkok), 13 December 1996.
43. See "A Great Leap Forward."
44. Poona Antaseeda, "Bangkok Fails WHO's 7 of 23 Indicators," *Bangkok Post,* 13 November 1997.
45. Transvestites and gays are openly accepted in Thailand, to a greater extent than any other country I know of. Thailand has had one openly gay prime minister and a couple of others who never officially came out of the closet. As for the *katoey,* they have their own culture in a way—some sociologists consider them to be a kind of third gender—which seems to include a particular affinity for volleyball. A controversy broke out several years ago when the national team declined to accept a couple of deserving *katoey* players, a sign that, although Thailand is more tolerant than most countries, it still has a way to go.

CHAPTER 2

1. See James Fahn, "Into the Underworld" and "Chemistry At Work in Bizarre Rock Formations," *Nation* (Bangkok), 1 April 1998. The chemical properties of limestone karsts endow them with all sorts of fantastic quirks, explains Dean Smart, a cave consultant with the Royal Forest Department. Rainfall may pick up small amounts of carbon dioxide in the atmosphere, or larger amounts when it hits the ground and mixes with decaying matter. It then turns into carbonic acid, which reacts quite rapidly with the calcium carbonate that makes up limestone. That's why you see so many small pits (known as "karren") and pinnacles in the rock on the karsts. They also tend to form indentations where they meet the sea, but these are not caused by the battering of the waves. Nor does the rock dissolve directly into the sea, since the ocean is already saturated with limestone. Rather, when a little bit of fresh water mixes with about 97 percent seawater, which often occurs when rain runs down the cliffs and meets the ocean, the compound becomes very aggressive, and that's what eats away at the base of the towers. Another strange feature is the "eucladiolith": a rocky, seemingly wind-blown protuberance often found at the mouths of caves. These are formed because algae tend to grow on the side of stalactites facing the light. As the algae absorb carbon dioxide from water dripping down the stalactite, the water compensates by depositing calcium, and so the eucladiolith grows out toward the light—almost like living rock.
2. Marque Rome, "Shooting to Kill," *Action Asia,* 1999.
3. Animal welfare activists are concerned the bird's population is declining due to the overcollection of the nests, and have sought to protect it through the Convention in International Trade in Endangered Species. But the move was blocked by Southeast Asian governments, which have benefited from the lucrative trade for hundreds of years.
4. Denis Gray, "Passion for Birds' Nest Soup Sparks Violence, Wildlife Loss," *Associated Press,* 5 March 2000.
5. Denis Gray, "Model Eco-Tourism Effort in Thailand Swamped by Greed, Violence," *Associated Press,* 10 July 1999.

6. Matthew Sherriff, "Thailand's Troubled Waters," *South China Morning Post* (Hong Kong), 10 November 1998.

7. James Fahn, "Paradox Week in Phuket," *Nation* (Bangkok), 6 December 1992.

8. "Visitors Exceed 10 Million Mark," *Bangkok Post,* 28 February 2002; and "3.4 Million Tourists Visit Thailand in 1987," *Xinhua News Agency,* 16 March 1988.

9. 1990 was Visit Malaysia Year, 1991 was Visit Indonesia Year, 1992 was Visit ASEAN Year, 1996 was Visit Myanmar (Burma) Year, 2000 was Visit Laos Year, 2002 will be Visit Vietnam Year, and so on.

10. For ASEAN tourism statistics, see the Web site at http://www.aseansec.org/view.asp?file=/economic/Summary.htm. No earnings figures were available, but estimates were made before 2000 based on the predicted number of arrivals. See "Tourism: Asean Target Is 36 Million Foreigners," *Bangkok Post,* 29 December 1999.

11. To be fair, Samut Prakan, an industrial center south of Bangkok, has also been declared a pollution control zone. One major reason resort areas have received so much attention is that the Thai government finds it much easier to tax tourists to pay for the cleanups. In Bangkok, meanwhile, where a much more ambitious citywide wastewater treatment project is underway, the government has yet to come up with the political will to make city-dwellers pay for treating their sewage.

12. The report in question was supported by UNESCO and is cited in a *Nation* (Bangkok) article from May 11, 1990.

13. Ing K, "TAT, This Is Not Party Time," *Nation* (Bangkok), 20 October 1997.

14. See the Web site at http://www.golfwar.org/.

15. "Vietnamese Peasants Attack Military Construction Team Over Land," *Deutsche Presse-Agentur,* 19 May 1996; and "The 'Wars' Against Golf Are Spreading," *Nation* (Bangkok), 4 June 1996. There was also controversy over a golf course planned in a protected forest just outside Ho Chi Minh City in southern Vietnam. See Jatuphol Rakthammachat, "Tourism: Saviour or Spoiler?" *Nation* (Bangkok), 31 January 1993.

16. James Fahn, "Going Against the Green," *Nation* (Bangkok), 29 April 1994; "Indonesia Students Seek Release of Evicted Farmers" (editorial), *United Press International,* 1 October 1993; and Jeremy Wagstaff, "Golf: Only Catering to the Rich in Indonesia," *Reuters,* 5 March 1993.

17. See Phil Davison, "Golf War," *Independent* (London), 16 April 1996; and "The 'Wars' Against Golf Are Spreading," *Nation* (Bangkok), 4 June 1996.

18. Sonni Efron, "Critics Take Swing At Japan's Golf Courses," *Los Angeles Times,* 7 July 1992; and Fahn, "Going Against the Green."

19. Much of the opposition to Japanese-backed real estate development in Hawaii has focused on the involvement of the Japanese mafia. A Japanese businessman testified before a Senate subcommittee in 1992 that as many as fifty "major" Hawaiian properties were actually bought with *yakuza* money.

20. These and other statistics have been provided by the Global Anti-Golf Movement (GAGM).

21. James Fahn, "Golf: The Money Game," *Nation* (Bangkok), 29 April 1993.

22. Being partly of Thai descent, Woods has naturally become a hero in the homeland of his mother, but he is not a Thai citizen. After he became famous, some of Thailand's fawning politicians suggested that special allowances be made to grant him citizenship. It was hypocrisy of the highest order, considering the country has enacted discriminatory laws against Thai women (but not Thai men) who marry foreigners.

23. John Hail, "Thai Golf Industry Out of the Rough," *United Press International,* 12 October 1994.

24. See the Web site at http://www.american.edu/TED/ASIAGOLF.HTM; and Fahn, "Golf: The Money Game."

25. Fahn, "Going Against the Green"; Howard French, "The Little White Ball That Put Japan in the Red," *New York Times,* 2 June 2000.

26. Fahn, "Going Against the Green."

27. In recent years, Thailand has slightly loosened the restrictions on ownership of land by foreign investors and corporations, which have been able to skirt the laws anyway by setting up cleverly designed joint ventures.

28. Fahn, "Golf: The Money Game"; French, "The Little White Ball."

29. Tony Walker, "China's Golf Plans Drive Into the Rough," *Financial Times* (London), 17 August 1993.

30. James Fahn, "An Uphill Victory," *Nation* (Bangkok), 26 August 1992. There have been other golf controversies in Malaysia as well, particularly over the development of a course on Pulau Redang. See Jatuphol Rakthammachat, "Troubled Waters," *Nation* (Bangkok), 18 July 1993.

31. Kultida Samabuddhi, "Grand Tourism Plan Might Be Unrealistic, Plodprasop Says," *Nation* (Bangkok), 16 February 2002.

32. James Fahn, "Nature Out of Bounds for Caravan," *Nation* (Bangkok), 26 October 1997.

33. Kamol Sukin and James Fahn, "Saving the Parks from a Flawed Forest Policy," *Nation* (Bangkok), 24 October 1993.

34. See Pennapa Hongthong and James Fahn, "Resorts Could Get Lease for 30 Years in National Parks," *Nation* (Bangkok), 23 October 1997, and "Green Lobby Attacks National Parks Proposal," *Nation* (Bangkok), 7 November 1997.

35. In the 1980s, a plan to build a cable car from Chiang Mai up to a famous temple atop Doi Suthep was stopped by local opposition. A long-running campaign by the Phuket Environmental Protection Club against the massive North Phuket resort project at Ta Chatrachai, which would have encroached on Nai Yang National Park, finally had the project scrapped in 1990. Around the same time, Ital-Thai, one of the country's biggest construction firms, sought permission to blast a tunnel through a limestone cliff at Ao Pai Plong, part of Nopparat Thara Beach in Phi Phi Islands National Park, so that it could build a hotel in the secluded cove. The company was eventually denied permission. But like many such development projects, it was never actually defeated. It just faded away, to be resuscitated another day.

36. James Fahn, "On the Edge," *Nation* (Bangkok), 28 March 1993.

37. Choosak Jirasakunthai, "Love Affair With Tourists Continues," *Nation* (Bangkok), 22 November 2000. Tourist arrivals in Phuket have been growing at an average yearly rate of 12.6 percent over the last decade, and international visitors have grown annually by 17.3 percent. See "Land Use and Tourism Seminar," *Nation* (Bangkok), 12 December 2001.

38. James Fahn, Klomjit Chandrapanya, and Jadet Na Bangthao, "Landmine: A Special Report," *Nation* (Bangkok), 25 May 1993.

39. Kamol Sukin, "Land Rights Issue Stirs Up a Storm," *Nation* (Bangkok), 15 October 1993.

40. Nantiya Tangwisutijit, "New Resort 'Poses Threat' to Wildlife," *Nation* (Bangkok), 9 December 1996; and Oy Kanjanavanit, "Grass Balls, Malayan Plover and California Palm," *Nation* (Bangkok), 13 December 1996. Khao Sam Roi Yot National Park, a world-class bird sanctuary, has been particularly beleaguered by shrimp farming, road building, and encroachment; see, for instance, "National Park for Sale," Letters to the Editor, *Nation* (Bangkok), 8 April 1994.

41. Pennapa Hongthong and Kamol Sukin, "Spectre of Insider Land Trading Looms," *Nation* (Bangkok), 5 May 1996.

42. See James Fahn and Klomjit Chandrapanya's "Si Chang Deeds Throw Doubt on Port Plans," *Nation* (Bangkok), 2 December 1993, and "On Shaky Ground," *Nation* (Bangkok), 6 December 1993.

43. Kamol Sukin, James Fahn, and Klomjit Chandrapanya, "Treasured Land," *Nation* (Bangkok), 22 November1993.

44. The RFD was able to make slightly more progress at Koh Samet in Rayong province, but there, too, it has become bogged down in judicial proceedings. In 1990, a crusading antilogging task force headed by future police chief Pratin Santiprapop actually arrested forty-six resort operators on charges of encroaching on Samet Island–Laem Ya Mountain National Park. Most of the cases are still pending because it has proved so difficult to determine whether the resorts were built before the park was declared. Even the three resort operators found guilty in 1991 of encroachment are still operating; meanwhile, cases against the owners wind on. For reference, see Pennapa Hongthong, "Encroachers Make Hay on Koh Samet," *Nation* (Bangkok), 2 November 1997.

45. "Crucial Proof in Kraby Property Dispute Missing," *Nation* (Bangkok), 5 November 1993; and Kamol Sukin, Klomjit Chandrapanya, and James Fahn, "The Latest Resort," *Nation* (Bangkok), 6 November 1993.

46. Environment Desk, "The Law of the Land vs. Encroachers," *Nation* (Bangkok), 23 November 1993.

47. See James Fahn, "Threat Emerges to Unspoilt Beaches of PM's Province," *Nation* (Bangkok), 10 April 1995; "The Road to Ruin," *Nation* (Bangkok), 25 April 1995; "Pichet, Others 'Wrongfully' Hold Haad Chaomai Land Documents," *Nation* (Bangkok), 20 June 1995; and "Shady Land Deals Threaten Coastal Wilderness," *Nation* (Bangkok), 12 February 1999.

48. See the Web site at http://www.amari.com/trangbeach/index.asp.

49. James Fahn, "Last Chance to Save Krabi," *Nation* (Bangkok), 1 December 1998.

50. David Gritten, "The Water's Hot This Time," *Los Angeles Times,* 13 February 1999.

51. Alan Cowell, "Novelist's Muse Lures Him to the Soul's Darkness," *New York Times,* 11 March 1999.

52. Jeffrey Ressner, "In the Swim Again," *Time,* 8 March 1999.

53. Ing K, "Phi Phi Islands Can Do Without Hollywood Promo," Letters to the Editor, *Nation* (Bangkok), 9 October 1998.

54. "Thailand okays filming of 'Survivor' on Tarutao Island," *Deutsche Presse Agentur,* 29 April 2002. Ironically, National Film Board secretary M. L. Sidhichai Jayanit promised that "the board will keep a close look on the filming process because we have learned a lesson from [Twentieth Century] Fox's filming of *The Beach,* whose production caused serious damage to the environment of Maya Bay."

55. Gritten, "The Water's Hot."

CHAPTER 3

1. Geoffrey Lean, "Explorers Hail 1990s as 'Golden Age of Discovery,'" *Independent* (London), 30 April 1995.

2. For an account of a trip across the Tonle Sap in a passenger ferry, see James Fahn, "A Floating World," *Nation* (Bangkok), 22 November 1992.

3. Milton Osborne, *River Road to China: The Search for the Source of the Mekong, 1866–73* (St. Leonards, New South Wales: Allen & Unwin, 1997).

4. James Fahn et al., "Changing Course," *Nation* (Bangkok), 29 October 1991, contains a series of articles about efforts to revive the cascade plan and its potential impact.

5. Walter Rainboth, "1,200 Different Fish Species in the Mekong Basin," *Catch and Culture: Mekong Fisheries Network Newsletter*, vol. 2, no. 1 (August 1996).

6. Jorgen Jensen, "One Million Tonnes of Fish in the Mekong?" *Catch and Culture: Mekong Fisheries Network Newsletter,* vol. 2, no. 1 (August 1996).

7. James Fahn and Ian Baird, "Death of the Dolphins," *Nation* (Bangkok), 7 July 1992.

8. "The Big Ones Get Away in Northern Thailand," *Deutsche Presse-Agentur,* 24 May 2001; and "Conservationists: Dams Threaten Giant Catfish of Mekong River," *Associated Press,* 12 May 1996. Another Mekong fish species reportedly under threat is the trey riel, which gave its name to the Cambodian currency, the riel. See Ron Moreau and Richard Ernsberger Jr., "Strangling the Mekong," *Newsweek,* 19 March 2001.

9. Malee Traisawasdichai, "A Dangerous Numbers Game," *Nation* (Bangkok), 22 May 1995.

10. Martin Fackler, "China's Plans to Develop Mekong Worry Its Southern Neighbors Downstream," *Associated Press,* 28 September 2001; and "Mekong Plan Approved," *Nation* (Bangkok), 31 January 2002.

11. "Chinese Dam May Threaten Food Source of Neighbors," *Associated Press,* 29 September 2001.

12. "China to Build Huge Power Station on Lancang-Mekong," *Xinhua News Agency,* 21 January 2002; "China Building Another World-Class Hydropower Station," *China Online,* 12 April 2001; "Mekong-Lancang Power Project Begins Amid Environmental Misgivings," *Interfax News Agency,* 25 January 2002.

13. "Dammed If You Don't," *Economist,* 18 November 1995.

14. Defined by the World Commission on Dams as those over fifty feet high or having a reservoir volume greater than 4 million cubic yards.

15. Malcolm Browne, "Dams for Water Supply Are Altering Earth's Orbit, Expert Says," *New York Times,* 3 March 1996.

16. The phenomenon is known as reservoir-induced seismicity. See, for example, James Fahn, "Reservoirs to Blame for Earthquakes, Egat Admits," *Nation* (Bangkok), 12 December 1995.

17. The World Bank and the World Conservation Union set up the WCD as a response to all the controversy surrounding big dams. Its report is on the Web site at www.damsreport.org.

18. "Unlike India or China, with a semblance of unity conditioned by a single landmass and the open plains of river systems, the intermediate region of Southeast Asia with confined river valleys, delta-plains and islands, is geographically segmented," explains Jeya Kathirithamby-Wells, a historian at the University of Malaya. "Due to the peculiar features of the environment, riverine configurations with dendritic patterns of human settlement and urbanization were the dominant characteristics of development." See Jeya Kathirithamby-Wells, "Socio-Political Structures and the Southeast Asian Ecosystem: An Historical Perspective Up to the Mid-Nineteenth Century," in Ole Bruun and Arne Kalland, eds., *Asian Perceptions of Nature: A Critical Approach* (Richmond, Surrey, U.K.: Curzon, 1995), 25–47.

19. Ibid.

20. James Fahn and Klomjit Chandrapanya, "Husbanding Nature by Royal Decree," *Nation* (Bangkok), 25 July 1993.

21. Environment Desk, "Are Big Dams the Right Solution?" *Nation* (Bangkok), 14 December 1993.

22. Sanitsuda Ekachai, "Residents Carry on Ancestors' Spirit: Struggle for Compensation Not Easy," *Bangkok Post,* 13 August 2000.

23. Editorial, "Shadow of Violence Looms Over Phrae Province," *Nation* (Bangkok), 29 November 1996.

24. James East, "More Activists Getting Killed in Thailand's 'Green Wars,'" *Straits Times,* 4 July 2001; and "Protect Those Who Protect a Better Life," *Bangkok Post,* 4 July 2001.

25. James Fahn and Pravit Rojanaphruk, "Pak Moon Puts Egat Back in the Firing Line," *Nation* (Bangkok), 14 March 1993.

26. James Fahn and Pravit Rojanaphruk, "Will Pak Mool Dam Be Protestors' Last Stand?" *Nation* (Bangkok), 6 March 1993.

27. Ibid.

28. Ibid.

29. "40 Hurt As Rivals Clash At Dam Site," *Nation* (Bangkok), 8 March 1993; and James Fahn and Pravit Rojanaphruk, "A Recipe for Disaster," *Nation* (Bangkok), 14 March 1993.

30. When it was first built, the World Trade Center was slated to use 50 MW of power during peak hours. At the time, that was more electricity than was consumed by all of Laos. See James Fahn et al., "Saving Energy, Saving Money, Saving Nature: A Special Report," *Nation* (Bangkok), 21 May 1991.

31. In addition to the report at http://www.damsreport.org, see "Pak Moon Dam a Failure in Every Respect, Says Report," *Bangkok Post,* 20 September 2000; and Moreau and Ernsberger, "Strangling the Mekong."

32. A Fisheries Department survey concluded that only 26 percent of fish species in the Mekong, and only those no longer than 30 centimeters, can climb the fish ladder. According to Dr. Tyson Roberts, an American expert on Mekong fishes and a long-standing critic of dams, fish ladders in the West were built to accommodate a limited number of species with very definite migratory patterns. "Can you imagine a female fish with half-a-billion eggs swimming up the ladder? As far as I know, no *plaa beuk,* the most important migratory species, has ever used it. Worse, the ladder does not allow fish to move downstream, and thus its life cycle cannot be completed." See Vasana Chinvarakorn, "Dam Decommissioning is the Answer," *Bangkok Post,* 25 July 2000.

33. David Nicholson-Lord, "Questioning Some of the Basic Assumptions of Development," *Bangkok Post,* 27 September 1998.

34. The women of the Chipko movement were essentially the first "tree-huggers." The name *Chipko* actually comes from a word meaning *embrace.*

35. Vasana Chinvarakorn, "Villagers Hail Return of Fish As Gates Re-Opened," *Bangkok Post,* 17 June 2001.

36. James Fahn, "Proposed Dam May Threaten Future of Local Teak Industry," *Nation* (Bangkok), 30 October 1995.

37. Chang Noi, "Dam's Benefits Less Than the Costs," *Nation* (Bangkok), 21 January 1998; and James Fahn and Pennapa Hongthong, "Doubts Cast Over Benefits of Kaeng Sua Ten Project," *Nation* (Bangkok), 6 February 1997.

38. Kultida Samabuddhi, "New Chief Irks Green Activists," *Bangkok Post,* 22 March 2002.

39. Moreau and Ernsberger, "Strangling the Mekong."

40. Ibid.

41. Statistics gathered from World Bank and CIA Web sites; for instance, see the Web site at http://www.cia.gov/cia/publications/factbook/geos/la.html.

42. In China, the valley of the Nu Jiang, as it is called there, is home to several unique indigenous groups, along with tigers, leopards, bears, deer, giant hawks, and rare pheasants. The Nu Jiang River Project, an environmental group, claims that 314 different medicinal plants have been discovered there, along with hundreds of different types of orchids. The group has a Web site at http://www.river.com/nujiang/.

43. James Fahn, "What Lurks in the Salween?" *Nation* (Bangkok), 17 July 1995.

44. James Fahn, "The Salween Under Attack Again," *Nation* (Bangkok), 26 January 1999.

45. James Fahn, "Dam Plan Calls for Major Relocation," *Nation* (Bangkok), 14 September 1996.

46. James Fahn, "The Negawatt Revolution Is Coming," *Nation* (Bangkok), 16 December 1992.

47. According to a 1993 World Bank study of Thailand's fuel options, reducing demand by 1 kilowatt-hour would cost only about 0.7 baht (roughly three U.S. cents) compared to spending from 1.0 to 1.4 baht (from four to six U.S. cents) for burning gas or coal, the next cheapest option.

48. EGAT's solution to the energy crunch was a proposal in 1992 to build half a dozen nuclear reactors. That didn't sit too well with either the budget crunchers at NEPO or the fallout-fearing Thai public. The proposal was indefinitely shelved, although EGAT still hopes to go nuclear eventually, and has an ongoing promotional campaign for it.

49. James Fahn, "Energy: Issue of Demand vs. Supply," *Nation* (Bangkok), 2 October 1992.

50. Nareerat Wiriyapong, "Lack of Funds Drains Energy-Saving Effort," *Nation* (Bangkok), 10 January 2001.

51. James Fahn, "Tower of Less Power," *Nation* (Bangkok), 12 November 1993.

52. In 1994, for instance, I had a conversation with Qu Geping, chairman of China's National Environmental Planning Committee in the National People's Congress, who expressed great interest in energy efficiency for his country, but noted it would be impossible to adopt until electricity prices there were reformed to reflect market levels.

53. Wiriyapong, "Lack of Funds."

54. UNESCAP, *State of the Environment in Asia and the Pacific* (New York: United Nations, 1995), 88; and "Raising Price of Water Will Curb Its Profligate Wastage," *Nation* (Bangkok), 13 August 1995; and Thana Poopat, "Agencies Blamed for Water Mismanagement," *Nation* (Bangkok), 19 February 1999.

55. James Fahn, "New Water Law Just a Drop in the Bucket," *Nation* (Bangkok), 12 December 1993.

56. Attempts to charge market prices for water supply are often tied up with privatization schemes, which in places such as Bolivia have drawn violent opposition. See, for instance, William Finnegan, "Leasing the Rain," *New Yorker,* 8 April 2002. Such schemes can undoubtedly be abused, especially if public monopolies are turned into private monopolies without proper oversight.

57. James Fahn, "Seeking a Truce in the Water Wars," *Nation* (Bangkok), 12 April 1999.

CHAPTER 4

1. Marwaan Macan-Markar, "Logging Bans Prove Ineffective—Report," *InterPress Service,* 13 November 2001.

2. Malaysian and Korean firms have both acquired logging concessions in the Amazon, and Thai firms are allegedly logging in South America and Africa. See "Asian Logging Firms Active in

the Amazon," *Kyodo, Nation* (Bangkok), 5 June 1996.These firms are responsible for only a minor portion of the annual 2.3 million hectares of deforestation in Brazil, but have drawn the ire of Brazilian environmentalists for their rapacious activities.

3. A copy of the 2001 report is available on the FAO Web site at http://www.fao.org/forestry/fo/sofo/sofo_e.asp. "As a consequence of trade liberalization and globalization, illegal logging and trade appear to be growing [in the Asia-Pacific region]," the report states. From 1990 to 1995, the region lost 16.3 million hectares of forest; Indonesia lost 5.4 million hectares; Malaysia, 2 million hectares; Burma, 1.9 million hectares; and Thailand, 1.6 million hectares. "[Southeast Asia] is losing forests five times faster than the global net annual forest loss of 0.2 percent experienced between 1990 and 2000," said the FAO's R. B. Singh.

4. A summary of Thailand's forest situation is available on the FAO Web site at http://www.fao.org/forestry/fo/country/index.jsp?geo_id=41&lang_id=1.

5. James Fahn, "The Future of Thai Forests," *Nation* (Bangkok), 13 January 1999.

6. James Fahn, "Creating a Buffer for Forests," *Nation* (Bangkok), 10 April 1999.

7. There is some scientific controversy among Western hydrologists over to what extent forests help regulate the flow of water. Certainly, increased water use for industry and intensive agriculture has also contributed to greater stress on water supplies. But wherever you go in Thailand, people confirm that once their forests were cut down, particularly in headwater regions, their rivers and streams tended to dry up during the hot season. It is virtually an article of faith among all parties that forests act as a "sponge," soaking up rainfall during the wet season, and releasing it in the dry season. See James Fahn, "Water Works," *Nation* (Bangkok), 22 June 1995.

8. Thai foresters were traditionally trained at the Phrae School of Forestry, in the northern part of the country, where they are schooled in British and Finnish methods of forestry management. See Pennapa Hongthong and James Fahn, "Education of Foresters Needs to Be Improved," *Nation* (Bangkok), 22 September 1996.

9. In Indonesia, for instance, following the end of Suharto's dictatorial rule, as power has been decentralized and a rough democracy has taken hold, the rate of deforestation has increased to 2 million hectares per year, up from an annual 1 million hectares in the 1980s and early 1990s. See "Corruption and Lawlessness Prime Culprits in Rape of Indonesian Forests," *Deutsche Presse-Agentur,* 22 February 2002. According to the FAO report, Indonesia officially has 105 million hectares of forest coverage.

10. Neil Englehart, *The Political Economy of Deforestation in Thailand and Indonesia,* prepared at Midwest Political Science Association Annual Meeting, 23 to 25 April, 1998. Englehart notes that modern liberal democracies are often criticized by greens as being unable to deal effectively with environmental problems, and he initially set out to show that "democracies are probably better on this score than the real alternative, [which is] authoritarian regimes." But he found the opposite to be true. For Indonesia and Thailand, the data is poor, but between 1981 and 1993 Thailand had lost an average of 1.93 percent of its forests every year, compared to 1.34 percent for Indonesia. Even after Thailand imposed a ban on the logging of state-owned forests in 1989, its deforestation rate still outstripped that of Indonesia's over the next five years, largely because of illegal logging activities.

11. For a good treatment of market-based environmental measures, see, for instance, Geoffrey Heal, *Nature and the Marketplace* (Washington, D.C.: Island Press, 2000); and Gretchen Daily and Katherine Ellison, *New Economy of Nature: The Quest to Make Conservation Profitable* (Washington D.C.: Island Press, 2002).

12. Malee Traisawasdichai, "Lao Dam Plan Wins Support of Environmentalist," *Nation* (Bangkok), 8 March 1996. The Foundation for Ecological Recovery responded to Rabinowitz's claims in an article by Veerawat Dheeraprasart, "Why the Nam Theun 2 Dam Won't Save Wildlife," *Watershed,* vol. 1, no. 3 (March–June, 1996).

13. Jake Brunner, Kirk Talbott, Chantal Elkin, *Logging Burma's Frontier Forests: Resources and the Regime* (Washington D.C.: World Resources Institute, 1996); the report is available on the World Resources Institute Web site at http://www.wri.org/ffi/burma/.

14. Rabinowitz made these comments in a telephone interview with me in 1995. He writes about his explorations in Burma in his new book, *Beyond the Last Village* (Washington D.C.: Island Press, 2001).

15. Les Line, "Indiana Jones Meets His Match in Burma Rabinowitz," *New York Times,* 3 August 1999.
16. "Border Logging Called 'Free For All,'" *Nation* (Bangkok), 18 September 1996; and "IMF Threatens Aid Cut Over Logging Issue," *United Press International,* 31 May 1996.
17. "Premier Threatens to Expel Forestry Monitor from Cambodia," *Deutsche Presse-Agentur,* 31 January 2001.
18. James Fahn, "Karenni Rebels Say Battle with Slorc Imminent," *Nation* (Bangkok), 30 June 1995.
19. Brunner, Talbott, and Elkin, *Logging Burma's Frontier Forests.*
20. Fiona Thompson, "Environment Will Not Survive Abuse By the Wealthy and Greedy," *Nation* (Bangkok), 7 October 1997.
21. Chang Noi, "Chavalit and the Salween Saga," *Nation* (Bangkok), 13 March 1998.
22. Ibid.; and William Barnes, "Thailand Finds Financial Ally in Old Foe," *South China Morning Post* (Hong Kong), 16 May 1997.
23. Paul Wedel, "Thai, Laotian Generals Discuss Border Cease-Fire," *United Press International,* 16 February 1988.
24. Paul Wedel, "Thailand Upgrades Ties with Burma's Military Regime," *United Press International,* 20 December 1988.
25. Tan Lian Choo, "Cabinet 'No' to Chavalit's Call to Re-open Checkpoints for Thai Loggers," *Straits Times* (Singapore), 5 May 1993.
26. Kamol Sukin, "Illegal Logs Flood Into Thailand," *Nation* (Bangkok), 18 December 1996.
27. William Barnes, "Crime Puts Cambodia in the Dock," *South China Morning Post,* 20 July 1996.
28. "Thai Prime Minister Linked to Logging in Cambodia," *Deutsche Presse-Agentur,* 16 January 1997.
29. "Slorc, Karenni Battle As Ceasefire Collapses," *Nation* (Bangkok), 1 July 1995; James Fahn and Yindee Lertcharoenchok, "1,000 Sent Fleeing as Burmese Army Continues Offensive," *Nation* (Bangkok), 2 July 1995; and James Fahn, "Burmese-Karenni Fighting Dislocates 3,000," *Nation* (Bangkok), 3 July 1995.
30. James Fahn, "All's Not Fair in Trade and War in Burma," *Nation* (Bangkok), 8 July 1995. See also James Fahn, "Over the Hills and Not So Far Away," *Nation* (Bangkok), 17 July 1995.
31. "Official Transferred as Logging Scam Exposed," *Nation* (Bangkok), 14 March 1997; Pennapa Hongthong and Nittayaporn Muangmit, "Forestry Police Zero in on Logging Firm," *Nation* (Bangkok), 21 March 1997; and Kamol Sukin, "Five Firms Given Logging Permits," *Nation* (Bangkok), 24 September 1997; "'Influential Figures' the Guiding Force Behind Thai-Burmese Border Trade," *Nation* (Bangkok), 29 September 1997; "More Effort Sought in Logging Probe," *Nation* (Bangkok), 17 February 1998; and the editorial "Logging Mafia Rules in Mae Hong Son," *Nation* (Bangkok), 20 February 1998.
32. Pennapa Hongthong and Piyanart Srivalo, "Seized Logs Net Over Bt60m in Auctions By FIO," *Nation* (Bangkok), 24 February 1998; and James Fahn and Pennapa Hongthong, "Forestry Agencies Still Take Big Profits from Woodland Destruction," *Nation* (Bangkok), 4 May 1996.
33. Chang Noi, "Chavalit and the Salween Saga," *Nation* (Bangkok), 13 March 1998.
34. "Bt5m Bribe Linked to 13,000 Logs," *Nation* (Bangkok), 20 February 1998; and Chaiyakorn Bai-Ngern, "Prawat Charged with Demanding Kickbacks," *Nation* (Bangkok), 26 February 1998.
35. "Chavalit Accused in Logging Case," *Nation* (Bangkok), 6 March 1998; and "Triphol Digs Up Dirt on Chavalit," *Nation* (Bangkok), 4 March 1998.
36. "Chavalit Dares Govt to Arrest Logging Culprits," *Nation* (Bangkok), 19 February 1998.
37. "Thai Deputy PM Going to Burma to Patch Up Relations," *Nation* (Bangkok), 1 March 2001.

CHAPTER 5

1. James Fahn, "The Spirit of Disobedience," *Nation* (Bangkok), 11 April 1996.
2. An account of the Karens's traditions and predicaments, and our trip into Thung Yai, can be found in James Fahn, "The End of Innocence," *Nation* (Bangkok), 6 February 1994, "The Law of

the Jungle," *Nation* (Bangkok), 8 February 1994, and "A Walk in the Woods with the Karen of Thung Yai," *Nation* (Bangkok), 27 February 1994.

3. This simmering conflict exploded into violence in late 1992 when five Border Police troops and seven villagers were killed in one bloody evening. Perhaps because of the Burmese Karen's notoriety in fighting a civil war, initial police reports of the event tended to portray the Karen as separatists or savage cultists or even Communists. To many of the Mae Chanta villagers, the violence was the result of a clumsy attempt to force modern Thai culture upon them.This bizarre and terrible incident was a major reason for our trip into Thung Yai to meet with the Karen. The lawyers and human rights workers sought to work with the villagers on how to gather information and use the law to defend themselves; the journalists, including myself, sought to ferret out what really lay behind the violence. It made for a fascinating story, but seems too tangential an issue to include here. Those interested in learning more about the episode can read an in-depth article I wrote about it, "The Law of the Jungle," *Nation* (Bangkok), 8 February 1994.

4. Making matters even more complicated, the World Bank put forward a project to research the tremendous biodiversity of Thung Yai Naresuan, among other places, with funding also coming from the Global Environmental Facility (GEF). The preinvestment study stated that although the project "does not foresee the use of donor funds" to resettle villagers, "the removal and resettlement of occupants of protected areas may be necessary at protected sites and should remain an option for the Royal Thai Government." Some of the more vocal Thai NGOs argued strongly against the 2-billion-baht project, contending it would benefit agroindustrial, pharmaceutical, and cosmetics companies at the expense of the Karen. The ensuing uproar caused so much controversy that the project was never carried out. Ironically, it is the Karen who could supply the knowledge to unlock the secrets of the forest, but some of this local wisdom may already be fading away. "There's a lot of medicine in the woods, but it's easier to ask the Border Police for drugs," N'der admits.

5. Among the eminent historical figures who espoused contemplation in nature were Shantideva in eighth-century India, Tibet's great yogi-saint Milarepa, and the Zen master Dogen.

6. Jeya Kathirithamby-Wells, "Socio-Political Structures and the Southeast Asian Ecosystem: An Historical Perspective up to the Mid-Nineteenth Century," in Ole Bruun and Arne Kalland, eds., *Asian Perceptions of Nature: A Critical Approach* (Richmond, Surrey, U.K.: Curzon, 1995).

7. Ibid.

8. Montri Umavijani, "Man and Nature As Reflected in Thai Literary Works," in *Man and Nature: A Cross-Cultural Perspective* (Bangkok: Chulalongkorn University Press: 1993), 74.

9. Ibid.

10. This anecdote was related to me by fellow *Nation* correspondent Ann Danaiya Usher.

11. Mont Redmond, "Nature: Human Outcry and Natural Outcome," *Nation* (Bangkok), 29 August 1993.

12. Roderick Nash, *Wilderness and the American Mind* (New Haven: Yale University Press, 1982).

13. James Fahn, "The Future of Thai Forests," *Nation* (Bangkok), 13 January 1999.

14. See Kritsana Kaewplang, "Stunted Growth," *Nation* (Bangkok), 15 February 1994; see also James Fahn, "Forest or Farm? The Great Eucalyptus Debate Rages On," *Nation* (Bangkok), 10 October 1993; and "'Cutting Eucalyptus Better Than Cutting Forest,'" *Nation* (Bangkok), 15 February 1994.

15. For an eloquent account of the RFD's shortcomings, see Belinda Stewart-Cox, "Dancing Around Dinosaurs," *Nation* (Bangkok), 23 September 1996. This article was part of a series marking the agency's one-hundredth anniversary. See also Kamol Sukin, "Politics Cited as Reason for Forestry Woes," *Nation* (Bangkok), 20 September 1996, and "Experts Decry Agency Efforts Over 100 Years," *Nation* (Bangkok), 21 September 1996.

16. Walakkamon Eamwiwatkit, "Park Rangers Left Out on a Limb," *Nation* (Bangkok), 19 September 1996.

17. A description of the importance of Huay Kha Khaeng, part of a series of conservation areas near the Thai-Burma border known as the Western Forest Complex, can be seen on the Web site at http://www.worldwildlife.org/wildworld/profiles/terrestrial/im/im0119_full.html.

18. A Web site recounting Seub's life and work can be seen at http://www.seub.or.th/SeubLife/seublife-eng.html.

19. Mark David Spence, *Dispossessing the Wilderness: Indian Removal and the Making of the National Parks* (New York: Oxford University Press, 1999).

20. Dr. Crespo made these comments during a personal interview at Columbia University in May 2001.

21. Chang Doi, "Solution to Land Mess At Hand," *Nation* (Bangkok), 20 April 1998.

22. Examples of the differing estimates can be found in Chang Noi, "Confront the Realities of Land Mess," *Nation* (Bangkok), 10 April 1998, which uses the estimate of 10 million forest dwellers in Thailand, and a response by Chang Doi, "Solution to Land Mess At Hand," *Nation* (Bangkok), 20 April 1998, which, by narrowing consideration to only the most vital forests, uses the 1995 RFD data to estimate the presence of 591,893 inhabitants: 102,230 in national parks, 53,695 in wildlife sanctuaries, and 435,968 in other crucial watershed areas.

23. The conflicts seem to be greatest in Asia, perhaps because of its greater population density, but encroachment by poor farmers is also a problem in Latin America. See for instance, Frank Clifford, "Which One: Food or the Forest?" *Los Angeles Times,* reprinted in the *Nation* (Bangkok), 10 August 1997.

24. Natedao Phatkul, "Broken Promises, Broken Lives," *Nation* (Bangkok), 26 September 1994. See also Walakkamon Eamwiwatkit, "Bt 30m Plan to Shift Villagers from Sanctuary Buffer Zone," *Nation* (Bangkok), 27 May 1994; and Natedao Phatkul, "The Soul Guardians," *Nation* (Bangkok), 27 July 1994, and "Resettlement Planned for Wildlife Sanctuary Dwellers," *Nation* (Bangkok), 23 December 1994.

25. Philip Hirsch, *The Politics of Environment in Southeast Asia: Resources and Resistance* (New York: Routledge, 1998).

26. Apichai made these comments at a talk given at the Foreign Correspondents Club of Thailand. A similar tolerance for criticism from environmental movements was evident in Eastern Europe during the final years of communism.

27. For a look at the debate over community forestry, forest dwellers, and protecting watersheds, see Oy Kanjanavanit, "The Integrity of Watershed Forest," *Nation* (Bangkok), 22 May 1995; James Fahn, "Nature's Rights and Human Rights," *Nation* (Bangkok), 24 May 1995, and "Changes Need on Draft Forest Bill," *Nation* (Bangkok), 12 July 1996; Pennapa Hongthong, "The Battle for Our Forest Resources," *Nation* (Bangkok), 7 October 1997; and Nantiya Tangwisutijit, "Finding a Solution to Our Watershed Problem," *Nation* (Bangkok), 13 October 1998.

28. Roy Ellen, "What Black Elk Left Unsaid," *Anthropology Today* 2, no. 6 (December 1986): 8–12.

29. Natedao Phatkul, "A Lasting Solution," *Nation* (Bangkok), 28 April 1995.

30. James Fahn, "Buffer Zones Strategy Mooted," *Nation* (Bangkok), 30 January 1997, and "Buffer Zones Hold Key to Conservation," *Nation* (Bangkok), 2 February 1997. Some buffer zone projects, such as that endorsed by the RFD for the area around Huay Kha Khaeng, are more controversial. See James Fahn, "In the Line of Fire," *Nation* (Bangkok), 18 February 1994, and "The Rarest of Creatures," *Nation* (Bangkok), 11 January 1996.

31. Fahn, "The Rarest of Creatures."

32. James Fahn, "River Nan: When a Holy Alliance Works," *Nation* (Bangkok), 16 June 1993.

CHAPTER 6

1. Comments made in the Australian television documentary *Thai Dugong.*

2. The keys to success in Chao Mai were related in an e-mail discussion with Jim Enright, who worked at Yadfon in the late 1990s, and currently works for the Mangrove Action Project (MAP).

3. For a better understanding of adaptive management, see Kai Lee, *Compass and Gyroscope* (Washington D.C.: Island Press, 1993), and C. S. Holling, *Adaptive Environmental Assessment and Management* (New York: John Wiley & Sons, 1978).

4. J. E. Cohen et al., "Estimates of Coastal Populations," *Science* 278, no. 5341 (14 November 1997): 1211–1212. This points out that the commonly quoted statistic that two-thirds of the

world's population lives near the coastline is incorrect. See also the Web site at http://www.sida.se/Sida/jsp/Crosslink.jsp?d=168&a=606.

5. Paul Pinet, *Invitation to Oceanography* (Sudbury, Mass.: Jones & Bartlett, 1998).

6. From a talk given to an IMMF workshop in May 1999.

7. "Net Benefits," *Economist,* 24 February 2001.

8. United Nations Food and Agriculture Organization, *State of the World's Fisheries and Aquaculture* (United Nations Food and Agriculture Organization, 2000), viewed on the Web site at http://www.fao.org/DOCREP/003/X8002E/x8002e04.htm#P1_6.

9. FAOSTAT, viewed on the Web site at http://apps.fao.org/page/collections?subset=fisheries.

10. John Virdin and David Schorr, *Hard Facts, Hidden Problems: A Review of Current Data on Fishing Subsidies,* World Wide Fund for Nature technical paper, October 2001.

11. This estimate was provided by MAP's Jim Enright.

12. See, for example, Mark Kurlansky, *Cod: A Biography of the Fish That Changed the World* (New York: Walker and Co., 1997), for an excellent discussion of how one fishery was decimated.

13. United Nations Food and Agriculture Organization Fisheries Department, *State of World Fisheries and Aquaculture* (United Nations Food and Agriculture Organization Fisheries Department, 1995), 29.

14. The use of modern pesticides and fertilizers on farms has made it increasingly difficult for farmers to cultivate fish. These chemical inputs often leach into the soil as persistent organic pollutants (POPs), poisoning the fish or contaminating them to the point where they become health risks for consumers.

15. Alfredo Quarto et al., "Choosing the Road to Sustainability," published in 1996 by MAP, available on the Web site at http://www.earthisland.org/map/rdstb.htm. See also M. Skaladany and C. Harriss, "On Global Pond: International Development and Commodity Chains in the Shrimp Industry," in P. McMichael, ed., *Food and Agrarian Orders in the World Economy* (Westport, Conn.: Greenwood Press, 1995).

16. Solon Barraclough and Andrea Finger-Stich, *Some Ecological and Social Implications of Commercial Shrimp Farming in Asia* (Geneva: United Nations Research Institute for Social Development, 1995), 6. The authors note that "a substantial part of this sum seems to be destined for intensification and expansion of shrimp ponds."

17. Mark Flaherty, Peter Vandergeest, and Paul Miller, "Rice Paddy or Shrimp Pond: Tough Decisions in Rural Thailand," *World Development* 27, no. 12 (1999): 2046.

18. Michael Phillips, "Management Strategies in Development of Sustainable Shrimp Culture" (Network of Aquaculture Centres in Asia-Pacific, 1999), 3. See the NACA Web site at http://www.agri-aqua.ait.ac.th/naca/aquaasia/aquaasia.htm.

19. Quarto et al., "Choosing the Road," 2.

20. John Forster, "Shrimp and Salmon Farming," *Science* 283 (29 January 1999): 639.

21. Rosamond Naylor et al., "Nature's Subsidies to Shrimp and Salmon Farming," *Science* 282 (30 October 1998): 883.

22. Flaherty et. al. "Rice Paddy or Shrimp Pond," p. 2047

23. Mark Flaherty and Peter Vandergeest, "Low-Salt Shrimp Aquaculture in Thailand," *Environmental Management* 22, no. 6 (1998): 818.

24. World Bank project brief, Thailand coastal resources management project, available on the Web site at http://www.worldbank.org/pics/pid/th34640.txt.

25. Anthony Charles et al., "Aquaculture Economics in Developing Countries" (United Nations Food and Agriculture Organization, November 1997), 12–13. Available on the Web site at http://www.fao.org/docrep/W7387E/W7387E00.htm.

26. The figures for the proportion of exported shrimp raised on farms seem to vary with different reports. See, for example, James Fahn, "Shrimp Export Dispute," *Nation* (Bangkok), 8 March 1996.

27. See Flaherty et al., "Rice Paddy or Shrimp Pond," 2048, and Michael Phillips, "Management Strategies in Development of Sustainable Shrimp Culture" (Network of Aquaculture Centres in Asia-Pacific, 1999), 3. See the NACA Web site at http://www.agri-aqua.ait.ac.th/naca/aquaasia/aquaasia.htm.

28. J. H. Primavera, "Tropical Shrimp Farming and Its Sustainability," in S. De Silva, ed., *Tropical Mariculture* (New York: Academic Press, 1998), 257–289.

29. Aphaluck Bhatiasevi, "Calls for Less Farm Use of Chemicals," *Bangkok Post,* 7 April 2000.

30. Quarto et al., "Choosing the Road," 6.

31. Brian Szuster et al., "Socio-Economic and Environmental Implications of Inland Shrimp Farming in the Chao Phraya Delta," in F. Molle and T. Srijantr, eds., *Perspectives on Social and Agricultural Change in the Chao Phraya Delta* (Bangkok: White Lotus, in press).

32. Ibid., 5. On the other hand, Naylor et al. ("Nature's Subsidies") say that "the life-span of intensive shrimp ponds in Asia rarely exceeds five to ten years."

33. Szuster et al., "Socio-Economic and Environmental Implications."

34. Stories about villagers who sell their assets and then use the proceeds in ill-advised ways are all too common in the developing world. NGO activists see this tendency as a major argument against the introduction of capitalist development and individualized property rights. Free-market economists view such spending as the result of individual choices; they argue that people should be allowed to use their money as they see fit. Where you come down on this debate essentially depends on your faith in liberalism.

35. Tunya Sukpanich, "The Unkindest Cut?" *Bangkok Post,* 24 January 1999.

36. Quarto et al., "Choosing the Road," 6.

37. In 1996, the World Bank proposed the establishment of a $50 million coastal resource management plan that courted controversy in its attempts to improve shrimp culture practice. According to Anthony Zola, a consultant who helped draft it, the original proposal "reflected community needs and was participatory in nature." It received initial approval from Thai authorities, but then was rewritten to include "more hardware" and has since languished in limbo.

CHAPTER 7

1. James Fahn, "Languishing in Limbo," *Nation* (Bangkok), 31 July 1994, and "Mon Hold to Safety As Deadline Passes for Return to Burma," *Nation* (Bangkok), 13 August 1994.

2. James Fahn, "Surprise Attack Seen As Slorc Attempt to Pressure Mon to Sign Ceasefire," *Nation* (Bangkok), 25 July 1994.

3. "PTT Pictures First Trans-National Regional Gas Pipeline," *Nation* (Bangkok), 22 September 1990.

4. For a detailed chronology of the pipeline dispute, see *Total Denial,* a report published by EarthRights International and the Southeast Asian Information Network in July, 1996. See also "Thailand Inks Historic Deal to Buy Gas from Rangoon," *Nation* (Bangkok), 3 February 1995; and Pichaya Changsorn, "Offshore Gas to Flow from PTT Field in Burma," *Nation* (Bangkok), 31 August 1994.

5. See, for instance, Maria Isabel Garcia, "Columbia-Ecuador: European Activists Protest Amazon Pipeline," *Inter Press Service,* 12 June 2002.

6. Danielle Knight, "Activists Urge Banks to Shun Peru Gas Project," *InterPress Service,* 13 May 2002.

7. James Fahn, "Thai-Burma Pipeline Raises Concern," *Nation* (Bangkok), 11 March 1994.

8. Total claimed that only a few kilometers of forest would require cutting and that, "since the [signing] of the Production Sharing Contract in 1992, no population has been moved to the best of our knowledge." An ad placed by EGAT in the *Bangkok Post* later claimed that "Myanmar [the new official name for Burma] has recently cleared the way by relocating a total of 11 Karen villages that would otherwise obstruct the passage of the gas resource development project." See EarthRights International and the Southeast Asian Information Network, *Total Denial.*

9. In April 1993, the military had burned down two camps—Democracy Village and Aung Tha Pye—just south of Ban I-Tong, displacing five hundred ethnic Tavoyan refugees. On February 28, 1994, Thai soldiers tore down a camp housing three hundred Burman refugees named Democracy Village, and the next day told the residents of the nearby Tanaosri camp that they also had to go. See James Fahn, "Pushed to the Edge," *Nation* (Bangkok), 23 March 1994, and "Tanaosri Camp Refugees Given 45 Days to Go Home," *Nation* (Bangkok), 6 May 1995.

10. See James Fahn, "Making Tracks," *Nation* (Bangkok), 22 March 1994. Eventually, Total and its contractors apparently decided against using the railroad to transport equipment and personnel, preferring to use barges instead. The railroad offered them little cost savings, and using it would have created a public relations nightmare. But critics claim that the *tatmadaw* uses the railroad to move its men and materiel, some of which is then used to protect the pipeline.

11. This was part of a policy to slowly squeeze the refugees until they returned to Burma. See for instance, "Authorities Seal Road to Mon Refugee Camp," *Bangkok Post,* 11 August 1994; "Soldiers Cut Off Water Supplies to Refugees," *Bangkok Post,* 12 August 1994; "NSC Blocks Access to Mon Refugee Camp," *Bangkok Post,* 19 August 1994; and James Fahn, "Mon Refugees Denied Access to Rice Depot," *Nation* (Bangkok), 1 September 1994.

12. "Army Denies Mon Fleeing Rights Abuses," *Nation* (Bangkok), 27 August 1994; "Generals Insist Mon Refugees Can't Stay," *Nation* (Bangkok), 25 August 1994; and "Mon Refugees Must Go Home: NSC Chief," *Bangkok Post,* 13 August 1994.

13. Pichayaporn Utomporn, "Marching Orders," *Nation* (Bangkok), 12 October 1993.

14. James Fahn, "Borderline Adventure," *Nation* (Bangkok), 26 August 1994.

15. See James Fahn, "Stuck in the Mud," *Nation* (Bangkok), 17 August 1994; and "New Deadline Set for Mon to Return to Burma," *Nation* (Bangkok), 20 August 1994.

16. James Fahn, "Mon Hold to Safety as Deadline Passes for Return to Burma," *Nation* (Bangkok), 13 August 1994.

17. James Fahn, "Repatriated Illegals Add to Strain on Camp," *Nation* (Bangkok), 13 August 1994.

18. "Amoco Decides to Call It Quits in Burma," *Nation* (Bangkok), 5 March 1994; and "Oil Company Put Under Pressure to Sever Links," *Nation* (Bangkok), 7 May 1996.

19. Unocal was able to keep its stake in the Yadana operation because of the grandfather clause and, in addition to its headquarters in Los Angeles, it set up another "joint headquarters" in Kuala Lumpur.

20. James Fahn, "Lawsuit Charges, Firms, Slorc with Abuse," *Nation* (Bangkok), 4 October 1996.

21. James Fahn, "Resistance Leader Predicts Gas Pipeline Disaster," *Nation* (Bangkok), 18 February 1995.

22. The honor of being the largest conservation corridor in Southeast Asia may now belong to a series of reserves set up in Cambodia's Cardamom Mountains. See Andrew Revkin, "On Minefields of Khmer Rouge, Wilderness Is Preserved," *New York Times,* 18 June 2002.

23. "Pipeline Firms Urged to Reveal Impact Data," *Bangkok Post,* 12 March 1995.

24. James Fahn, "Fears of Environmental Damage in Natural Gas Pipeline Project," *Nation* (Bangkok), 23 February 1996.

25. Pennapa Hongthong, "Pipeline Fears Remain,"*Nation* (Bangkok), 6 December 1996.

26. Pennapa Hongthong, "Pipeline Re-Routed for Crabs' Comfort," *Nation* (Bangkok), 7 December 1996.

27. Pennapa Hongthong, "Gas Line Impact Report Rejected," *Nation* (Bangkok), 7 February 1997; Kamol Sukin, "A Bt100m Question Confronts Gas Project," *Nation* (Bangkok), 10 March 1997.

28. Piyanart Srivalo, "NEB to Go Ahead with Controversial Pipeline," *Nation* (Bangkok), 25 April 1997.

29. Nantiya Tangwisutijit and Pennapa Hongthong, "PTT Accused of Misinterpreting Gas Sales Agreement," *Nation* (Bangkok), 8 January 1998.

30. Nantiya Tangwisutijit, "Rival Groups Square off Over Gas Pipeline," *Nation* (Bangkok), 21 January 1998; and James Fahn, "Tension Rises as Protestors Block Workers," *Nation* (Bangkok), 16 January 1998.

31. Pennapa Honthong and Nantiya Tangwisutijit, "PTT Ordered to Suspend Construction of Pipeline," *Nation* (Bangkok), 9 January 1998.

32. Watcharapong Thongrung and Pennapa Hongthong, "Ratchaburi Power Plant in for Delay," *Nation* (Bangkok), 19 January 1998. It was also becoming clear that following the onset of the

regional financial crisis, Thailand's energy demand had stagnated, so the electricity from the gas wasn't going to be needed for a while yet anyway.

33. James Fahn and Pennapa Hongthong, "PTT Has Badly Misled the Public," *Nation* (Bangkok), 24 February 1998.

34. James Fahn and Pennapa Hongthong, "No Action By Yadana Committee," *Nation* (Bangkok), 26 February 1998.

35. Wayne Arnold, "A Gas Pipeline to World Outside," *New York Times,* 26 October 2001. Meanwhile, a $400 million gas pipeline from Indonesia to Singapore was completed in 2001, but other pipeline projects connecting the two countries are reportedly struggling to come up with the necessary financing.

36. Jed Greer, "From Denial to Undeniability: Unocal and Atrocities in Burma," *Nation* (Bangkok), 27 September 2000.

37. "Unocal to Face Suit on Human Rights," *Reuters,* 12 June 2002.

38. PTTEP, a listed subsidiary of the PTT that's in charge of exploration and production also has a stake in Yadana.

CHAPTER 8

1. Andrew Pollack, "Victims of Mercury-Contaminated Fish Cheer Government Edict, but Scars Remain," *New York Times,* 23 August 1997.

2. Following the passage of its new Constitution in 1997, Thailand did pass its own version of the Freedom of Information Act, another sign of the relatively open nature of its society. The ability to exercise this right to access information, however, is still somewhat piecemeal.

3. The government had previously established a small toxic waste treatment facility in Bang Khuntien as a pilot project. One businessman in the hazardous waste industry described it to me as a "fake."

4. James Fahn, "Deadly Toxins Found Off Eastern Seaboard," *Nation* (Bangkok), 12 January 1996.

5. The U.S. National Academy of Sciences has established a committee to investigate the mercury toxicity issue; see Robert Saar, "New Efforts to Uncover Dangers of Mercury," *New York Times,* 2 November 1999.

6. "FDA Warns Women Not to Eat Some Fish," *Associated Press,* 13 January 2001. There have been allegations that tuna was left off the list of potentially harmful fish as a result of pressure from the fishing industry; see Marian Burros, "Second Thoughts on Mercury in Fish," *New York Times,* 13 March 2002.

7. James Fahn, "Ingestion of Toxin Linked to Infertility," *Nation* (Bangkok), 27 June 1997.

8. James Fahn, "High Levels of Toxin Found in Gulf Fish," *Nation* (Bangkok), 13 January 1996.

9. James Fahn, "Gulf Mercury Levels on the Rise Near Rayong Industrial Estate," *Nation* (Bangkok), 25 January 1996.

10. "Rayong Trembles in Fear of Doomsday," *Bangkok Post,* 12 January 1996.

11. Fahn, "Total Adds to Mercury Problems in the Gulf," *Nation* (Bangkok), 19 June 1996.

12. James Fahn, "Unocal Admits to Gulf Discharges of Mercury," *Nation* (Bangkok), 18 June 1996.

13. A consultant working for Unocal later reported that the amount dumped by all the natural gas platforms in the Gulf was 400 kilograms; see Frank Lombard, "Gulf's Mercury Content Seen at Dangerous Level," *Nation* (Bangkok), 10 January 1997. But Unocal reported in 1998 that it had reduced its annual mercury discharge to 24 kilograms; see Nantiya Tangwisutijit, "To Eat or Not to Eat, Fish Eaters Wonder," *Nation* (Bangkok), 1 August 1998.

14. James Fahn, "Unocal to Dump Waste Via Deep-Well Injection," *Nation* (Bangkok), 9 July 1996.

15. James Fahn, "High Levels of Cadmium Discovered," *Nation* (Bangkok), 28 October 1996.

16. James Fahn, "Paint Chemical Induces Sex Change in Shellfish," *The Nation,* 4 April 1997.

17. James Fahn, "Total Admits Dumping Mercury," *Nation* (Bangkok), 10 September 1996.

18. James Fahn, "Mercury Levels on the Rise in Gulf," *Nation* (Bangkok), 9 October 1996.

19. In 1998, Tetra Tech reported that 6 percent of the fish it had collected at the Unocal platforms had mercury levels above the WHO standard; see Tangwisutijit, "To Eat or Not to Eat."

20. "Mercury Levels Low in Gulf of Thailand," Letters to the Editor, *Nation* (Bangkok), 10 October 1996.

21. See the May 1997 issue of the *PTIT Focus*; see also James Fahn, "Experts to Study Level of Mercury Present in Gulf," *Nation* (Bangkok), 27 June 1997.

22. James Fahn, "Experts to Study Level of Mercury Present in Gulf," *Nation* (Bangkok), 27 June 1997.

23. Herbert Windom and Gary Cranmer, "Lack of Observed Impacts of Gas Production of Bongkok Field, Thailand on Marine Biota," *Marine Pollution Bulletin* 36, no. 10 (1998): 799–807.

24. James Fahn, "Aromatics Output Grows Despite Problems with Mercury Removal Unit," *Nation* (Bangkok), 25 September 1997.

25. James Fahn, "Amid State of Decay, Crime Pays," *Nation* (Bangkok), 15 November 1997; see also Anchalee Kongrut, "Pollution Takes its Toll on Marine Life," *Bangkok Post,* 21 January 1999.

26. James Fahn, "Rayong and Chon Buri Seafood May Be Unsafe," *Nation* (Bangkok), 24 November 1997. Other chemical contaminants have also reportedly damaged marine life along the Eastern Seaboard; see Kongrut, "Pollution Takes its Toll."

27. James Fahn, "Mercury Situation in Gulf Appears Stable," *Nation* (Bangkok), 29 April 1999.

28. James Fahn, "River Mercury Findings Stump Experts," *Nation* (Bangkok), 3 May 1999, and "Govt Ignoring Mercury Peril," *Nation* (Bangkok), 9 May 1999.

29. Ard Chana, director of Thailand's Mineral Fuels Division, said he saw mercury absorption equipment on a Malaysian rig off Sarawak. Ir Hussein Rahmat, an official with the state-owned oil firm Petronas, confirmed that there is a small amount of mercury in Malaysian gas, but claimed there is none in the produced water, so mercury is not being discharged out to sea. Rahmat added that the gas also contains naturally occurring radioactive materials (known as "NORMs"). All in all, it's hard to know what to make of the situation because, despite being wealthier than Thailand, Malaysia places greater restrictions on the flow of information. See James Fahn, "Toxins Found in VN Offshore Gas Field," *Nation* (Bangkok), 13 September 1996.

30. Joe Cochrane, "Raping Borneo," *Newsweek,* 10 September 2001.

31. M. Hungspreugs et al., "Preliminary Investigation on the Mercury Distribution in the Mekong Estuary and Adjacent Sea," paper presented at the International Workshop on the Mekong Delta held in Chiang Rai, Thailand, 23–27 February 1998. The European Union–sponsored research team was led by Daniel Cossa of the French marine research institute IFREMER and Manuwadi Hungspreugs, a professor of marine science at Chulalongkorn University. Two sample sites showed elevated mercury levels: a site off Vung Tau had mercury levels of 4.14 ppb (and another in the Bassac River near Can Tho showed a level of 1.58 ppb). Also see James Fahn, "Toxin Threatens Mekong Delta," *Nation* (Bangkok), 26 April 1999. Unfortunately, this article contains an error in the units used (although the conclusions remain the same) that was corrected at a later date. A corrected version of the article can be seen on the Web site at http://www.geocities.com/jdfahn/mercuryVietnam.htm.

32. "Economic Man, Cleaner Planet," *Economist,* 29 September 2001.

CHAPTER 9

1. Pasuk Pongpaichit and Chris Baker, authors of *Thailand's Boom and Bust* (Chiang Mai, Thailand: Silkworm Books, 1998), presented these figures at a talk at Columbia University in October, 2001.

2. According to Tahir Qadri, an environmental specialist at the Asian Development Bank; see "Asean Launches Plan Against Haze, Officials Highlight Problems," *Agence France Presse,* 14 August 2001.

3. James Fahn, "Skies Darken Over SE Asia," *Nation* (Bangkok), 3 October 1997.

4. "An Asian Pea-Souper," *Economist,* 27 September 1997.

5. James Fahn, "Skies Darken Over SE Asia," *Nation* (Bangkok), 3 October 1997.

6. Ibid.

7. Ibid.

8. Ibid.

9. Rustam Abrus, chairman of the center for land and forest fire control in Riau province on Indonesia's Sumatra island, said loopholes in environmental laws made it difficult to punish business-

men who were responsible for forest fires. He noted that eleven investors were eventually prosecuted but were freed by the court, and that although the law stipulates that plantation owners are responsible for fires in their areas, it does not specify sanctions. "It's very difficult to prove their crimes. Even if we use articles in the criminal code then we'll need evidence, which is difficult to find," he said. See "Asean Launches Plan Against Haze, Officials Highlight Problems," *Agence France Presse,* 14 August 2001.

10. James Fahn, "Asean Ministers Tackle Air, Marine Pollution," *Nation* (Bangkok), 8 January 1997.

11. "$4.4bn Up in Smoke Due to 'Poor Govt Policy'," *Jakarta Post,* May 30, 1998.

12. James Fahn, "Skies Darken Over SE Asia," *Nation* (Bangkok), 3 October 1997.

13. Kamarulzaman Salleh, "Issues Affecting Asean 'No Longer Internal,'" *New Straits Times* (Kuala Lumpur), 23 July 2000.

14. ASEAN member states signed an agreement to control trans-boundary haze pollution in June 2002. See "ASEAN Members Sign Agreement to Control Haze Pollution," BBC, 11 June 2002.

15. James Fahn and Pennapa Hongthong, "Asean Endorses Turtle Protection Agreements," *Nation* (Bangkok), 13 September 1997.

16. Daniel Esty, "Bridging the Trade-Environment Divide," *Journal of Economic Perspectives* 15, no. 3 (summer 2001): 113.

17. "Thailand to Propose Asean Retaliation Against 'Greenies,'" *Xinhua News Agency,* 29 February 1992.

18. Things did not go as smoothly for Baird himself. Several years later, he was expelled from Thailand and, despite being married to a Thai woman, was considered *persona non grata.* He moved to Laos, where he worked to protect the few remaining freshwater dolphins living in the Mekong River.

19. Earth Island seems to be the most active, or at least the most effective, group pushing the U.S. government to impose trade sanctions in defense of conservation. In 1994, the U.S. slapped trade sanctions worth an estimated $22 million against Taiwan for trafficking in endangered tiger and rhinoceros parts. Taiwan complained, but its diplomatic isolation (it was not a member of the GATT/WTO at the time) meant that it had little recourse. The trade action may have been just the kind of political cover that Taiwan's government was seeking because it almost immediately announced a six-year, $227 million crackdown campaign against the trade in wildlife parts. See the editorial "Regulating Trade: Cheap Labour, Cheap Lives?" *Nation* (Bangkok), 18 April 1994. For other stories on the wildlife trade, see my "Chinese Take-Out," *Nation* (Bangkok), 12 September 1993, and "Vietnam's War on Wildlife," *Nation* (Bangkok), 8 September 1993; see also Jatuphol Rakthammachat, "Skinning the Cats," *Nation* (Bangkok), 15 October 1993.

20. James Fahn, "Wild Shrimp Embargo By U.S. Now in Force," *Nation* (Bangkok), 2 May 1996; and "Thailand Welcomes U.S. Lifting Sea Shrimp Import Ban," *Xinhua News Agency,* 19 November 1996.

21. Editorial, "The Trouble with U.S. Laws and Thai Turtles," *Nation* (Bangkok), 14 March 1996.

22. Walkkamon Eamwiwatkit, "Fishermen Test Device to Save Turtles from Nets," *Nation* (Bangkok), 11 September 1996.

23. See the Southeast Asian Fisheries Development Center Web site at http://www.seafdec.org.

24. Peter Fugazzotto, Associate Director of the Sea Turtle Restoration Project, in an e-mail interview, 24 April 2000.

25. "U.S. Set to Raise Profile of Environmental Issues in Foreign Policy Dealings," *Nation* (Bangkok), 12 April 1996.

26. Cited in "A. R. Dizon, Victory on Technicality Not Enough," *BusinessWorld* (Manila), 10 November 1998.

27. James Fahn, "Trade-Labor Backers State Case," *Nation* (Bangkok), 10 December 1996.

28. Kyle Bagwell and Robert Staiger, "The WTO as a Mechanism for Securing Market Access Property Rights: Implications for Global Labor and Environmental Issues," *Journal of Economic Perspectives* 15, no. 3 (summer 2001): 70.

29. Eric Neumayer, *Greening Trade and Investment* (London: Earthscan Publications, 2001), 151.

30. See James Fahn, "Trade-Labor Backers State Case," *Nation* (Bangkok), 10 December 1996; "Greens Place Trading Rules Under Spotlight," *Nation* (Bangkok), 11 December 1996; and "The WTO Battle That Had to Happen," *Nation* (Bangkok), 7 May 1999.

31. Fahn, "The WTO Battle."

32. Dai Qing made these comments at a talk given at Columbia University in November 2001.

33. See, for instance, Arik Levinson, "Environmental Regulations and Industry Location: International and Domestic Evidence," in *Fair Trade and Harmonization: Prerequisites for Free Trade?* Jagdish Bhagwati and Robert Hudec, eds. (Cambridge, Mass.: MIT Press, 1996), 429–459.

34. There are obvious exceptions however, for instance the two huge power plants being built just over the border in Mexico to supply power to California. See Tim Weiner, "U.S. Will Get Power, and Pollution, from Mexico," 17 September 2002.

35. Neumayer, *Greening Trade,* 55–57.

36. Esty, "Bridging the Trade-Environment Divide."

37. See the Woodrow Wilson International Center for Scholars Web site at http://wwics.si.edu/tef/measurm.htm. Some of the most important MEAs that could conflict with the WTO are the Convention on International Trade in Endangered Species, the CBD's Biosafety Protocol, the Montreal Protocol on Substances that Deplete the Ozone Layer, the Basel Convention on the Transboundary Movement of Hazardous Materials, the FCCC's Kyoto Protocol, and the Agreement on Persistent Organic Pollutants.

38. James Fahn, "WTO Could Undermine Environmental Treaties," *Nation* (Bangkok), 7 December 1996.

39. See Joel Trachtman, *European Journal of International Law,* vol. 10, no. 1, viewed on the Web site at http://www.ejil.org/journal/Vol10/No1/sr4–02.html#b.

40. Ibid.

41. "Victory on Technicality Not Enough," *BusinessWorld* (Manila), 10 November 1998.

42. Hakan Nordstrom and Scott Vaughan, *Special Studies: Trade and Environment* (Geneva: World Trade Organization, 1999). See also, "Embracing Greenery," *Economist,* 9 October 1999.

43. Edward Tang, "He Moves Like a Cool, Analytical Chess Player," *Straits Times* (Singapore), 1 August 1999.

44. Dirk Beveridge, "Environment, Labor Issues Strangle Trade Talks at APEC Summit," *Associated Press,* 14 November 2000.

45. David Pilling and Ted Bardacke, "Genetic Pirates Walk the Plank," *Financial Times,* 9 January 1999; and Pennapa Hongthong, "Potency Pill Ready to Take on Viagra," *Nation* (Bangkok), 24 March 1999.

46. See the Web site at http://www.newscientist.com/hottopics/biodiversity/biodiversityupdate.jsp.

47. James Fahn, "Trade Greed Gets in Way of a Green Treaty," *Nation* (Bangkok), 3 June 1999.

48. Peter Stoett and Shane Mulligan, "A Global Bioprospecting Regime: Partnership or Piracy?" *International Journal of Canadian Institute of International Affairs,* vol. 55, no. 2 (March 2000).

49. Pilling and Bardacke, "Genetic Pirates."

50. James Fahn, "Trade Greed Gets in Way of a Green Treaty," *Nation* (Bangkok), 3 June 1999.

51. James Fahn, "Confusion on Property Rights Rules Clouds Battle Against 'Biopirates,'" *Nation* (Bangkok), 19 March 1997.

52. "Sankyo to Produce Drug Material in Thailand," *Jiji Press,* 3 November 1985.

53. Carlos Correa, "Access to Genetic Resources and Intellectual Property Rights," p. 6, background study paper no. 8, FAO Commission on Genetic Resources for Food and Agriculture, April 1999.

54. James Fahn, "Tiny Island Nation Leads Southeast Asia in Potentially Rich Field," *Nation* (Bangkok), 19 March 1997.

55. See Pilling and Bardacke, "Genetic Pirates." The ethical issues raised by patenting forms of life have even become part of the debate over patients' rights, thanks to the efforts of a fascinating individual named John Moore. A surveying engineer who worked on the Alaska pipeline,

founded a college in Guadalajara and ran a worm farm in Washington state, Moore suffered from a rare and potentially fatal disease known as hairy cell leukemia. Near death in 1976, he checked into the UCLA Medical Center where doctors removed his dangerously swollen spleen. Amazed by his subsequently rapid recovery, his physician, David Golde, studied Moore's removed spleen and found unique blood cells that produced a protein which stimulates the growth of white blood cells to help fight infections. The "Mo" cell line was later patented by the regents of the University of California; Golde and one of his colleagues were named the inventors. When he learned of the patent, Moore filed a lawsuit seeking a share of the profits and claimed that his doctors had never informed him of their research. "Without my knowledge or consent, the doctors and the research institutions used a part of me for their own gain," he told the *Seattle Post-Intelligencer.* "They stole something from me." Moore won one of his appeals, but following an outcry from scientists and drug firms, who claimed the ruling would undermine biomedical research, the U.S. Supreme Court ruled against him in 1990. In 1996, the disease returned and five years later Moore passed away. See "John Moore, 56, Sued to Share Profits from His Cells," *Los Angeles Times* (obituaries), 13 October 2001.

56. Stoett and Mulligan, "A Global Bioprospecting Regime."

57. Fahn, "Confusion on Property Rights."

58. Pilling and Bardacke, "Genetic Pirates."

59. Andrew Revkin, "Biologists Sought a Treaty; Now They Fault It," *New York Times,* 7 May 2002.

60. "Biotech Conference Urges Adoption of International Protocol on GM Food," *Agence France Presse,* 12 July 2001.

60a. Bill Joy, "Why the Future Doesn't Need Us," *Wired,* Issue 8.04, April 2000.

61. Neumayer, *Greening Trade.*

62. Michele Landsberg, "Canadians Winners in Seattle, the Sequel," *Toronto Star,* 6 February 2000.

63. James Fahn, "Greens Raise Concern Over Property Rights," *Nation* (Bangkok), 12 December 1996.

64. See Revkin, "Biologists Sought a Treaty."

65. See Stoett and Mulligan, "A Global Bioprospecting Regime."

66. Ibid.

67. David Simpson et al., "Valuing Biodiversity for Use in Pharmaceutical Research," *Journal of Political Economy* 104, no. 1 (1996): 163–185; also see Table 1 in Gordon Rausser and Arthur Small, "Valuing Research Leads: Bioprospecting and the Conservation of Genetic Resources," *Journal of Political Economy* 108, no. 1 (February 2000): 173–206.

68. Rausser and Small, "Valuing Research Leads: Bioprospecting and the Conservation of Genetic Resources," *Journal of Political Economy,* February 2000.

69. Revkin, "Biologists Sought a Treaty."

70. For a summary of IPCC's assessment on the projected impacts of climate change, see the Web site at http://www.ipcc.ch/pub/wg2SPMfinal.pdf.

71. "World Bank: World's Poor Will Suffer Most From Climate Change," *Deutsche Presse-Agentur,* 14 June 2001.

72. Ploenpote Atthakor, "Ministry Pledges to Strive for Aid Access," *Bangkok Post,* 16 September 1999.

73. From a presentation given by Richard Bradley at Columbia University in October 2001.

74. The only possible exception, says Bradley, is whale oil, and even that may still be lighting some lamps in northern Norway.

75. James Fahn, "Global Game of Blind Man's Bluff," *Nation* (Bangkok), 30 March 1995.

76. James Fahn, "Japan Moots Joint Greenhouse Gas Project," *Nation* (Bangkok), 6 December 1997.

77. Peter Du Pont, former director of the Thailand branch of the International Institute for Energy Conservation (IIEC), supplied this information in an e-mail interview on April 6, 2000.

78. James Fahn, "Global Warming: 'Sink or Swim,'" *Nation* (Bangkok), 12 November 1991.

79. Annual Energy Outlook 1999, published by the Energy Information Agency, U.S. Department of Energy, viewed on the Web site at http://www.eia.doe.gov/oiaf/archive/aeo99/results.html#report.

80. As Thailand's official statement to the Kyoto conference noted, "The vast disparities amongst developing countries, specifically regarding levels of greenhouse gas emissions and industrial development must also be acknowledged. An approach that classifies developing nations into least-developed countries, countries with middle income and upper income levels, should be adopted to enable the application of differentiated policies and measures for the prevention of global warming that corresponds to a developing country's capabilities and national circumstances. In addition, a pragmatic time-frame for commitments should be adopted that is appropriate to each level." See James Fahn, "Thailand Presents Case At Climate Summit," *Nation* (Bangkok), 10 December 1997.

81. James Fahn, "Optimism Surrounds Probable Treaty on Climate Change," *Nation* (Bangkok), 9 December 1997.

82. James Fahn, "Carbon Emissions For Sale," *Nation* (Bangkok), 22 December 1997.

83. For a Web site with good information on Lovelock, Margulis, and Gaia, see http://www.magna.com.au/~prfbrown/gaia_jim.html.

84. Daniel Altman, "Just How Far Can Trading of Emissions Be Extended?" *New York Times,* 31 May 2002.

85. Tim Forsyth, Fellow in Environment and Development at the Institute of Development Studies, made the comments in an e-mail interview on 14 April 2000.

86. See the UNDP Web site at http://www.undp.org/gef/portf/funds.htm.

87. Saritdet Marukatat and Kultida Samabuddhi, "Debt Swap Agreement Rejected," *Bangkok Post,* 2 March 2002.

88. Pennapa Hongthong, "Forest Fund Was a 'Kyoto Ploy,'" *Nation* (Bangkok), 5 March 2002.

CHAPTER 10

1. James Fahn, "An American in Iraq," *Nation* (Bangkok), 1991; see also my "Across the Kurdish Heartland," *Nation* (Bangkok), 24 May 1991.

2. "Destination Chiang Mai: A Flight That Wasn't to Be," *Nation* (Bangkok), 24 February 1991.

3. See Transparency International's Web site at http://www.transparency.org/cpi/2001/cpi2001.html.

4. Steiner gave a talk on "Accountability and the Environment: The Need for a Joint Initiative of Public, Private and Civil Society Sectors" at the 9th International Anti-Corruption Conference. Viewed on the Web site at http://www.transparency.org/iacc/9th_iacc/papers/day3/ws8/d3ws8_asteiner.html.

5. See, for instance, Walakkamon Eamwiwatkit, "Environmental Awareness at Higher Level Than Expected," *Nation* (Bangkok), 26 January 1996, and "Survey Finds General Concern on Environment," *Nation* (Bangkok), 21 April 1995.

6. Daniel Goleman, "The Group and the Self: New Focus on a Cultural Rift," *New York Times,* 25 December 1990. In the article, anthropologist Raoul Narrol notes that communitarian societies have lower rates of homicide, suicide, juvenile delinquency, divorce, child abuse, and alcoholism, but they also have lower economic productivity, and grow more individualistic as they become richer, Japan being a good example. Harry C. Triandis, a psychologist of Greek origin at the University of Illinois provides examples of the distinct patterns of behavior in these cultures: "A telephone operator in Greece will be rude and unhelpful until the caller establishes his identity, while American operators tend to be uniformly friendly to everyone," he said. "In countries like Pakistan, Peru, and Singapore, people take it for granted that those at the top of the social hierarchy are very different from those at the bottom and that they should keep their distance. In countries like the United States, Australia, and the Netherlands, such an attitude is seen as offensively anti-egalitarian."

7. Thomas Friedman, "In Bangkok, Who Is Strong Enough to Fix the Boom Blight?" *New York Times,* 21 March 1996.

8. Jones's comments and others on Southeast Asian governance can be found in Hoi-kowk Wong and Hon Chan, eds., *Handbook of Comparative Public Administration in the Asia-Pacific*

Basin (New York: M. Dekker, 1999). Malaysia also has an Auditor-General's Office that exhibits less corruption than Thailand's. But the country certainly has its share of government scandals, and it is not considered as clean as Singapore. According to Khai Leong Ho of the National University of Singapore, examples of corruption, administrative abuse, and incompetence have increased in recent years.

9. Personal communication, circa 1996.

10. Ian Buruma made these remarks at a talk he gave at Columbia University's East Asian Institute on 4 December 2001.

11. Steven Erlanger, "A Fast Fire, and a Thai Slum's Slow Poisoning," *New York Times,* 21 March 1991.

12. Ibid.

13. A discussion of the impacts of rapid economic growth in Thailand, including an article by Bell, can be found in Michael Parnwell, ed., *Uneven Development in Thailand* (Aldershot, England: Avebury, 1996).

14. Poona Antaseeda, "Unlikely Heroine Raises Hopes for the Poor," *Bangkok Post,* 28 October 2001.

15. Anjira Assavanonda, "Victims of Chemical Fire Fighting for Redress 10 Years On," *Bangkok Post,* 28 February 2001; and Poona Antaseeda, "Justice Against the Odds," *Bangkok Post,* 28 October 2001.

16. Wasant Techawongtham, "Judicial Attitudes Need a Shake-Up," *Bangkok Post,* 7 August 1997.

17. Walakkamon Eamwiwatkit, "Port Fire Victims Still Picking Up the Pieces," *Nation* (Bangkok), 4 March 1996.

18. Anchalee Kongrut, "Victims Face Uphill Battle in Court," *Bangkok Post,* 14 May 2001.

19. James Fahn, "We're Not Just Another Asian Tragedy," *Nation* (Bangkok), 24 May 1992.

20. William Callahan makes this point in *Imagining Democracy : Reading "The Events of May" in Thailand* (Singapore: Institute of Southeast Asian Studies, 1998).

21. Scot Donaldson, "On the Wrong Side of Ratchadamnoen Ave," *Nation* (Bangkok), 24 May 1992.

22. Barrington Moore, *Social Origins of Dictatorship and Democracy: Lord and Peasant in the Making of the Modern World* (Boston: Beacon Press, 1993).

23. Ian Buruma points out another thing most of these countries have in common: "The U.S. client states in the '60s and '70s are precisely those that have become democracies, because the U.S. can pressure the generals to ease control." (Indonesia was not exactly a client state, but it certainly developed close economic and military ties to the United States.) Indeed, following the 1991 coup, the United States cut off its military aid to Thailand, although the effect seemed minimal compared to domestic political pressures. It remains to be seen how Asian countries with ruling Communist parties will evolve politically.

24. James Fahn, "Ecology and Democracy Closely Tied," *Nation* (Bangkok), 2 June 1992.

25. Baker and Pasuk made these remarks at a talk at Columbia University in October 2001.

26. Thongchai Winichakul, "Thai Democracy in Public Memory: Monuments of Democracy and Their Narratives" (paper presented at the 7th International Conference on Thai Studies, at the University of Amsterdam, the Netherlands, 4–8 July 1999).

EPILOGUE

1. See *Sustainable Solutions: Building Assets for Empowerment and Sustainable Development* (Ford Foundation, 2002).

2. Antonio G. M. La Vina, Gretchen Hoff, and Anne Marie DeRose, *The Success and Failure of Johannesburg: A Story of Many Summits,* working paper, World Resources Institute, September 2002.

INDEX